Logic in Asia: Studia Logica Library

Logic in Asia: Studia Logica Library

This book series promotes the advance of scientific research within the field of logic in Asian countries. It strengthens the collaboration between researchers based in Asia with researchers across the international scientific community and offers a platform for presenting the results of their collaborations. One of the most prominent features of contemporary logic is its interdisciplinary character, combining mathematics, philosophy, modern computer science, and even the cognitive and social sciences. The aim of this book series is to provide a forum for current logic research, reflecting this trend in the field's development.

The series accepts books on any topic concerning logic in the broadest sense, i.e., books on contemporary formal logic, its applications and its relations to other disciplines. It accepts monographs and thematically coherent volumes addressing important developments in logic and presenting significant contributions to logical research. In addition, research works on the history of logical ideas, especially on the traditions in China and India, are welcome contributions.

The scope of the book series includes but is not limited to the following:

- Monographs written by researchers in Asian countries.
- Proceedings of conferences held in Asia, or edited by Asian researchers.
- Anthologies edited by researchers in Asia.
- Research works by scholars from other regions of the world, which fit the goal of "Logic in Asia".

The series discourages the submission of manuscripts that contain reprints of previously published material and/or manuscripts that are less than 165 pages/ 90,000 words in length.

Please also visit our webpage: http://tsinghualogic.net/logic-in-asia/background/

Relation with Studia Logica Library

This series is part of the Studia Logica Library, and is also connected to the journal Studia Logica. This connection does not imply any dependence on the Editorial Office of Studia Logica in terms of editorial operations, though the series maintains cooperative ties to the journal.

This book series is also a sister series to Trends in Logic and Outstanding Contributions to Logic.

For inquiries and to submit proposals, authors can contact the editors-in-chief Fenrong Liu at fenrong@tsinghua.edu.cn or Hiroakira Ono at ono@jaist.ac.jp.

More information about this series at http://www.springer.com/series/13080

Fenrong Liu · Hiroakira Ono · Junhua Yu
Editors

Knowledge, Proof and Dynamics

The Fourth Asian Workshop on Philosophical Logic

Springer

Editors
Fenrong Liu
Department of Philosophy
Tsinghua University
Beijing, China

Hiroakira Ono
Japan Advanced Institute of Science and Technology
Kanazawa, Japan

Junhua Yu
Department of Philosophy
Tsinghua University
Beijing, China

ISSN 2364-4613 ISSN 2364-4621 (electronic)
Logic in Asia: Studia Logica Library
ISBN 978-981-15-2220-8 ISBN 978-981-15-2221-5 (eBook)
https://doi.org/10.1007/978-981-15-2221-5

This Springer imprint is published by the registered company Springer Nature Singapore Pte Ltd.
The registered company address is: 152 Beach Road, #21-01/04 Gateway East, Singapore 189721, Singapore

Preface

We are glad to see that all the papers are finally ready for this volume. We have chosen the title *Knowledge, Proof and Dynamics* with the idea in mind that each paper can be labeled with at least two words from the title. This collection consists of ten papers, which are arranged in two parts. The first part includes six papers that were presented at the 4th Asian Workshop on Philosophical Logic (AWPL2018) organized by Tsinghua University in Beijing, October 20–21, 2018. The four papers in the second part were later invited particularly for the volume. First of all, we would like to express our most profound appreciation to all the authors for being willing to contribute. Without their effort, this could never have happened.

In the last two decades, there has been a fast-growing community of logicians in Asian countries. In order to provide a platform for researchers to publish their latest works, the book series *Logic in Asia* was launched in 2014, with six volumes appeared by now. At the same time, a series of Asian Workshop on Philosophical Logic was started and held biyearly, to strengthen the interaction between logicians and philosophers. Both the book series and the workshop have witnessed a new and vibrant generation of logicians joining the community. They are carrying out important research of their own, but also in collaboration with international colleagues. This volume reflects the latest research in this respect. In what follows, we provide a quick introduction to each paper. Whenever possible, we try to find some common thread between them. Nevertheless, we warmly invite the audience to read the papers for further details.

The paper "Logics for Knowability Paradox with a Non-normal Possibility Operator" by Youan Su and Katsuhiko Sano studies the knowability paradox in terms of sequent calculus. They propose Hilbert-style axiomatization and Gentzen-style sequent calculus for an intuitionistic modal logic, with and without the knowability axiom, and prove logical properties of the corresponding formal systems. As the title suggests, "Refutation Systems: An Overview and Some Applications to Philosophical Logics" by Valentin Goranko, Gabriele Pulcini, and Tomasz Skura is an overview of refutation systems. The paper gives a detailed development of the theory of refutation systems and its essential ideas. In particular, it discusses their applications to philosophical logics. Against the background of modal quantification logic, "De Re, De Dicto, and Binding Modalities" by Melvin Fitting carries on the tradition of justification logic with possible world semantics. The paper continues the study of binding modalities for the system K. It shows that binding modalities provide a natural intuition for Kripke's nonstandard axiomatization, and can make a better interpretation regarding the distinction between *de re* and *de dicto*. "Unary and Two-Variable Interval Logics" by Kamal Lodaya is a technical paper. It shows that over finite word models, the interval logic of Halpern and Shoham 1991 is expressively complete for two-variable logic with betweenness relations introduced in his earlier work with Krebs and others in 2016. The paper proves that satisfiability of formulae can be checked in polynomial space. "A Logical Characterization of the Continuous Bar Induction" by Makoto Fujiwara and Tatsuji Kawai is in the area of logical foundations of mathematics. The paper contains two main results: (1) The continuous bar induction is equivalent to the monotone bar induction restricted to Π^0_2 bars. (2) The continuous bar induction for bounded Σ^0_2 side-predicates together with the lesser limited principle of omniscience is equivalent to the classical bar induction for bounded Σ^0_2 side-predicates.

The interaction between game theory and logic has a long and fruitful history, which we will not spell out the details here. The latest research in social network logic has a natural connection with games, in particular, those in graph theory. We are delighted to include several papers on this topic. "Graph Games and Logic Design" by Johan van Benthem and Fenrong Liu is a position paper. It discusses old and new graph games and argues that there is a tight connection between how the game is being played and how the matching logic is being set up. A back-and-forth methodology is highly needed for designing new graph games, as well as creating new logics as a way of analyzing existing graph games. "The Modal Logics of the Poison Game" by Francesca Zaffora Blando, Krzysztof Mierzewski, and Carlos Areces investigates the well-known poison game in graph theory from the perspective of modal logic. They introduce new model-changing operators to describe changes in the relational structures. Besides, they consider three memory logics, their fit with the game, their expressive power, and complexity of the model-checking problem. "Solution Complexity of Local Variants of Sabotage Game" by Tianwei Zhang is a case study on the respective graph game for three main local variants of sabotage modal logic. The paper analyzes the solution complexity for each game, with an attempt to identify the parameters of games and logic that crucially affect complexity. "Local Fact Change Logic" by Declan Thompson introduces a new operator allowing for the valuation change at a particular state in a model. "Local fact change" has a close connection with Boolean games that have been studied by Michael Wooldridge and his colleagues. The paper studies some properties of the new dynamic logic, its expressive power, and also shows it is undecidable. "Influence in Different Network Structures" by Yunqi Xue, Rohit Parikh, and Mihai Gociu studies possible conflicting influence in social networks. They define a new notion of the expected influence. Using this notion, they illustrate how agents make strategic decisions when they face different influence that suggests conflicting actions. The paper includes simulations of the strategic influence emerging in various realistic networks.

We feel very grateful to the experts below for their support. They spent time reading the papers and provided valuable comments for the authors to improve their work.

Alexandru Baltag (University of Amsterdam)
Davide Grossi (University of Groningen)
Minghui Ma (Sun Yat-sen University)
Chuangjie Xu (Ludwig Maximilians Universität München)
Chin-mu Yang (National Taiwan University)

Finally, we would like to thank Leana Li and Fiona Wu at Springer Beijing office for their assistance and patience, and Lei Li at Tsinghua University for his help with preparing documents for this volume.

Beijing, China — Fenrong Liu
Kanazawa, Japan — Hiroakira Ono
Beijing, China — Junhua Yu
December 2019

Contents

Contributed Papers

The Modal Logics of the Poison Game

Francesca Zaffora Blando, Krzysztof Mierzewski and Carlos Areces

Abstract The poison game is a two-player zero-sum game played on directed graphs, first introduced by Duchet and Meyniel (1993) in the context of graph theory, where one of the players travels along the edges of a graph, while the other modifies the underlying structure by marking vertices. In this paper, we investigate the poison game from the perspective of modal logic, as a natural case study of the use of modal languages equipped with model-changing operators to describe evolving relational structures. In particular, to model the poison game, we consider three memory logics of decreasing expressive power but increasing fit with the game. We begin with $\mathcal{ML}_\emptyset$, the basic memory logic restricted to the initial class of models with an empty memory (see Areces et al. 2011). We then identify two fragments of $\mathcal{ML}_\emptyset$, which we respectively, denote as $\mathcal{PML}$ and $\mathcal{PSL}$, and whose modal operators capture operations on models that mimic more closely the moves of both players. We show that these logics form a chain in expressive power with $\mathcal{PSL} < \mathcal{PML} < \mathcal{ML}_\emptyset$, and we introduce suitable notions of bisimulation for the two new logics presented in this paper. We then show that model checking for both $\mathcal{PML}$ and $\mathcal{PSL}$ is PSPACE-complete. The construction also establishes that determining the existence of a winning strategy in the poison game is PSPACE-hard. We conclude by proving that $\mathcal{PML}$, while strictly less expressive than $\mathcal{ML}_\emptyset$, nonetheless has an undecidable satisfiability problem.

This chapter is in its final form and it is not submitted to publication anywhere else.

F. Zaffora Blando (✉) · K. Mierzewski
Department of Philosophy, Stanford University and Logical Dynamics Lab, Center for the Study of Language and Information, Stanford, California, USA
e-mail: fzaffora@stanford.edu

C. Areces
Facultad de Matemática, Astronomía, Física y Computación, Universidad Nacional de Córdoba and CONICET, Córdoba, Argentina

F. Liu et al. (eds.), *Knowledge, Proof and Dynamics*,
Logic in Asia: Studia Logica Library,
https://doi.org/10.1007/978-981-15-2221-5_1

1 Introduction

Logic and games have gone hand in hand for quite some time. As noted by Hodges (2013), one can already find connections between logic and argumentation games in Aristotle's work on syllogisms. Nowadays, logical games are used in a multitude of settings. There are games for model comparison (Fraïssé 1954; Ehrenfeucht 1961), argumentation and dialogue (Lorenzen 1955), model checking (Hintikka 1973), as well as for building models for a given formula (Hodges 2006).

A more recent strand of research tackles the opposite direction: in addition to using game-theoretic tools in logic, one can also focus on the use of logical languages for analysing games (see van Benthem 2014). For instance, games of imperfect information may be naturally modelled within epistemic or doxastic logic, and certain common computational tasks might be 'gamifiable', which then facilitates their analysis from the perspective of modal logic. An example of this latter application is the *sabotage game* introduced by van Benthem (2005), a two-player zero-sum game played on graphs. In a sabotage game, one player tries to get from one vertex to another fixed set of vertices, while the other player tries to prevent this by deleting edges in the graph. In other words, a sabotage game is a version of the *graph reachability* problem involving an edge-deleting (or sabotaging) player, whose goal is rendering the target vertices inaccessible.

In order to model this game, van Benthem (2005) introduces a modal calculus, *sabotage modal logic*, which differs from the basic modal language, in that it is equipped with a transition-deleting modality which modifies the underlying model. This means that sabotage modal logic, as opposed to standard modal logic, can express changes of transition systems, on top of the usual properties of static models.[1]

In this paper, we consider another two-player zero-sum game played on graphs called the *poison game*—first introduced by Duchet and Meyniel (1993)—from a modal perspective. While the sabotage game is a logic-inspired graph game, the poison game comes from graph theory: studying the poison game is therefore a good way of testing whether methods from modal logic can be fruitfully extended to a wider domain.

Let (G, R) be a directed graph (with no double edges) and $s \in G$ a distinguished *starting vertex*. The poison game proceeds as follows: the two players—Traveller and Poisoner—alternate their moves and, at each step, choose a vertex that is a successor of the vertex previously chosen by the other player. The game begins at the starting vertex s.[2] Poisoner makes the first move by choosing a successor s' of s (that is, a point $s' \in G$ such that $(s, s') \in R$), which she then poisons with a poison that affects exclusively Traveller. This means that Poisoner's move renders vertex s' inaccessible to Traveller, but not to Poisoner. Then, Traveller has to choose a non-poisoned successor of s', and so on. The winning conditions are as follows.

[1]For a detailed discussion of sabotage modal logic and other relation-changing logics, see, for instance, (Löding and Rohde 2003a, b; Aucher et al. 2015, 2017; Areces et al. 2015).

[2]Here we treat the starting position as being given. In the original formulation of the game (Duchet and Meyniel 1993), Traveller makes the first move by *choosing* a starting vertex $s \in G$.

Traveller wins the poison game if either (i) she manages to keep choosing non-poisoned successors no matter what vertices Poisoner selects, or (ii) she begins her turn at a vertex with no successors, or (iii) Poisoner begins her turn at a vertex with no successors. If at least one of (i)–(iii) obtains, Traveller *survives* the game. Poisoner, on the other hand, wins the poison game if she manages to poison a vertex that (a) has at least one successor and (b) all of whose successors have already been poisoned (we then say that Traveller gets poisoned, as she would be forced to move to a poisoned vertex in the next round).

The landscape of modal logics is rich and varied. When designing a modal system to model a graph game like the poison game, one is immediately confronted with the question of which language is best suited for capturing the game, and of which minimal requirements a logic should meet in order to qualify as a plausible contender. For instance, it should be possible, within the logic, to talk about the moves of both players and to express, at least approximately, the existence of winning strategies.

In this modal setting, where we evaluate formulas at individual vertices in the graph, we are particularly interested in describing games from a local perspective. In the poison game, the players construct a path through the graph, moving from vertex to vertex: modal languages are singularly well suited to describe the current stage of the game, step-by-step, by capturing local (and possibly dynamic) properties of the graph as seen from the vantage point of the vertex being currently occupied. From this perspective, a good fit between the logic and the game also means that the logic should not be excessively expressive: i.e. it should not be possible to express (too many) global properties that have no natural counterpart in the poison game.

Since, by poisoning a vertex in a graph, Poisoner is basically marking that vertex, signalling in this way that it is no longer accessible for Traveller, memory logics (Areces 2007; Mera 2009) seem a natural starting point, for they have the ability to store states into a memory. In particular, in this paper we consider three memory logics of decreasing expressive power but increasing fit with the poison game. The first one is the basic memory logic restricted to the initial class of models with an empty memory, which we denote as $\mathcal{ML}_\emptyset$ (see Areces et al. 2011). The other two are syntactic fragments of $\mathcal{ML}_\emptyset$, which we, respectively, denote as $\mathcal{PML}$ and $\mathcal{PSL}$.

We analyse the expressive power of these languages, define notions of bisimulation that are appropriate for $\mathcal{PML}$ and $\mathcal{PSL}$, respectively, and prove that $\mathcal{PSL}$ is strictly less expressive than $\mathcal{PML}$ and $\mathcal{PML}$ is strictly less expressive than $\mathcal{ML}_\emptyset$. We also show that model checking for both $\mathcal{PML}$ and $\mathcal{PSL}$ is PSPACE-complete. In establishing this, we also provide a lower bound on the complexity of the poison game itself: determining the existence of a winning strategy for Traveller is shown to be PSPACE-hard. It is known that the satisfiability problem for $\mathcal{ML}_\emptyset$ is undecidable. Here, we prove that the satisfiability problem for $\mathcal{PML}$ is undecidable, too. We leave it as an open question whether $\mathcal{PSL}$ satisfiability is decidable or not.

The present paper extends and improves on some previous work (Mierzewski and Zaffora Blando 2016) by the first two authors. The logic $\mathcal{PML}$ was also independently investigated by Grossi and Rey (2019) in the context of abstract argumentation theory (specifically, to study *credulously admissible arguments*[3]).

[3] See, for instance, (Vreeswijk and Prakken 2000).

2 $\mathcal{ML}_\emptyset$

The simplest memory logic extends the basic modal language with two operators: ⓚ and ⓡ. A model for this logic is a tuple $\mathcal{M} = (W, R, V, M)$, where (W, R, V) is a standard relational structure, and $M \subseteq W$ is the *memory* of the model. The semantics of the new operators is then given by

$$\mathcal{M}, w \models ⓚ \quad \text{iff } w \in M,$$
$$\mathcal{M}, w \models ⓡ\varphi \quad \text{iff } \mathcal{M}[w], w \models \varphi \quad \text{where } \mathcal{M}[w] = (W, R, V, M \cup \{w\}).$$

The ⓚ operator allows to check whether the current state has been memorised, while the ⓡ operator elicits the memorisation of the current state and the subsequent evaluation of φ there.

To model the poison game, it is reasonable to focus on the class of initial models where $M = \emptyset$, as no vertices are poisoned at the beginning of the game. We will refer to the memory logic restricted to this class of initial models as $\mathcal{ML}_\emptyset$. In this language, we can use formulas of the form $\Diamond$ⓡφ to model Poisoner's moves and formulas of the form $\Diamond\neg$ⓚ to model Traveller's moves.

The logic $\mathcal{ML}_\emptyset$ is very expressive. For instance, we can express the property that Traveller can survive at least n rounds of a poison game by means of the following inductive scheme:

$$\rho_1 := \Box ⓡ(\Box\bot \vee \Diamond\neg ⓚ)$$

$$\rho_n := \Box ⓡ(\Box\bot \vee \Diamond(\neg ⓚ \wedge \rho_{n-1})).$$

We can also express the property that the point of evaluation has a successor that has itself as its only successor via the formula $\Diamond$ⓡ($\Diamond$ⓚ $\wedge$ $\Box$ⓚ). This implies that $\mathcal{ML}_\emptyset$ does not have the tree model property: the formula $\Diamond$ⓡ($\Diamond$ⓚ $\wedge$ $\Box$ⓚ) cannot be satisfied at the root of a tree.

The property of having n non-poisoned successors, on the other hand, is not expressible within $\mathcal{ML}_\emptyset$. To see why, consider the models $\mathcal{M} = (W, R, V, \emptyset)$ and $\mathcal{N} = (W', R', V', \emptyset)$ below, where $V(q) = V'(q) = \emptyset$ for all proposition letters q.

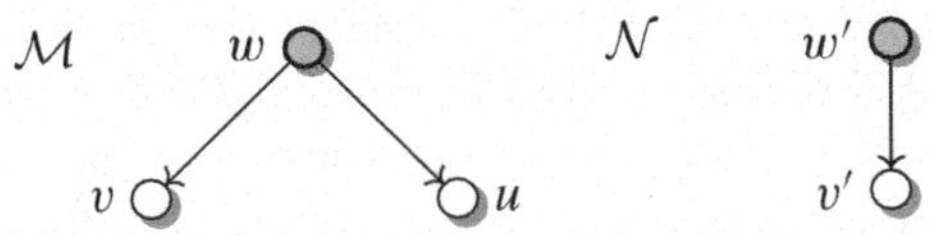

Note that it is possible to define a notion of bisimulation that is suitable for $\mathcal{ML}_\emptyset$ (see [Definition 3.4] Areces et al. 2011). Now, although state w from model $\mathcal{M}$ has two non-poisoned successors, while state w' from model $\mathcal{N}$ only has one, $(\mathcal{M}, w)$ and $(\mathcal{N}, w')$ are $\mathcal{ML}_\emptyset$ bisimilar. The basic intuition behind this observation is that, in order to mark the successors of w and w' via the ⓡ operator, one has to first move there via the standard $\Diamond$ modality. But then, from the perspective of states v, u and v', models $\mathcal{M}$ and $\mathcal{N}$ are completely indistinguishable.

A first worry raised by the use of $\mathcal{ML}_\emptyset$ to model the poison game is that the $\textcircled{r}$ operator does not, by itself, naturally correspond to Poisoner's moves, for it allows one to memorise the current state, while Poisoner must always poison a *successor* of the current state. In other words, $\mathcal{ML}_\emptyset$ seems to be too expressive for the purpose of faithfully modelling the poison game.

In addition, $\mathcal{ML}_\emptyset$ is not computationally well behaved:

Theorem 2.1 (Areces et al. 2011) *The satisfiability problem for $\mathcal{ML}_\emptyset$ is undecidable.*

In light of these considerations, a natural question is whether we can find a logic that is closer to the poison game and with a better computational behaviour.

3 $\mathcal{PML}$

As our first attempt, we shall focus on the fragment of $\mathcal{ML}_\emptyset$ that extends the basic modal language with two operators, $\textcircled{p}$ and $\langle \mathtt{p} \rangle$, respectively, defined as $\textcircled{p} \leftrightarrow \textcircled{k}$ and $\langle \mathtt{p} \rangle \varphi \leftrightarrow \Diamond \textcircled{r} \varphi$. We will refer to this logic as $\mathcal{PML}$.[4]

Expressive power. Since the $\langle \mathtt{p} \rangle$ operator forces the poisoning move to occur one step ahead of the point of evaluation, the following formula is a simple validity of $\mathcal{PML}$:

$$\langle \mathtt{p} \rangle \textcircled{p} \leftrightarrow \Diamond \top.$$

It is also worth noting that $\langle \mathtt{p} \rangle$ and $\Diamond$ agree on formulas that do not contain $\textcircled{p}$: i.e. $\langle \mathtt{p} \rangle \varphi \leftrightarrow \Diamond \varphi$ is valid when φ is a formula not containing $\textcircled{p}$. This no longer holds in the presence of $\textcircled{p}$. For instance, consider the formula $\langle \mathtt{p} \rangle \textcircled{p} \leftrightarrow \Diamond \textcircled{p}$ and the model $\mathcal{M} = (W, R, V, \emptyset)$ below, where $V(q) = \emptyset$ for all proposition letters q.

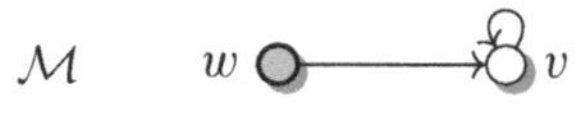

Here, we have that $\mathcal{M}, w \models \langle \mathtt{p} \rangle \textcircled{p}$, but $\mathcal{M}, w \not\models \Diamond \textcircled{p}$.

As in the case of $\mathcal{ML}_\emptyset$, the following inductive scheme allows to express the property that Traveller can survive at least n rounds of a poison game:

$$\rho_1 := [\mathtt{p}](\Box\bot \vee \Diamond \neg \textcircled{p})$$

$$\rho_n := [\mathtt{p}](\Box\bot \vee \Diamond(\neg \textcircled{p} \wedge \rho_{n-1})).$$

---

[4] We use $\textcircled{p}$ instead of $\textcircled{k}$ simply because the former is more suggestive of the *poison* game. Also note that $\mathcal{PML}$ models are of course exactly the same as $\mathcal{ML}_\emptyset$ models, except that we will write P instead of M to denote the set of poisoned states. $\mathcal{PML}$ stands for *poison memory logic*.

Note that the $\mathcal{ML}_\emptyset$ formula $\Diamond ⓡ(\Diamond ⓚ \wedge \Box ⓚ)$ from the previous section—which, as we saw, forces non-tree models in $\mathcal{ML}_\emptyset$—falls within $\mathcal{PML}$ (we can write it as $\langle \mathrm{p} \rangle(\Diamond ⓟ \wedge \Box ⓟ)$). Hence, $\mathcal{PML}$ lacks the tree model property, too.

It is possible to define a notion of bisimulation that fits $\mathcal{PML}$ by adding the following clauses to the standard ones for the basic modal language. Let Z below denote a bisimulation between models $\mathcal{M} = (W, R, V, P)$ and $\mathcal{N} = (W', R', V', P')$:

Non-empty: there are $w \in W$ and $w' \in W'$ with $(P, w)Z(P', w')$;
Agree: if $(S, u)Z(S', u')$, then

(1) $\mathcal{M}, u \models q$ if and only if $\mathcal{N}, u' \models q$ for any proposition letter q, and
(2) $u \in S$ if and only if $u' \in S'$;

Zig$_{\langle \mathrm{p} \rangle}$: if $(S, u)Z(S', u')$ and there exists $v \in W$ with $(u, v) \in R$, then there exists $v' \in W'$ with $(u', v') \in R'$ and $(S \cup \{v\}, v)Z(S' \cup \{v'\}, v')$;
Zag$_{\langle \mathrm{p} \rangle}$: if $(S, u)Z(S', u')$ and there exists $v' \in W'$ with $(u', v') \in R'$, then there exists $v \in W$ with $(u, v) \in R$ and $(S \cup \{v\}, v)Z(S' \cup \{v'\}, v')$.

In the **Agree**, **Zig**$_{\langle \mathrm{p} \rangle}$ and **Zag**$_{\langle \mathrm{p} \rangle}$ clauses above, we use S and S' instead of P and P' because the $\langle \mathrm{p} \rangle$ modality allows to add new states to the initial memories of the models $\mathcal{M}$ and $\mathcal{N}$.

This notion of $\mathcal{PML}$ bisimulation is correct, in that it implies $\mathcal{PML}$ equivalence:

Proposition 3.1 *Let $\mathcal{M} = (W, R, V, P)$ and $\mathcal{N} = (W', R', V', P')$ be two $\mathcal{PML}$ models, $w \in W$ and $w' \in W'$. If Z is a bisimulation linking (P, w) and (P', w'), then, for any $\varphi \in \mathcal{PML}$,*

$$\mathcal{M}, w \models \varphi \textit{ iff } \mathcal{N}, w' \models \varphi.$$

Proof The proof is by induction on φ and the only non-standard cases are the ones involving ⓟ and $\langle \mathrm{p} \rangle$. First of all, we have that

$$\begin{aligned} \mathcal{M}, w \models ⓟ &\text{ iff } w \in P && \text{by the semantics of ⓟ,}\\ &\text{ iff } w' \in P' && \text{by the **Agree** condition,}\\ &\text{ iff } \mathcal{N}, w' \models ⓟ && \text{by the semantics of ⓟ.} \end{aligned}$$

Now, suppose $\mathcal{M}, w \models \langle \mathrm{p} \rangle\varphi$. Then, there is $v \in W$ with $(w, v) \in R$ and $\mathcal{M}[v], v \models \varphi$. Since $(P, w)Z(P', w')$ and there is $v \in W$ with $(w, v) \in R$, the **Zig**$_{\langle \mathrm{p} \rangle}$ condition gives us that there exists $v' \in W'$ with $(w', v') \in R'$ and $(P \cup \{v\}, v)Z(P' \cup \{v'\}, v')$. Now, since $(P \cup \{v\}, v)Z(P' \cup \{v'\}, v')$ and $\mathcal{M}[v], v \models \varphi$, the induction hypothesis implies that $\mathcal{N}[v'], v' \models \varphi$. Hence, $\mathcal{N}, w' \models \langle \mathrm{p} \rangle\varphi$. The other direction is analogous, except that it relies on the **Zag**$_{\langle \mathrm{p} \rangle}$ condition. □

Now, how does $\mathcal{PML}$ compare with $\mathcal{ML}_\emptyset$ in terms of expressive power? We will show that $\mathcal{PML}$ is strictly less expressive than $\mathcal{ML}_\emptyset$.

Definition 3.2 Let $\mathcal{L}$ and $\mathcal{L}'$ be two logics. We say that $\mathcal{L}'$ is *at least as expressive as* $\mathcal{L}$ (in symbols, $\mathcal{L} \leq \mathcal{L}'$) if there is a translation $T : \mathcal{L} \to \mathcal{L}'$ such that, for every model $\mathcal{M}$, every w in $\mathcal{M}$ and every $\varphi \in \mathcal{L}$,

$$\mathcal{M}, w \models_{\mathcal{L}} \varphi \text{ iff } \mathcal{M}, w \models_{\mathcal{L}'} T(\varphi),$$

where $\mathcal{M}$ is seen as an $\mathcal{L}$ model on the left-hand side and as an $\mathcal{L}'$ model on the right-hand side, and, in each case, we use the appropriate semantic relation ($\models_{\mathcal{L}}$ and $\models_{\mathcal{L}'}$, respectively). Logic $\mathcal{L}'$ is *strictly more expressive* than logic $\mathcal{L}$ (in symbols, $\mathcal{L} < \mathcal{L}'$) if $\mathcal{L} \leq \mathcal{L}'$ but $\mathcal{L}' \not\leq \mathcal{L}$.

We will appeal to the notion of $\mathcal{PML}$ bisimulation defined above to show that $\mathcal{PML}$ is indeed strictly less expressive than $\mathcal{ML}_\emptyset$:

Proposition 3.3 $\mathcal{PML} < \mathcal{ML}_\emptyset$.

Proof Since $\mathcal{PML}$ is a syntactic fragment of $\mathcal{ML}_\emptyset$, we trivially have that $\mathcal{PML} \leq \mathcal{ML}_\emptyset$. To show that $\mathcal{ML}_\emptyset \not\leq \mathcal{PML}$, consider the models $\mathcal{M} = (W, R, V, \emptyset)$ and $\mathcal{N} = (W', R', V', \emptyset)$ below, where $V(q) = V(q') = \emptyset$ for all proposition letters q.

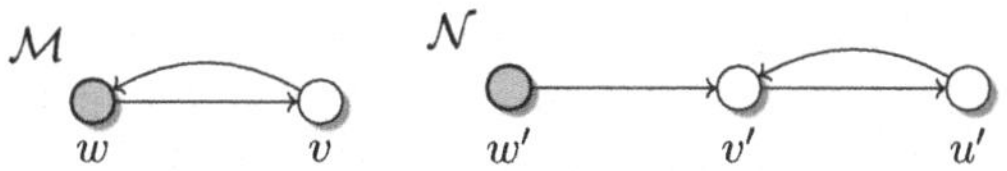

We have that $(\mathcal{M}, w)$ and $(\mathcal{N}, w')$ are $\mathcal{PML}$ bisimilar. However, $\mathcal{M}, w \models_{\mathcal{ML}_\emptyset}$ ⓡ$\Diamond\Diamond$ⓚ, while $\mathcal{N}, w' \not\models_{\mathcal{ML}_\emptyset}$ ⓡ$\Diamond\Diamond$ⓚ. □

Since it is a fragment of $\mathcal{ML}_\emptyset$, the logic $\mathcal{PML}$ is evidently a fragment of first-order logic (Areces 2007; Mera 2009). A direct translation into first-order logic is given below.

Proposition 3.4 *There is an effective meaning-preserving translation from* $\mathcal{PML}$ *into first-order logic.*

Proof We define a translation ST_y^X from $\mathcal{PML}$ formulas to first-order formulas, where y is a variable and X a finite set of variables:

$$\begin{aligned}
\mathsf{ST}_y^X(p) &= Py \\
\mathsf{ST}_y^X(\neg\varphi) &= \neg\mathsf{ST}_y^X(\varphi) \\
\mathsf{ST}_y^X(\varphi \vee \psi) &= \mathsf{ST}_y^X(\varphi) \vee \mathsf{ST}_y^X(\psi) \\
\mathsf{ST}_y^X(\Diamond\varphi) &= \exists z\big(Ryz \wedge \mathsf{ST}_z^X(\varphi)\big) \\
\mathsf{ST}_y^X(\langle \mathrm{p} \rangle\varphi) &= \exists z\big(Ryz \wedge \mathsf{ST}_z^{X\cup\{z\}}(\varphi)\big) \\
\mathsf{ST}_y^X(ⓟ) &= \bigvee_{x \in X} y = x.
\end{aligned}$$

Given a pointed model $(\mathcal{M}, w)$ and a finite set $D = \{d_1, \ldots, d_n\}$ of states in $\mathcal{M}$, it is easy to prove by induction on $\mathcal{PML}$ formulas φ and sets of variables $X = \{x_1, \ldots, x_n\}$ that

$$(\mathcal{M}, w) \models \varphi \text{ iff } \mathcal{M} \models \mathsf{ST}_y^X(\varphi)[y := w, x_1 := d_1, \ldots, x_n := d_n].$$

(Note that the set X in the above translation is used to keep track of the states that have already been poisoned.) □

Model checking. We now show that model checking for $\mathcal{PML}$ is PSPACE-complete. That it is at most PSPACE is established by the first-order translation given in Proposition 3.4, which results in a polynomial increase in length and can be done in polynomial time (alternatively, recall that $\mathcal{PML}$ is a fragment of the basic memory logic $\mathcal{ML}_{\emptyset}$). For the lower bound, we provide a polynomial-time reduction from the true quantified Boolean formula (QBF) problem. Given a QBF φ, we build a graph in which Traveller has a winning strategy in the poison game if and only if φ is true, and for which the existence of a winning strategy is expressible by a $\mathcal{PML}$ formula (where the size of the graph and of the formula increase only linearly in the size of φ). This also gives a reduction of QBF to the problem of determining the existence of a winning strategy in the poison game itself, and it thus provides a lower complexity bound for the game. This method of reduction by games for modal logics originates in Löding and Rohde (2003a) and Rohde (2005), where it was used in the study of the sabotage game: the method is worth noting, as it can yield simple and useful model checking results in dynamic modal logics more generally (see Benthem et al. 2019 for another application).

Theorem 3.5 *Model checking for $\mathcal{PML}$ is* PSPACE*-complete.*

Proof We reduce QBF to model checking for $\mathcal{PML}$. Recall that a *fully quantified Boolean formula* (QBF) is one of the form

$$Q_1 x_1 \ldots Q_n x_n \bigwedge_{1 \leq i \leq m} C_i,$$

where each $Q_j \in \{\exists, \forall\}$ and each C_i is a disjunction of literals $\pm x_j$ $(j \leq n)$. Without loss of generality, we can assume that $Q_1 = \exists$. The QBF problem consists in determining whether the formula is true when the quantifiers range over truth-value assignments to the variables x_j. For each given QBF φ, we construct a pointed model $(\mathcal{M}_\varphi, s)$ and a formula Γ_φ such that φ is true if and only if $\mathcal{M}_\varphi, s \models \Gamma_\varphi$. The model $\mathcal{M}_\varphi$ is constructed from basic modules as depicted in Fig. 1. We begin with the *initial module* (Fig. 1a), which contains the evaluation point s and corresponds to the initial quantifier $\exists x_1$. We then concatenate subsequent modules—each either a $\forall x_j$*-module* (Fig. 1c) or a $\exists x_j$*-module* (Fig. 1b)—for consecutive variables x_j, in the order corresponding to the order of quantifiers Q_j in φ.

Concatenating modules here simply amounts to identifying the last row of vertices in each module (with labels $a(x_j), x_j, \neg x_j$...) with the first row of vertices of the next

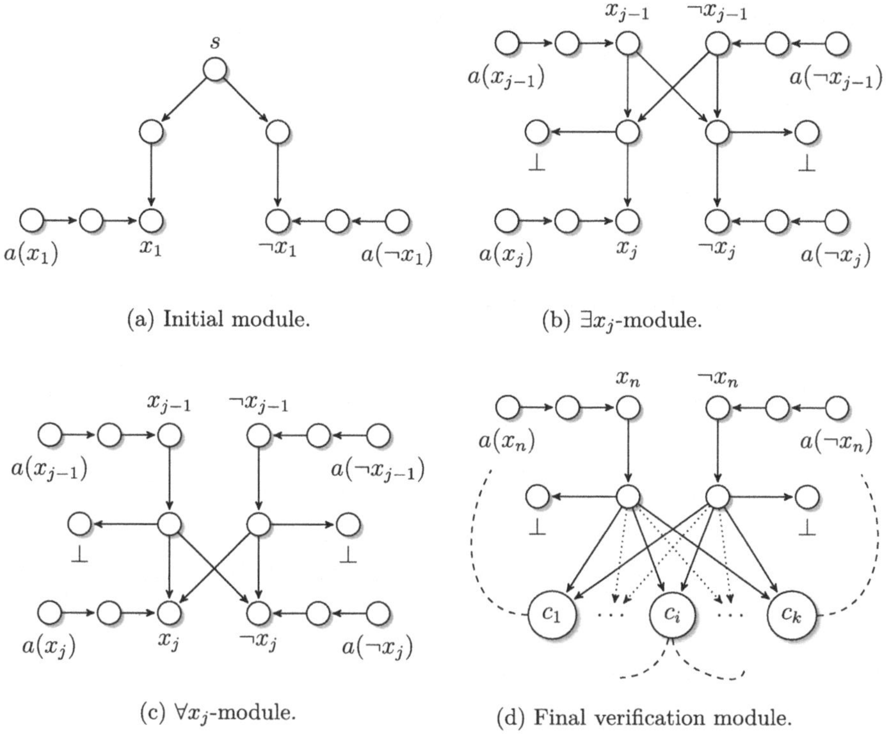

Fig. 1 The four basic modules. In (b), (c) and (d), the top nodes labelled by x_{j-1} and $\neg x_{j-1}$ (x_n and $\neg x_n$ in (d)) are the end nodes of the previous module. In (d), each clause vertex c_i has an outgoing edge to all and only the vertices $a(\ell)$, where ℓ is a literal $\pm x_i$ that makes clause C_i true

module: that is, for the top nodes of the j-th module we take the end nodes of the $(j-1)$-th module. Once all n quantifier modules have been added in this fashion, we append the *final verification module* (Fig. 1d) in the same way. Each clause C_i in φ is associated with exactly one node c_i in the verification module—its *clause vertex*. Each clause vertex c_i has an outgoing edge to all and only the vertices $a(\ell)$, where $\ell = \pm x_j$ is a literal that makes C_i true.

Consider now the poison game played on this graph. Traveller begins the game at s and makes the first move. By choosing to go left or right, she forces Poisoner to poison exactly one of the vertices labelled x_1 or $\neg x_1$. The same holds at each $\exists x_j$-module: Traveller has the power to determine which of x_j or $\neg x_j$ will be poisoned. Similarly, at $\forall x_j$-modules, Traveller makes the only available first move, after which it is Poisoner who chooses which one between x_j and $\neg x_j$ to poison (note that Poisoner never selects the endpoints, marked by $\bot$, since endpoints in the graph are winning positions for Traveller).

Observation Traveller *has a winning strategy for the poison game on* $(\mathcal{M}_\varphi, s)$ *if and only if the initial* QBF *formula* φ *is true.*

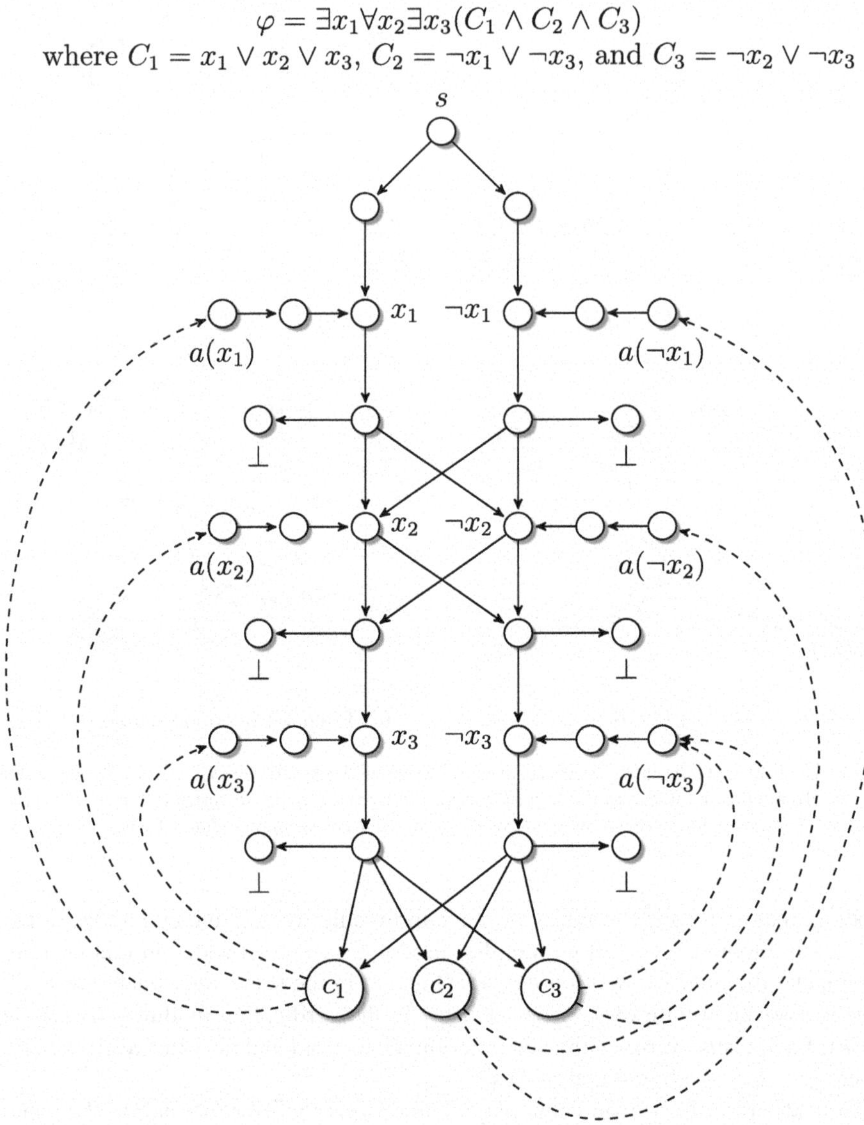

Fig. 2 An example. At each $\exists x_j$-module, Traveller forces Poisoner to poison a node labelled by the literal $\pm x_j$ of Traveller's choice. Each $\forall x_j$-module, on the other hand, leaves the choice to Poisoner. In the final verification module, Poisoner forces Traveller into some clause node c_i. For each literal $\pm x_j$ that makes the corresponding clause C_i true, Traveller can move to the node $a(\pm x_j)$ above. In this example, Traveller has a winning strategy which ensures that she can reach an endpoint $\perp$, where she wins

In this setting, Traveller winning the poison game is equivalent to the $\exists$-player winning the *formula game* on QBF formulas (see Sipser 2012). A useful heuristic for thinking about the game is as follows. Each passage through the graph all the way until the final verification module corresponds to selecting a valuation: for each j, exactly one of the vertices marked with x_j or $\neg x_j$ is poisoned. The non-poisoned vertices correspond to the selected valuation. Traveller aims at a final valuation that makes the QBF formula true, while Poisoner tries to make the formula false. If Traveller wants $\pm x_j$ to be true at a $\exists x_j$-module, she forces Poisoner to pass through $\mp x_j$, so that the vertex $\pm x_j$ is spared. Similarly, if Poisoner wishes to make $\pm x_j$ false at a $\forall x_j$-module, she makes sure to poison it. At the verification module, Poisoner selects a clause node c_i: at this point, either there is some vertex $a(\pm x_j)$ accessible from c_i for some non-poisoned $\pm x_j$ that makes C_i true, or $\pm x_j$ is poisoned for every accessible vertex $a(\pm x_j)$. In the former case, Traveller can move from c_i to such a vertex $a(\pm x_j)$, where $\pm x_j$ is not poisoned. After any three more moves in the game, Traveller then finds herself, at the beginning of her turn, at a point that sees an endpoint (marked by $\bot$ on the graph). She travels to the endpoint and wins the game. If, on the other hand, $\pm x_j$ is poisoned for every $\pm x_j$ that makes C_i true, Traveller is compelled to move to some $a(\pm x_j)$ where the vertex $\pm x_j$ is poisoned. At her next turn, Traveller loses the game, since she is forced to move to a poisoned vertex.

Lastly, we make sure that the existence of a winning strategy is expressible by a $\mathcal{PML}$ formula. Recall the formula ρ_k expressing survival for at least k rounds, given by the scheme

$$\rho_1 := [\mathrm{p}](\Box\bot \vee \Diamond\neg \textcircled{p})$$

$$\rho_k := [\mathrm{p}](\Box\bot \vee \Diamond(\neg \textcircled{p} \wedge \rho_{k-1})).$$

Given a QBF φ with n variables, take the $\mathcal{PML}$ formula

$$\Gamma_\varphi = \Diamond(\neg \textcircled{p} \wedge \rho_{n+3}).$$

It is easy to see that Traveller has a winning strategy if and only if $\mathcal{M}_\varphi, s \models \Gamma_\varphi$. The formula states that Traveller can make a first move to a non-poisoned point, after which she can survive for $n+3$ rounds in the poison game. This is equivalent to Traveller winning the poison game. Note that it takes exactly n rounds of the game to reach the last row of vertices from which the clause nodes are accessible. The $(n+1)$-st round starts with Poisoner selecting a clause vertex in the final verification module, and Traveller responding by selecting some accessible vertex $a(\ell)$. If all accessible points $a(\ell)$ have the vertex ℓ poisoned, then Traveller loses in the next, $(n+2)$-nd round, and Γ_φ fails. Otherwise, Traveller survives for one more round (and so Γ_φ holds), and now she has a guaranteed victory by moving to an endpoint in the $(n+3)$-rd round. The reduction is polynomial in the size of φ. Note that the size of the model $\mathcal{M}_\varphi$ grows linearly in the number of variables, and so does the size of Γ_φ: when φ has n variables and m clauses, we have that $|\mathcal{M}_\varphi| \leq \alpha(n+1)+m$, where α is the maximum size of a quantifier module, and the number of edges is bounded above by $\beta(n+1)+m(n+2)$, where β is the maximum number of edges is a non-final module. $\Box$

The construction just given reduces the truth of a QBF not only to the truth of a $\mathcal{PML}$ formula in a model, but also to the existence of a winning strategy in a poison game. Thus, the argument above gives us an immediate corollary:

Proposition 3.6 *The existence of a winning strategy for* Traveller *in the poison game is* PSPACE-*hard.*

Thus, what falls out of this simple analysis of model checking is a lower bound on the complexity of testing for the existence of a winning strategy in the poison game. In fact, Zhang (2019) has independently shown that this problem is PSPACE-complete (Fig. 2).

Satisfiability. We conclude this section by considering the satisfiability problem for $\mathcal{PML}$. In spite of being strictly less expressive than $\mathcal{ML}_{\emptyset}$ (Proposition 3.3), the satisfiability problem for $\mathcal{PML}$ is undecidable, just like the one for $\mathcal{ML}_{\emptyset}$.

We begin by showing that $\mathcal{PML}$ does not have the finite model property. To do so, we make use of the 'spy-point technique', first introduced by Blackburn and Seligman (1995) in the context of *hybrid logics*.

Theorem 3.7 *$\mathcal{PML}$ lacks the finite model property.*

Proof Consider the following formulas:

$$\begin{array}{rl}
(\mathit{Back}) & q \wedge \neg s \wedge \Diamond\top \wedge \Box(\neg q \wedge s \wedge \Diamond\top \wedge \Box(\neg q \wedge \neg s \wedge \Box\neg q)) \wedge [\mathrm{p}]\Box\Diamond\textcircled{p} \wedge [\mathrm{p}]\Box\Box(s \rightarrow \textcircled{p}) \\
(\mathit{Spy}) & [\mathrm{p}][\mathrm{p}][\mathrm{p}](\neg s \rightarrow \Diamond(s \wedge \textcircled{p} \wedge \Diamond(\neg s \wedge \textcircled{p} \wedge \Box(\neg s \rightarrow \neg\textcircled{p})))) \\
(\mathit{Irr}) & \Box[\mathrm{p}]\Box\neg\textcircled{p} \\
(\mathit{Succ}) & \Box\Box\Diamond\neg s \\
(\mathit{No\text{-}3cyc}) & [\mathrm{p}][\mathrm{p}][\mathrm{p}](\neg s \rightarrow [\mathrm{p}](\neg s \rightarrow \Box(\neg s \rightarrow \neg\textcircled{p}))) \\
(\mathit{Trans}) & [\mathrm{p}][\mathrm{p}]\Box(\neg s \rightarrow [\mathrm{p}](\neg s \rightarrow \Diamond(s \wedge \textcircled{p} \wedge \Diamond(\neg s \wedge \textcircled{p} \wedge \Diamond(\neg s \wedge \textcircled{p})))))).
\end{array}$$

Now, let *Inf* be the formula (*Back* $\wedge$ *Spy* $\wedge$ *Irr* $\wedge$ *Succ* $\wedge$ *No-3cyc* $\wedge$ *Trans*). We are going to show that, for any $\mathcal{PML}$ model $\mathcal{M} = (W, R, V, \emptyset)$ and any $w \in W$, if $\mathcal{M}, w \models$ *Inf*, then W must be infinite. So, suppose that $\mathcal{M}, w \models$ *Inf*. Then, by (*Back*), we know that w has a successor, call it v, that satisfies s. Any such successor of w will behave as a spy point. Now, (*Back*) ensures that w does not see itself, and that no spy point can access w. Moreover, (*Back*) guarantees that (i) spy points have at least one successor and that no spy point can see itself, (ii) all successors of a spy point can see this spy point back, and (iii) all successors of a spy point see exactly one spy point. (*Spy*) then ensures that all successors of a successor of a spy point see this spy point back and are directly seen by it. By (*Irr*), all successors of a spy point are irreflexive and, by (*Succ*), all successors of a spy point see a non-spy point (which cannot be w). (*Back*), (*Spy*), (*Irr*) and (*No-3cyc*) together ensure that there are no 2-cycles and no 3-cycles among the successors of a spy point. Finally, (*Trans*)—together with (*Back*), (*Spy*), (*Irr*), (*Succ*) and (*No-3cyc*)—guarantees that the relation R is transitive on the set of successors of a spy point. Now, consider the spy point v: it follows from the above reasoning that its set of successors is an unbounded strict partial order, which entails that W is infinite. Lastly, the model depicted in Fig. 3 establishes that (*Inf*) is indeed satisfiable. □

Fig. 3 A model for *Inf*

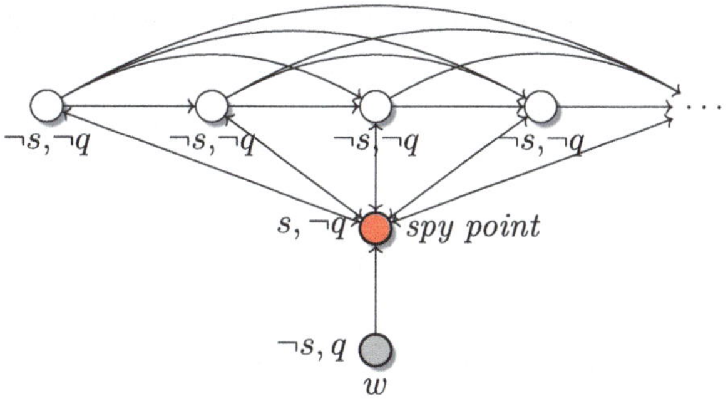

To prove undecidability, we will encode the $\omega \times \omega$ tiling problem within our language. In doing so, we shall once again employ a spy-point argument. To streamline the presentation of the proof, we will at first help ourselves to three accessibility relations: R_s, R_u and R_r—where R_s will be used to model transitions from/to the spy point to/from the grid points, R_u for upward transitions along the grid, and R_r for rightward transitions along the grid. We will thus show that the satisfiability problem for $\mathcal{PML}(\Diamond_s, \Diamond_u, \Diamond_r)$ is undecidable. After presenting this argument, we will then explain how the proof of Theorem 3.8 can be turned into a proof of the fact that the satisfiability problem for $\mathcal{PML}$, where only one accessibility relation is available, is undecidable, too.

Theorem 3.8 *The satisfiability problem for $\mathcal{PML}(\Diamond_s, \Diamond_u, \Diamond_r)$ is undecidable.*

Proof Let $\mathcal{T} = \{T_1, \ldots, T_n\}$ be a finite set of tile types. Given a tile type T_i, $u(T_i)$, $r(T_i)$, $d(T_i)$ and $l(T_i)$ will represent the colours of the upper, right, lower and left edges of T_i, respectively. For each tile type T_i, we fix a proposition letter t_i that is going to encode T_i. We now define a formula $\varphi_{\mathcal{T}}$ such that $\varphi_{\mathcal{T}}$ is satisfiable if and only if $\mathcal{T}$ tiles $\omega \times \omega$. Consider the following formulas:

(*Back*) $q \wedge \neg s \wedge \Diamond_s \top \wedge \Box_s(\neg q \wedge s \wedge \Diamond_s \top \wedge \Box_s(\neg q \wedge \neg s)) \wedge [\mathrm{p}]_s \Box_s \Diamond_s (s \wedge \textcircled{p}) \wedge$
$[\mathrm{p}]_s \Box_s \Box_s (s \wedge \textcircled{p})$

(*Spy*) $[\mathrm{p}]_s [\mathrm{p}]_s [\mathrm{p}]_u (\neg q \wedge \neg s \wedge \Diamond_s (s \wedge \textcircled{p} \wedge \Diamond_s (\textcircled{p} \wedge \Box_u \neg \textcircled{p} \wedge \Box_r \neg \textcircled{p})))$
$[\mathrm{p}]_s [\mathrm{p}]_s [\mathrm{p}]_r (\neg q \wedge \neg s \wedge \Diamond_s (s \wedge \textcircled{p} \wedge \Diamond_s (\textcircled{p} \wedge \Box_r \neg \textcircled{p} \wedge \Box_u \neg \textcircled{p})))$

(*Grid*) $\Box_s \Box_s (\Diamond_u \top \wedge \Diamond_r \top)$

(*Func*) $\Box_s [\mathrm{p}]_s [\mathrm{p}]_u \Diamond_s (s \wedge \Diamond_s (\textcircled{p} \wedge \Diamond_u \textcircled{p} \wedge \Box_u \textcircled{p}))$
$\Box_s [\mathrm{p}]_s [\mathrm{p}]_r \Diamond_s (s \wedge \Diamond_s (\textcircled{p} \wedge \Diamond_r \textcircled{p} \wedge \Box_r \textcircled{p}))$

$$(UR\text{-}no\text{-}Cycle) \quad \Box_s[\mathrm{p}]_s[\mathrm{p}]_u\Box_r\neg\circledS \wedge \Box_s[\mathrm{p}]_s[\mathrm{p}]_r\Box_u\neg\circledS$$

$$(URU\text{-}no\text{-}Cycle) \quad \Box_s[\mathrm{p}]_s[\mathrm{p}]_u[\mathrm{p}]_r\Box_u\neg\circledS$$

$$(Confl) \quad \Box_s[\mathrm{p}]_s\langle\mathrm{p}\rangle_u\langle\mathrm{p}\rangle_r\Diamond_s\Diamond_s(\circledS \wedge \Box_r\neg\circledS \wedge \Diamond_u\circledS \wedge \Diamond_r\Diamond_u(\circledS \wedge \Box_r\neg\circledS))$$

$$(Unique) \quad \Box_s\Box_s\Big(\bigvee_{1\le i\le n} t_i \wedge \bigwedge_{1\le i<j\le n}(t_i \to \neg t_j)\Big)$$

$$(Vert) \quad \Box_s\Box_s \bigwedge_{1\le i\le n}\Big(t_i \to \Diamond_u \bigvee_{1\le j\le n, u(T_i)=d(T_j)} t_j\Big)$$

$$(Horiz) \quad \Box_s\Box_s \bigwedge_{1\le i\le n}\Big(t_i \to \Diamond_r \bigvee_{1\le j\le n, r(T_i)=l(T_j)} t_j\Big).$$

Now, let $\varphi_{\mathcal{T}}$ be the conjunction of all of the formulas above.

($\Rightarrow$) Suppose that $\mathcal{M}, w \models \varphi_{\mathcal{T}}$, for some $\mathcal{PML}(\Diamond_s, \Diamond_u, \Diamond_r)$ model $\mathcal{M} = (W, R_s, R_u, R_r, V, \emptyset)$ and $w \in W$. The formula (*Back*) and the two (*Spy*) formulas ensure that w has a successor that is a spy point via the relation R_s. They also guarantee that (1) w is not a successor of the spy point; (2) w is not a successor of any successor of the spy point; (3) R_s, R_u and R_r are irreflexive; and (4) R_u and R_r are asymmetric. The points that are R_s-accessible from the spy point represent the tiles. The formula (*Grid*)—together with (*Back*) and (*Spy*)—ensures that every tile (i.e. every point accessible from the spy point) has a tile above it, via the R_u relation, and a tile to its right, via the R_r relation. The two (*Func*) formulas—together with (*Back*) and (*Spy*)—on the other hand, guarantee that R_u and R_r are functional: namely, that every tile has at most one tile above it and at most one tile to its right. So, (*Grid*) and (*Func*) together ensure that every tile has exactly one tile above it and exactly one tile to its right. Now, (*UR-no-Cycle*) ensures that no tile can be both above/below and to the left/right of another tile, while (*URU-no-Cycle*) forbids cycles following successive steps of the R_u, R_r and R_u relations, in this order. Together with (*UR-no-Cycle*) and (*URU-no-Cycle*), (*Conf*) then ensures that the tiles are arranged in a grid. Finally, (*Unique*) guarantees that every tile has a unique type, while (*Vert*) and (*Horiz*) ensure that the colours of the tiles match appropriately. It then follows that $\mathcal{M}$ yields a tiling of $\omega \times \omega$ (Fig. 4).

($\Leftarrow$) For the other direction, suppose that $f : \omega \times \omega \to \mathcal{T}$ is a tiling of $\omega \times \omega$. Let $\mathcal{M}$ be the $\mathcal{PML}(\Diamond_s, \Diamond_u, \Diamond_r)$ model $(\omega \times \omega \cup \{w, v\}, R_s, R_u, R_r, V, \emptyset)$, where the accessibility relations R_s, R_u and R_r are given by

$$R_s = \{(w, v)\} \cup \{(v, x), (x, v) : x \in \omega \times \omega\}$$
$$R_u = \{((n, m), (n, m + 1)) : n, m \in \omega\}$$
$$R_r = \{((n, m), (n + 1, m)) : n, m \in \omega\}$$

and the valuation V is defined as

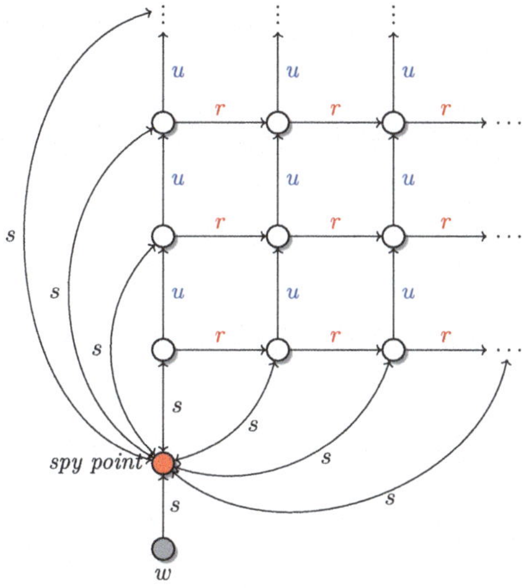

Fig. 4 A model for φ_T: a $\omega \times \omega$ grid with a spy point

$$\begin{aligned} V(q) &= \{w\} \\ V(s) &= \{v\} \\ V(t_i) &= \{x : x \in \omega \times \omega \text{ and } f(x) = T_i\} \text{ for all } 1 \leq i \leq n \\ V(p) &= \emptyset \text{ for all other proposition letters } p. \end{aligned}$$

The above specification of $\mathcal{M}$ ensures that v is a spy point. By construction, we then have that $\mathcal{M}, w \models \varphi_T$. □

So, $\mathcal{PML}(\Diamond_s, \Diamond_u, \Diamond_{r.})$ is undecidable. The additional modalities render the encoding less cumbersome, but, as a matter of fact, they are not required for undecidability: an argument analogous to the one given by Hoffmann (2015) can be employed to adapt our proof of Theorem 3.8 to the original $\mathcal{PML}$ language with only one accessibility relation R. The basic idea for such a modification consists in using proposition letters u and r to appropriately encode the relations R_u and R_r.

Theorem 3.9 *The satisfiability problem for $\mathcal{PML}$ is undecidable.*

Proof Consider the set of proposition letters $V := \{0, 1, 2, q, s, r, u\}$. For any $\ell \in V$, let $\mathsf{X}_\ell := \ell \wedge \bigwedge_{v \in V \setminus \{\ell\}} \neg v$ be the formula stating that ℓ holds and all other atomic propositions in V are false. For $i \in \{0, 1, 2\}$, let $\mathsf{n}(i) = i + 1 \pmod 3$ and $\mathsf{e}(i) = i + 2 \pmod 3$. Now, consider the following formulas, where $i \in \{0, 1, 2\}$ and $a \in \{r, u\}$:

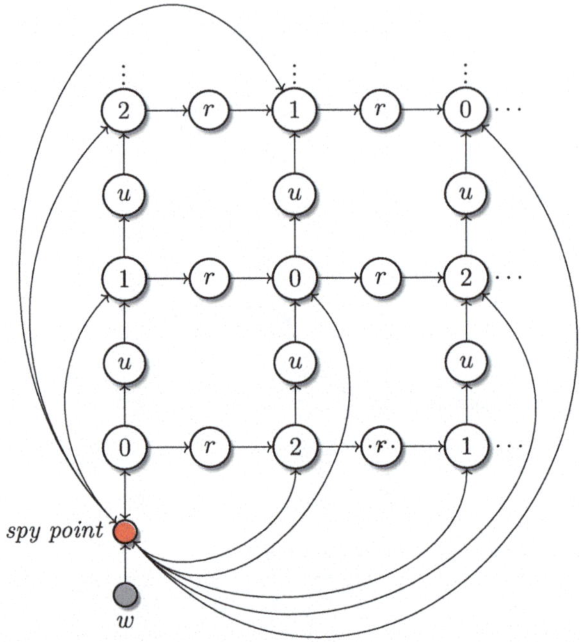

Fig. 5 Encoding the grid with a single accessibility relation

$$
\begin{array}{rl}
(\mathit{Grid}_1) & \mathsf{X}_q \wedge \Diamond\top \wedge \Box(\mathsf{X}_s \wedge \Diamond\top \wedge \Box(\mathsf{X}_0 \vee \mathsf{X}_1 \vee \mathsf{X}_2)) \\
(\mathit{Grid}_2) & \Box\Box(\Diamond\mathsf{X}_r \wedge \Diamond\mathsf{X}_u \wedge \Box(\mathsf{X}_r \vee \mathsf{X}_u \vee \mathsf{X}_s)) \\
(\beta_i) & \Box\Box\Big(i \to \Box((r \to \Diamond\mathsf{e}(i) \wedge \Box\mathsf{X}_{\mathsf{e}(i)}) \wedge (u \to \Diamond\mathsf{n}(i) \wedge \Box\mathsf{X}_{\mathsf{n}(i)}))\Big) \\
(\mathit{Back}) & [\mathtt{p}]\Box(\Diamond(s \wedge \textcircled{p}) \wedge \Box(s \to \textcircled{p})) \\
(\mathit{Spy}) & [\mathtt{p}][\mathtt{p}][\mathtt{p}]\Big(a \to [\mathtt{p}]\Diamond(s \wedge \textcircled{p} \wedge \Diamond(\textcircled{p} \wedge \Box(a \to \neg\textcircled{p})))\Big) \\
(\mathit{Func}) & \Box[\mathtt{p}][\mathtt{p}]\Big(a \to [\mathtt{p}]\Diamond(s \wedge \Diamond(\textcircled{p} \wedge \Diamond(\textcircled{p} \wedge a \wedge \Diamond\textcircled{p}) \wedge \Box(a \to \textcircled{p} \wedge \Box\textcircled{p})))\Big).
\end{array}
$$

We then also add the formulas (*UR-no-Cycle**), (*Conf**), (*Unique**), (*Vert**) and (*Horiz**), each of which is obtained by a simple translation scheme which replaces every multimodal $\mathcal{PML}(\Diamond_s, \Diamond_r, \Diamond_u)$ formula φ by a standard $\mathcal{PML}$ formula φ^*. The translation $\varphi \mapsto \varphi^*$ leaves Boolean formulas unchanged and, otherwise, is defined as follows:

$$
\begin{array}{lll}
(\Box_s\varphi)^* & = \Box\varphi^* & \\
([\mathtt{p}]_s\varphi)^* & = [\mathtt{p}]\varphi^* & \\
(\Box_a\varphi)^* & = \Box(a \to \Box\varphi^*) & \text{for } a \in \{r, u\} \\
([\mathtt{p}]_a\varphi)^* & = [\mathtt{p}](a \to [\mathtt{p}]\varphi^*) & \text{for } a \in \{r, u\}.
\end{array}
$$

The conjunction of the above formulas is satisfiable if and only if the corresponding instance of the tiling problem has a solution. The formula forces a grid-like structure, as depicted in Fig. 5. Each point in the valuation of $0, 1, 2$ is assigned a tile type: the type of each such point i is required to match the type of its rightward successor $i + 2 \pmod 3$ and of its upward successor $i + 1 \pmod 3$. The rest of the argument is then analogous to the undecidability proof for $\mathcal{PML}(\Diamond_s, \Diamond_r, \Diamond_u)$. □

4 $\mathcal{PSL}$

Besides undecidability, another reason why one might be unhappy with $\mathcal{PML}$ as a language for modelling the poison game concerns the $\textcircled{p}$ operator. When used in conjunction with $\Diamond$, $\textcircled{p}$ allows to talk about Traveller's moves. However, $\textcircled{p}$ by itself does not have a counterpart in the poison game: neither player is allowed to roam the graph without poisoning, simply to check whether certain vertices are poisoned or not.

To overcome these difficulties, we will now consider the fragment of $\mathcal{PML}$ (and, a fortiori, $\mathcal{ML}_\emptyset$) that does not feature the basic $\Diamond$ modality, but which includes $\langle \mathtt{p} \rangle$ and the operator $\langle \mathtt{t} \rangle$, defined as $\langle \mathtt{t} \rangle \varphi \leftrightarrow \Diamond(\neg \textcircled{p} \wedge \varphi)$. As before, the $\langle \mathtt{p} \rangle$ operator captures Poisoner's moves. The $\langle \mathtt{t} \rangle$ modality, on the other hand, captures Traveller's *safe* moves: i.e. moves along edges that do not lead to a poisoned state. We will refer to this logic as $\mathcal{PSL}$.

Expressive power. Even though there are no modalities that prompt the explicit deletion of edges, $\mathcal{PSL}$ is rather close to sabotage modal logic in spirit.[5] This is because we could also think of $\langle \mathtt{t} \rangle$ as the modality associated with a second graph relation, one that corresponds to Traveller's 'safe accessibility' relation and which shrinks over time, as more and more states get poisoned. Under this interpretation, the poison modality $\langle \mathtt{p} \rangle$ behaves analogously to the sabotage modality, in that poisoning moves result in the deletion of links from the safe accessibility relation, just as sabotaging moves trigger the deletion of links from the basic graph relation. The crucial difference between sabotage modal logic and $\mathcal{PSL}$, however, is that, while sabotaging only allows to remove one link at a time, poisoning prompts the deletion of *all* safe links leading to the poisoned state.

It is worth noting that $\langle \mathtt{p} \rangle$ and $\langle \mathtt{t} \rangle$ agree on atomic propositions and Boolean formulas, since here we restrict attention to the class of initial models where $P = \emptyset$. However, this is clearly not the case for arbitrary formulas. For a simple example, consider the formula $\langle \mathtt{p} \rangle \langle \mathtt{t} \rangle q \leftrightarrow \langle \mathtt{t} \rangle \langle \mathtt{t} \rangle q$, and the model $\mathcal{M} = (W, R, V, \emptyset)$ below, where $V(q) = \{v\}$. We then have that $\mathcal{M}, w \models \langle \mathtt{t} \rangle \langle \mathtt{t} \rangle q$, but $\mathcal{M}, w \not\models \langle \mathtt{p} \rangle \langle \mathtt{t} \rangle q$.

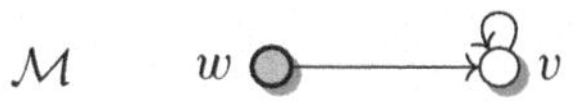

[5]This is why the logic is called $\mathcal{PSL}$, which is an abbreviation for *poison sabotage logic*.

Like $\mathcal{ML}_\emptyset$ and $\mathcal{PML}$, $\mathcal{PSL}$ can express that Traveller has a strategy for surviving at least n rounds of a poison game via the inductive scheme below:

$$\rho_1 := [\mathtt{p}]([\mathtt{p}]\bot \vee \langle \mathtt{t}\rangle\top)$$

$$\rho_n := [\mathtt{p}]([\mathtt{p}]\bot \vee \langle \mathtt{t}\rangle\rho_{n-1}).$$

We can also express the property that (i) the current state has a safely accessible successor that has itself as its only safely accessible successor, and that (ii) every state that is safely accessible from the current state is either an endpoint or has itself as its only safely accessible successor via the formula $\langle \mathtt{t}\rangle\langle \mathtt{t}\rangle\top \wedge [\mathtt{p}][\mathtt{t}]\bot$. It then follows that the tree model property fails for $\mathcal{PSL}$, too.

Now, note that the notion of $\mathcal{PML}$ bisimulation defined in Sect. 3 can be easily adapted to the case of $\mathcal{PSL}$ by replacing the standard clauses for $\Diamond$ with the following clauses for $\langle \mathtt{t}\rangle$:

Zig$_{\langle \mathtt{t}\rangle}$: if $(S, u)Z(S', u')$ and there exists $v \in W$ with $(u, v) \in R$ and $v \notin S$, then there exists $v' \in W'$ with $(u', v') \in R'$ and $v' \notin S'$, and $(S, v)Z(S', v')$;
Zag$_{\langle \mathtt{t}\rangle}$: if $(S, u)Z(S', u')$ and there exists $v' \in W'$ with $(u', v') \in R'$ and $v' \notin S'$, then there exists $v \in W$ with $(u, v) \in R$ and $v \notin S$, and $(S, v)Z(S', v')$.

Next, we make sure that the above definition is correct:

Proposition 4.1 *Let $\mathcal{M} = (W, R, V, P)$ and $\mathcal{N} = (W', R', V', P')$ be two $\mathcal{PSL}$ models, $w \in W$ and $w' \in W'$. If Z is a bisimulation linking (P, w) and (P', w'), then, for any $\varphi \in \mathcal{PSL}$,*

$$\mathcal{M}, w \models \varphi \text{ iff } \mathcal{N}, w' \models \varphi.$$

Proof The only case that requires checking is the one involving $\langle \mathtt{t}\rangle$. Suppose that $\mathcal{M}, w \models \langle \mathtt{t}\rangle\varphi$. Then, by the semantics of $\langle \mathtt{t}\rangle$, there is $v \in W$ with $(w, v) \in R$, $v \notin P$ and $\mathcal{M}, v \models \varphi$. By the **Zig**$_{\langle \mathtt{t}\rangle}$ condition, we then have that there exists $v' \in W'$ with $(w', v') \in R'$, $v' \notin P'$ and $(P, v)Z(P', v')$. By the induction hypothesis, we can then conclude that $\mathcal{N}, v' \models \varphi$. Hence, $\mathcal{N}, w' \models \langle \mathtt{t}\rangle\varphi$. The other direction is analogous, except that it instead relies on the **Zag**$_{\langle \mathtt{t}\rangle}$ condition. □

At this point, it is natural to ask how $\mathcal{PSL}$ compares with $\mathcal{PML}$ in terms of expressivity. Using the above notion of $\mathcal{PSL}$ bisimulation, we can show that $\mathcal{PSL}$ is strictly less expressive than $\mathcal{PML}$:

Proposition 4.2 $\mathcal{PSL} < \mathcal{PML}$.

Proof Since $\mathcal{PSL}$ is a syntactic fragment of $\mathcal{PML}$, we trivially have that $\mathcal{PSL} \leq \mathcal{PML}$. To show that $\mathcal{PML} \not\leq \mathcal{PSL}$, consider the two infinite binary trees $\mathcal{M} = (W, R, V, \emptyset)$ and $\mathcal{N} = (W', R', V', \emptyset)$ shown below, where $V(q) = V'(q) = \emptyset$ for all proposition letters q.

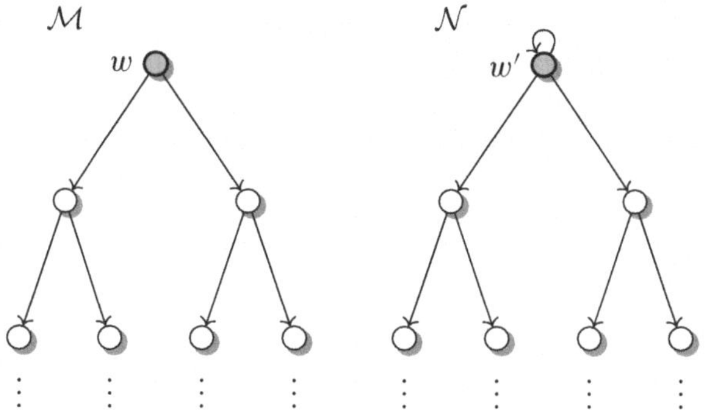

The two pointed models $(\mathcal{M}, w)$ and $(\mathcal{N}, w')$ are $\mathcal{PSL}$ bisimilar. However, we have that $\mathcal{N}, w' \models_{\mathcal{PML}} \langle \mathtt{p} \rangle \Diamond \textcircled{p}$, while $\mathcal{M}, w \not\models_{\mathcal{PML}} \langle \mathtt{p} \rangle \Diamond \textcircled{p}$. □

Just as in the case of $\mathcal{PML}$, we can define a direct translation ST_y^X from $\mathcal{PSL}$ formulas to first-order formulas, where, once again, y is a variable and X a finite set of variables. We only give the clause for the $\langle \mathtt{t} \rangle$ operator:

$$\mathsf{ST}_y^X(\langle \mathtt{t} \rangle \varphi) = \exists z \left(Ryz \wedge \bigwedge_{x \in X} \neg(z = x) \wedge \mathsf{ST}_z^X(\varphi) \right).$$

Model checking. The model-checking complexity of $\mathcal{PSL}$ can be analysed in the exact same way as that of $\mathcal{PML}$. The standard translation into first-order logic (or simply the fact that $\mathcal{PSL}$ is a fragment of $\mathcal{ML}_{\emptyset}$) provides a PSPACE upper bound, while the lower bound can be shown via the same reduction from the true QBF problem. For the latter, it is sufficient to note that $\mathcal{PSL}$ can express survival for at least n rounds.

Theorem 4.3 *Model checking for $\mathcal{PSL}$ is* PSPACE*-complete.*

Proof By the same reduction as in Theorem 3.5: given a QBF φ with n variables, construct the pointed model $(\mathcal{M}_\varphi, s)$. As we saw, $\mathcal{PSL}$ can express, through the formula ρ_n given above, the existence of a survival strategy for Traveller for at least n rounds. Then, Traveller has winning strategy in the poison game on $(\mathcal{M}_\varphi, s)$ if and only if $\mathcal{M}_\varphi, s \models \langle \mathtt{t} \rangle \rho_{n+3}$. □

The logic $\mathcal{PSL}$ appears to be particularly well suited for talking about the poison game. It allows to describe each stage of the game from the local perspective of the vertex currently occupied by the players, and quantification is restricted so as to precisely match the possible moves of the players: the $\langle \mathtt{t} \rangle$ and $\langle \mathsf{p} \rangle$ modalities capture exactly statements of the form 'there is an available move by Traveller/Poisoner resulting in φ'. Note, in particular, the elegant way in which $\mathcal{PSL}$ expresses n-round survival for the two players: with every application of a modal operator corresponding to a player's move, the language allows to talk about the game steps with no frills. Thus, $\mathcal{PSL}$ seems to have a strong claim for being the most appropriate logic, at least among the memory logics discussed in this paper, for modelling the poison game from a local, stepwise perspective.

These considerations render the decidability question for $\mathcal{PSL}$ especially poignant. The question is open: should $\mathcal{PSL}$ satisfiability be decidable, we would then have a logic that strikes a very pleasing balance between close fit with the original game, expressivity and good computational behaviour.

5 Conclusion

In this paper, we studied three logics for modelling the poison game, moving from the memory logic $\mathcal{ML}_{\emptyset}$ to two fragments thereof: $\mathcal{PML}$ and $\mathcal{PSL}$. We showed that these logics form a chain in expressive power, with $\mathcal{PSL} < \mathcal{PML} < \mathcal{ML}_{\emptyset}$, and we introduced suitable notions of bisimulation for the two new logics presented in this paper. We proved that $\mathcal{PML}$, while strictly less expressive than $\mathcal{ML}_{\emptyset}$, has a PSPACE-complete model-checking problem and an undecidable satisfiability problem. We also showed that model checking for $\mathcal{PSL}$, a more sabotage-style logic for describing the poison game, is PSPACE-complete, and we concluded by identifying a natural open question: namely, whether the satisfiability problem for $\mathcal{PSL}$ is decidable.

Our results indicate that methods from modal logic are indeed well suited for modelling graph games such as the poison game and the sabotage game—and, more broadly, 'evolving' relational structures. They also suggest the adoption of a useful technical perspective for this wider purpose: the systematic use of memory logics and, more generally, of techniques from hybrid logics (such as the spy-point method), in addition to game-reduction techniques for analysing model-checking complexity (Löding and Rohde 2003a). Lastly, our findings invite a broad methodological question concerning the emerging study of modal logics for graph games: which logics are the natural candidates for studying a given class of games, and how do such design choices affect significant properties of these logics?

Acknowledgements We would like to thank Johan van Benthem for many helpful suggestions and inspiring discussions, the organisers and participants of the 4th Asian Workshop on Philosophical Logic (AWPL 2018) held at Tsinghua University, as well as the anonymous referees for their feedback and valuable comments.

References

Areces, C. 2007. Hybrid logics: the old and the new. In the *Proceedings of LogKCA-07*, 15–29, San Sebastian.

Areces, C., D. Figueira, S. Figueira, and S. Mera. 2011. The expressive power of memory logics. *Review of Symbolic Logic* 4 (2): 290–318.

Areces, C., R. Fervari, and G. Hoffmann. 2015. Relation-changing modal operators. *Logic Journal of the IGPL* 23: 601–627.

Aucher, G., J. van Benthem, and D. Grossi. 2015. Sabotage modal logic: some model and proof theoretic aspects. In the *Proceedings of the 5th International Workshop on Logic, Rationality and Interaction (LORI '15)*, 1–13, Taipei.

Aucher, G., J. van Benthem, and D. Grossi. 2017. Modal logics of sabotage revisited. *Journal of Logic and Computation* 28 (2): 269–303.

van Benthem, J. 2005. An Essay on Sabotage and Obstruction. In *Mechanizing Mathematical Reasoning*, vol. 2605, ed. D. Hutter, and W. Stephan, 268–276., Lecture Notes in Computer Science.

van Benthem, J. 2014. *Logic in Games*. Cambridge, MA: The MIT Press.

Benthem, J., K. Mierzewski, and F. Zaffora Blando. 2019. The modal logic of stepwise removal. Under review.

Blackburn, P., and J. Seligman. 1995. Hybrid languages. *Journal of Logic, Language and Information, Special issue on decompositions of first-order logic* 4 (3): 251–272.

Duchet, P., and H. Meyniel. 1993. Kernels in directed graph: a poison game. *Discrete Mathematics* 115: 273–276.

Ehrenfeucht, A. 1961. An application of games to the completeness problem for formalized theories. *Fundamenta Mathematicae* 49: 129–141.

Fraïssé, R. 1954. Sur quelques classifications des systèmes de relations. *Publications des Sciences de l'Université de l'Algérie, Série* A1: 35–182.

Grossi, D., and S. Rey. 2019. Credulous acceptability, poison games and modal logic. To appear in the *Proceedings of SYSMICS 2019*.

Hintikka, J. 1973. *Logic, language-games and information: Kantian themes in the philosophy of logic*. Oxford: Clarendon Press.

Hodges, W. 2006. *Building models by games*. Mineola, NY: Dover Publications.

Hodges, W. 2013. Logic and Games. *The Stanford Encyclopedia of Philosophy*, Zalta Edward, N. (ed.) Spring 2013th ed.

Hoffmann, G. 2015. Undecidability of a Very Simple Modal Logic with Binding. Preprint.

Löding, C., and P. Rohde. 2003a. Model checking and satisfiability for sabotage modal logic. In , *Proceedings of the 23rd Conference on the Foundations of Software Technology and Theoretical Computer Science (FSTTCS 2003)*, Pandya, P.K., and J.. Radhakrishnan, eds. Mumbai, India, December 15-17, 2003, Lecture Notes in Computer Science 2914, 302–313, Springer.

Löding, C., and P. Rohde. 2003b. Solving the Sabotage Game is PSPACE-hard. In *Proceedings of the 28th International Symposium on the Mathematical Foundations of Computer Science (MFCS 2003)*, Rovan, B., and Vojtás, P. eds. Bratislava, Slovakia, August 25-29, 2003, Lecture Notes in Computer Science 2747, 531–540, Springer.

Lorenzen, P. 1955. *Einführung in die Operative Logik und Mathematik*. Berlin: Springer.

Mera, S. 2009. Modal Memory Logics. Ph.D. Diss., Universidad de Buenos Aires, Buenos Aires, Argentina and Université Henri Poincaré, Nancy, France.

Mierzewski, K., and F. Zaffora Blando. 2016. The Modal Logic(s) of Poison Games. Manuscript, Stanford University.

Rohde, P. 2005. On Games and Logics over Dynamically Changing Structures. Ph.D. Dissertation, RWTH Aachen University, Aachen, Germany.

Sipser, M. 2012. *Introduction to the theory of computation*, 3rd ed. Boston: Cengage Learning.

Vreeswijk, G., and H. Prakken. 2000. Credulous and Sceptical Argument Games for Preferred Semantics. In the *Proceedings of the 7th European Workshop on Logic for Artificial Intelligence (JELIA '00), LNAI*, 239–253.

Zhang, T. 2019. Solution Complexity of Local Variants of Sabotage Game. To appear in the *Proceedings of the Workshop on Logics for the Formation and Dynamics of Social Norms (LFDSN 2019)*, Zheijang University, Hangzhou, China, May 4-5.

A Logical Characterization of the Continuous Bar Induction

Makoto Fujiwara and Tatsuji Kawai

Abstract The continuous bar induction (c-BI) is an instance of the monotone bar induction (MBI) which is constructively equivalent to the statement that every pointwise continuous function from $\mathbb{N}^{\mathbb{N}}$ to $\mathbb{N}$ is induced by an inductively generated neighborhood function. In this paper, we give a simple logical characterization of c-BI that c-BI is equivalent to the restriction of MBI to Π_1^0 bars over intuitionistic elementary analysis. We also show that the difference between c-BI and the classical bar induction (without monotonicity on the bar) can be captured by the lesser limited principle of omniscience (LLPO) in the special case when the side-predicates of bar induction are restricted to bounded Σ_2^0 predicates, respectively.

Keywords Bar induction · Intuitionistic mathematics · Neighborhood function · Constructive reverse mathematics · LLPO

MSC2010: 03F55 · 03F35 · 03B30 · 03B20

1 Introduction

Inductively generated neighborhood functions provide a fundamental notion of continuity on Baire space in Brouwer's intuitionism; in particular they play an important role in the theory of choice sequences (cf. Section 4.8 and Chapter 12 of Troelstra and van Dalen 1988 and Kreisel and Troelstra 1970). Based on this notion, in Kawai

This chapter is in its final form and it is not submitted to publication anywhere else.

M. Fujiwara (✉)
School of Science and Technology, Meiji University, 1-1-1 Higashi-Mita, Tama-ku, Kawasaki-shi, Kanagawa 214-8571, Japan
e-mail: makotofujiwara@meiji.ac.jp

T. Kawai
School of Information Science, Japan Advanced Institute of Science and Technology, 1-1 Asahidai, Nomi, Ishikawa 923-1292, Japan
e-mail: tatsuji.kawai@jaist.ac.jp

F. Liu et al. (eds.), *Knowledge, Proof and Dynamics*,
Logic in Asia: Studia Logica Library,
https://doi.org/10.1007/978-981-15-2221-5_2

(2019), the second author formulated the following principle of continuity on Baire space:

(UCB) Every pointwise continuous function from $\mathbb{N}^{\mathbb{N}}$ to $\mathbb{N}$ is induced by an inductively generated neighborhood function,

and showed that UCB is constructively equivalent to a version on the monotone bar induction called continuous bar induction c-BI, in which the bar predicate is required to be a c-bar.[1] Here, a bar P is a *c-bar* if

$$\forall a\,[P(a) \leftrightarrow \forall b\gamma(a) = \gamma(a * b)]$$

for some function γ over the finite sequences of natural numbers. As explained there, the principles UCB can be considered as a generalization of the uniform continuity principle:

(UC) Every pointwise continuous function from $\{0, 1\}^{\mathbb{N}}$ to $\mathbb{N}$ is uniformly continuous,

which is known to be equivalent to the fan theorem for c-sets c-FAN (see Berger 2006). Hence, the equivalence of UCB and c-BI can be considered as a generalization of the equivalence between UC and c-FAN to the setting of Baire space.

Although c-BI is equivalent to a mathematical statement like UCB that is quite natural from the intuitionistic point of view, the formulation of c-BI as presented above is awkward and difficult to deal with. The purpose of this paper is to give a simple characterization of c-BI in terms of the logical complexity of the bar. Specifically, in the spirit of constructive reverse mathematics Ishihara (2005), we show that c-BI is equivalent to Π^0_1-MBI, which is the restriction of the monotone bar induction MBI to Π^0_1 bars. We also show that c-BI admits a similar characterization as that of c-FAN based on the notion of c-set in the sense of Berger (2006); see Theorem 3.2. Furthermore, it is also natural to ask the relation between c-BI and Π^0_1-BI, which is a principle obtained from Π^0_1-MBI by omitting the monotonicity condition on bars. In the previous work Kawai (2019), the second author showed that Π^0_1-BI implies the lesser limited principle of omniscience LLPO, and hence it is not compatible with Brouwer's intuitionism, whereas its monotone counterpart Π^0_1-MBI (equivalently c-BI) is. Thus, c-BI does not imply Π^0_1-BI constructively. From a viewpoint of constructive reverse mathematics, a natural question is whether c-BI together with LLPO is equivalent to Π^0_1-BI. In fact, this is the case when we impose a restriction on the side-predicates to the respective bar inductions. Specifically, we show that c-BI + LLPO becomes equivalent to Π^0_1-BI when the side-predicates of both bar inductions are restricted to bounded Σ^0_2 predicates; see Corollary 4.8.

[1]The equivalence is proved in Heyting arithmetic in all finite types HA^ω extended with the class of inductively generated neighborhood functions and some constructive choice principles.

2 Preliminary

We work (except Remark 3.5) in the theory of intuitionistic elementary analysis EL, which is an extension of the Heyting arithmetic HA with variables and quantifiers for functions between natural numbers. See Section 3.6 of Troelstra and van Dalen (1988) for detailed information about EL. As far as this paper is concerned, we only need mathematical induction on arithmetical formulas (i.e., formulas without quantifiers for functions). We assume a fixed bijective coding of finite sequences of natural numbers, and identify a finite sequence with its code. A function with several arguments is regarded as a unary function that acts on the arguments coded into a single finite sequence.

We use the following notations: $\mathbb{N}$ and $\mathbb{N}^{\mathbb{N}}$ denote the sorts for natural numbers and functions between natural numbers respectively; $\mathbb{N}^*$ denotes the class of finite sequences of natural numbers. The letters x, y, i, k, n range over $\mathbb{N}$; a, b, c range over $\mathbb{N}^*$; α, β, γ range over $\mathbb{N}^{\mathbb{N}}$. We write $\langle x_0, \ldots, x_{n-1} \rangle$ for a finite sequence with kth ($k < n$) element x_k; in particular $\langle\rangle$ denotes the empty sequence. For a finite sequence a, $|a|$ denotes its length and a_i denotes its ith element if $i < |a|$. For an infinite sequence α and a natural number n, $\overline{\alpha}n$ denotes $\langle \alpha(0), \ldots, \alpha(n-1) \rangle$. For finite sequences a and b, $a * b$ denotes the concatenation of a followed by b. We write $a \sqsubseteq b$ to mean that a is an initial segment of b.

We recall some notions related to bar induction (see e.g. Section 4.8 of Troelstra and van Dalen 1988). Let P be a predicate on $\mathbb{N}^*$. Then

- P is a *bar* if $\forall \alpha \exists n P(\overline{\alpha}n)$;
- P is *monotone* if $\forall a\, [P(a) \to \forall b P(a * b)]$;
- P is *inductive* if $\forall a\, [\forall x P(a * \langle x \rangle) \to P(a)]$.

The *monotone bar induction* (MBI) is the following principle: if P and Q are predicates on $\mathbb{N}^*$ such that

1. P is a bar;
2. P is monotone;
3. $\forall a\, [P(a) \to Q(a)]$;
4. Q is inductive,

then $Q(\langle\rangle)$. The predicate Q is called a *side-predicate*. The *unrestricted bar induction* BI is a statement similar to MBI, but without the condition that the bar be monotone. In this paper, we only deal with bar induction for the lowest type where the bar P and the side-predicate Q are predicates on the finite sequences of natural numbers.

If P and Q are restricted to some formula classes Γ and Γ', respectively, then the corresponding principle is called (Γ, Γ')-BI. For example, (Π^0_1, Π^0_1)-BI denotes a fragment of BI in which the bar P and the side-predicate Q are Π^0_1 predicates, i.e., of the form $\forall y A(a, y)$ for some quantifier-free formula A. In addition, Γ-BI denotes a fragment of BI in which only the bar P is a predicate in Γ (there is no restriction on the side-predicate Q.) Similarly, we introduce fragments (Γ, Γ')-MBI and Γ-MBI of MBI as for BI.

3 A Logical Characterization of c-BI

Definition 3.1 (cf. Berger 2006) A predicate P on $\mathbb{N}^*$ is a *c-set* if

$$\forall a\,[P(a) \leftrightarrow \forall b \gamma(a * b) = 0] \tag{1}$$

for some function γ over the finite sequences of natural numbers. The principle ĉ-BI is an instance of MBI in which the bar is required to be a c-set.

The goal of this section is to show the following theorem.

Theorem 3.2 *The following are pairwise equivalent:*

1. c-BI;
2. ĉ-BI;
3. Π^0_1-MBI *which denotes the fragment of* MBI *with a bar* P *being* Π^0_1.

Theorem 3.2 would be trivial if the three notions, c-bar, c-set bar, and monotone Π^0_1 bar, are pairwise equivalent, which we still do not know (cf. Remark 3.6). However, the following two lemmas suffice to show that the three versions of bar induction in Theorem 3.2 are equivalent.

Lemma 3.3 *Every c-set bar is a c-bar.*

Proof Let P be a c-set bar. Then P satisfies $\forall \alpha \exists n P(\overline{\alpha}n)$ and $\forall a[P(a) \leftrightarrow \forall b \gamma(a * b) = 0]$. It suffices to show that

$$\forall a\,[P(a) \leftrightarrow \forall b \gamma(a) = \gamma(a * b)]\,.$$

The direction $(\rightarrow)$ is obvious. For the direction $(\leftarrow)$, fix $a \in \mathbb{N}^*$ and assume $\forall b \gamma(a) = \gamma(a * b)$. Since P is a bar, there exists an n such that $P(\overline{\hat{a}}n)$, where $\hat{a}$ denotes an infinite sequence (namely, a function between $\mathbb{N}$) obtained from a by appending the infinite sequence of 0s. Either $n < |a|$ or $n \geq |a|$. In the former case, we have $P(a)$ by the monotonicity on P. In the latter case, there is a finite sequence $c \equiv \langle 0, \ldots, 0 \rangle$ such that $\gamma(a * c) = 0$. Then, $\gamma(a * b) = \gamma(a) = \gamma(a * c) = 0$ for all b, and hence $P(a)$. □

Lemma 3.4 *Let P be a monotone Π^0_1 bar. Then there is a c-set bar P' which is contained in P.*

Proof We can assume that P satisfies $P(a) \leftrightarrow \forall b \forall x \beta(a * b, x) = 0$ for some β. Define $\gamma \colon \mathbb{N}^* \rightarrow \{0, 1\}$ by

$$\gamma(\langle\rangle) = 0$$
$$\gamma(a * \langle x \rangle) = \beta(a, x)$$

and define a c-set P' by $P'(a) \overset{\text{def}}{\Longleftrightarrow} \forall b \gamma(a * b) = 0$. Then, for any a, we have

$$\begin{aligned} P(a) &\leftrightarrow \forall b \forall x \beta(a * b, x) = 0 \\ &\leftrightarrow \forall b \forall x \gamma(a * b * \langle x \rangle) = 0 \\ &\leftrightarrow \forall x \forall b \gamma(a * \langle x \rangle * b) = 0 \\ &\leftrightarrow \forall x P'(a * \langle x \rangle). \end{aligned}$$

Since P' is monotone by its definition, P' is contained in P. Since P is a bar, for any infinite sequence α, there exists n such that $P(\overline{\alpha}n)$, and hence $P'(\overline{\alpha}(n+1))$. □

Proof of Theorem 3.2 Note that c-BI is an instance of Π^0_1-MBI, and hence ($3 \to 1$). ($1 \to 2$) follows from Lemma 3.3. ($2 \to 3$) follows from Lemma 3.4. □

Remark 3.5 The proof of Lemma 3.4 reveals that for each monotone Π^0_1 subset P, there exists a c-set $P' \subseteq P$ such that

$$P(a) \leftrightarrow \forall x P'(a * \langle x \rangle), \tag{2}$$

and then P is a bar if and only if P' is a bar. On the other hand, for a non-trivial bar P (namely P is not the set of all sequences of natural numbers), such a c-set bar P' is never equivalent to P in $\mathsf{EL} + \Pi^0_1$-MBI: If $P(a) \to P'(a)$, then $\forall x P'(a * \langle x \rangle) \to P'(a)$ by (2). Applying Π^0_1-MBI, we have $P'(\langle\rangle)$, which leads to $P(\langle\rangle)$ and hence $\forall a P(a)$ by the monotonicity of P.

Remark 3.6 At the first glance, the reader may think that it requires almost nothing to show that ĉ-BI implies Π^0_1-MBI because the definition of c-set is somewhat similar to that of monotone Π^0_1 set. However, the situation is not so simple as what is expected. The fact that b might be an empty sequence in the condition (1) of Definition 3.1 causes difficulties when we try to show that a monotone Π^0_1-set is a c-set, and even that a monotone Π^0_1 bar is a c-set bar as suggested by Remark 3.5. Thus, it is still open whether a monotone Π^0_1 bar is a c-set bar. If it is true, since a c-bar is a monotone Π^0_1 bar, we have that the three notions, c-bar, c-set bar, and monotone Π^0_1 bar, are pairwise equivalent.

4 LLPO and Fragments of Bar Induction

In this section, we analyze the relation between c-BI and Π^0_1-BI. In general, BI implies some classical principles. For example, Σ^0_1-BI implies the limited principle of omniscience:

$$\text{(LPO)} \qquad \exists x \alpha(x) \neq 0 \vee \neg \exists x \alpha(x) \neq 0$$

(see Exercise 4.8.11 in Troelstra and van Dalen 1988), and Π^0_1-BI implies the lesser limited principle of omniscience:

$$\text{(LLPO)} \qquad \neg\left[\exists x \alpha(x) \neq 0 \wedge \exists x \beta(x) \neq 0\right] \to \left[\forall x \alpha(x) = 0 \vee \forall x \beta(x) = 0\right].$$

(see Proposition 7.1 in Kawai 2019). Since LLPO (and hence also LPO) contradicts the weak continuity axiom WC-N (see Section 4.6.5 of Troelstra and van Dalen 1988), Π^0_1-BI is incompatible with Brouwer's intuitionism, and hence c-BI does not imply Π^0_1-BI. On the other hand, in the presence of the weak limited principle of omniscience:

$$\text{(WLPO)} \qquad \forall x \alpha(x) = 0 \vee \neg \forall x \alpha(x) = 0,$$

Π^0_1 bar is decidable, and hence the decidable bar induction (which is a consequence of Π^0_1-MBI, namely c-BI, see Theorem 4A in Howard and Kreisel 1966) implies Π^0_1-BI. Moreover, it is known in the study of the arithmetical hierarchy of logical principles (see e.g. Akama et al. 2004; Fujiwara 2015) that WLPO is strictly stronger than LLPO. Therefore, from the equivalence between c-BI and Π^0_1-MBI, it is natural to ask whether c-BI + LLPO is equivalent to Π^0_1-BI. In what follows, we show that this is actually the case if we impose a suitable restriction on the side-predicates of bar induction, respectively.

By a careful inspection of the proof of Proposition 7.1 in Kawai (2019), one can see that $(\Pi^0_1, \mathrm{B}\Sigma^0_2)$-BI suffices to derive LLPO. Here, a predicate $P(x)$ is $\mathrm{B}\Sigma^0_2$ if it is of the form $P(x) \equiv \exists y \leq \bar{n}\, Q(x, y)$ for some numeral $\bar{n}$ and a Π^0_1 predicate Q (cf. Akama et al. 2004). Note that the class of $\mathrm{B}\Sigma^0_2$ sets is identical with that of Π^0_1 sets in classical (only in the presence of LLPO as shown in Corollary 4.6 below) arithmetic but this is not the case in intuitionistic arithmetic. On the other hand, its monotone counterpart $(\Pi^0_1, \mathrm{B}\Sigma^0_2)$-MBI (equivalently $(\mathrm{c}, \mathrm{B}\Sigma^0_2)$-BI) does not imply LLPO. In the following, we show that LLPO captures the difference between $(\Pi^0_1, \mathrm{B}\Sigma^0_2)$-BI and $(\Pi^0_1, \mathrm{B}\Sigma^0_2)$-MBI, while we state our theorem in seemingly the strongest form.

Theorem 4.1 *The following are pairwise equivalent*:

1. (Full, $\mathrm{B}\Sigma^0_2$)-BI *which denotes the fragment of* BI *with the side-predicate Q being* $\mathrm{B}\Sigma^0_2$;
2. $(\Pi^0_1, \mathrm{B}\Sigma^0_2)$-BI;
3. (Full, $\mathrm{B}\Sigma^0_2$)-MBI + LLPO *where* (Full, $\mathrm{B}\Sigma^0_2$)-MBI *denotes the fragment of* MBI *with the side-predicate Q being* $\mathrm{B}\Sigma^0_2$;
4. (Π^0_1, Π^0_1)-MBI + LLPO;
5. (c, Π^0_1)-BI + LLPO *where* (c, Π^0_1)-BI *denotes the fragment of* c-BI *with the side-predicate Q being* Π^0_1.

The crucial part of the proof is to show that (Π^0_1, Π^0_1)-MBI + LLPO implies (Full, $\mathrm{B}\Sigma^0_2$)-BI. The basic idea for this comes from the one for the full versions: the full monotone bar induction MBI together with the full numerical constant domain axiom CD^0 implies the full unrestricted bar induction BI (see Lemma 8 in Fujiwara 2019). Recall that the *numerical constant domain axiom* is the following principle

$$(\mathrm{CD}^0) \qquad \forall x\, (C \vee D(x)) \rightarrow C \vee \forall x D(x)$$

where x does not occur free in C. The superscript 0 indicates that the universal quantifiers in CD^0 are for natural numbers. We write (Π^0_1, Π^0_1) - CD^0 for the fragment of CD^0 with the predicates C and D being both Π^0_1.

Proposition 4.2 (Π^0_1, Π^0_1) - CD^0 *is equivalent to* LLPO.

Proof First, assume (Π^0_1, Π^0_1) - CD^0. Fix α and β, and suppose

$$\neg\,[\exists x \alpha(x) \neq 0 \wedge \exists x \beta(x) \neq 0]\,.$$

Put $C \equiv \forall x \alpha(x) = 0$ and $D(x) \equiv \beta(x) = 0$. Then, for any x, either $D(x)$ or $\neg D(x)$. In the latter case, $\exists x \alpha(x) \neq 0$ implies a contradiction. Thus, we have C and so $C \vee D(x)$. In the former case, we trivially have $C \vee D(x)$. Abstracting x, we obtain $\forall x\,(C \vee D(x))$. By (Π^0_1, Π^0_1) - CD^0, we have $\forall x \alpha(x) = 0 \vee \forall x \beta(x) = 0$.

Next, assume LLPO. Let C and $D(x)$ be Π^0_1 predicates where x does not occur free in C. We may write $C \equiv \forall y \alpha(y) = 0$ and $D(x) \equiv \forall y \beta(x, y) = 0$ for some α and β. Now, suppose that $\forall x\,(C \vee D(x))$. If $\exists y \alpha(y) \neq 0 \wedge \exists x, y \beta(x, y) \neq 0$, then we clearly have a contradiction. Thus, by LLPO, we have $\forall x \alpha(x) = 0 \vee \forall x, y \beta(x, y) = 0$, i.e. $C \vee \forall x D(x)$. □

Next, we shall present several consequences of LLPO. Before that, we recall the following fact which is provable in a weak fragment of EL.

Lemma 4.3 $\forall n\,[\forall i \leq n \exists x \alpha(i, x) \neq 0 \rightarrow \exists x \forall i \leq n \exists y \leq x \alpha(i, y) \neq 0]$.

Proof See e.g. Lemma 2.2.15 in Fujiwara (2015). □

Lemma 4.4 LLPO *implies* $\forall n\,[\neg \forall i \leq n \exists x \alpha(i, x) \neq 0 \rightarrow \exists i \leq n \forall x \alpha(i, x) = 0]$.

Proof Assume LLPO. Put

$$P(n) \equiv \neg \forall i \leq n \exists x \alpha(i, x) \neq 0 \rightarrow \exists i \leq n \forall x \alpha(i, x) = 0.$$

We show that $\forall n P(n)$ by induction on n. $P(0)$ is obvious. For the inductive case, assume $P(n)$, and suppose $\neg \forall i \leq n + 1 \exists x \alpha(i, x) \neq 0$. Define β and γ by

$$\beta(x) \stackrel{\text{def}}{=} \prod_{i=0}^{n} \sum_{k=0}^{x} \alpha(i, k),$$
$$\gamma(x) \stackrel{\text{def}}{=} \alpha(n + 1, x).$$

Note that $\beta(x) \neq 0$ if and only if $\forall i \leq n \exists k \leq x \alpha(i, k) \neq 0$.

If $\exists x \beta(x) \neq 0 \wedge \exists x \gamma(x) \neq 0$, then we have $\forall i \leq n + 1 \exists x \alpha(i, x) \neq 0$, a contradiction. Thus, $\forall x \beta(x) = 0 \vee \forall x \gamma(x) = 0$ by LLPO. We have two cases:

Case $\forall x \beta(x) = 0$: Suppose $\forall i \leq n \exists x \alpha(i, x) \neq 0$. Then by Lemma 4.3, there exists x such that $\forall i \leq n \exists y \leq x \alpha(i, y) \neq 0$. Since $\beta(x) = 0$, there exists $i \leq n$ such that $\forall y \leq x \alpha(i, y) = 0$, a contradiction. Thus, $\neg \forall i \leq n \exists x \alpha(i, x) \neq 0$. By induction hypothesis, we have $\exists i \leq n \forall x \alpha(i, x) = 0$. Hence, $\exists i \leq n + 1 \forall x \alpha(i, x) = 0$.

Case $\forall x \gamma(x) = 0$: Obviously, we have $\exists i \leq n + 1 \forall x \alpha(i, x) = 0$.

Thus, in either case we have $\exists i \leq n + 1 \forall x \alpha(i, x) = 0$. Therefore, $P(n + 1)$. □

Corollary 4.5 LLPO *implies*

$$\forall n\,[\forall x \exists i \leq n \forall y \leq x \alpha(i, y) = 0 \leftrightarrow \exists i \leq n \forall x \alpha(i, x) = 0]\,.$$

Proof The implication from the right to the left is trivial. The converse direction follows from the preceding two lemmas. □

Corollary 4.6 *Under the assumption of* LLPO*, every* $\mathrm{B}\Sigma^0_2$ *predicate is equivalent to a* Π^0_1 *predicate.*

We now prove Theorem 4.1.

Proof of Theorem 4.1 It is trivial that $(1 \to 2)$, $(1 \to 3)$ and $(3 \to 4)$. The equivalence between 4 and 5 follows from (the proof of) Theorem 3.2. On the other hand, since $(\Pi^0_1, \mathrm{B}\Sigma^0_2)$-BI implies LLPO (see Proposition 7.1 in Kawai 2019), we have $(2 \to 4)$. Therefore, it suffices to show $(4 \to 1)$, namely, (Π^0_1, Π^0_1)-MBI + LLPO implies $(\text{Full}, \mathrm{B}\Sigma^0_2)$-BI.

Let P be a bar, a side-predicate $Q \in \mathrm{B}\Sigma^0_2$ be inductive and $P(a)$ imply $Q(a)$ for all $a \in \mathbb{N}^*$. By Corollary 4.6, there is a Π^0_1 predicate Q' equivalent to Q. Note that Q' is a bar and is inductive by our assumptions. We must show that $Q'(\langle\rangle)$. Define a predicate Q'' on $\mathbb{N}^*$ by

$$Q''(a) \overset{\text{def}}{\iff} \exists b \sqsubseteq a Q'(b).$$

Note that Q'' is equivalent to some Π^0_1 predicate by Corollary 4.5. In addition, $Q' \subseteq Q''$ and $Q'(\langle\rangle) \leftrightarrow Q''(\langle\rangle)$. Since Q' is a bar, Q'' is a monotone bar. To see that Q'' is inductive, suppose $\forall x Q''(a * \langle x\rangle)$. For each x, using the decidability of the length of finite sequences, we have that $\exists b \sqsubseteq a * \langle x\rangle Q'(b)$ implies $\exists b \sqsubseteq a Q'(b) \vee Q'(a * \langle x\rangle)$. Thus, $\forall x \left[Q''(a) \vee Q'(a * \langle x\rangle)\right]$. Then, by Proposition 4.2, we have $Q''(a) \vee \forall x Q'(a * \langle x\rangle)$. Since Q' is inductive, we have $Q''(a) \vee Q'(a)$, which implies $Q''(a)$. Therefore, applying (Π^0_1, Π^0_1)-MBI, we conclude $Q''(\langle\rangle)$, which is equivalent to $Q'(\langle\rangle)$. □

Remark 4.7 While (Π^0_1, Full)-BI (even $(\Pi^0_1, \mathrm{B}\Sigma^0_2)$-BI) derives LLPO, (Π^0_0, Full)-BI does not derive LLPO. This is because each instance of (Π^0_0, Full)-BI is an instance of the decidable bar induction, which is derivable from MBI (see Theorem 4A in Howard and Kreisel 1966).

Corollary 4.8 $(\Pi^0_1, \mathrm{B}\Sigma^0_2)$-BI *is equivalent to* $(\mathrm{c}, \mathrm{B}\Sigma^0_2)$-BI + LLPO*, where* $(\mathrm{c}, \mathrm{B}\Sigma^0_2)$-BI *denotes the fragment of* c-BI *with the side-predicate* Q *being* $\mathrm{B}\Sigma^0_2$.

Proof Immediate from Theorem 4.1 and Corollary 4.6. □

Acknowledgements We thank Josef Berger for pointing out a crucial error in an earlier draft of this paper. A part of this work had been carried out at Mathematisches Institut, LMU, München in August 2017. We are grateful to Helmut Schwichtenberg and Chuangjie Xu for their invitation and hospitality for our visit. The visit of the first author was supported by the Core-to-Core Program (A. Advanced Research Networks) of Japan Society for the Promotion of Science. This work is also supported by Waseda University Grant for Special Research Projects 2018K-461.

References

Akama, Yohji, Stefano Berardi, Susumu Hayashi, and Ulrich Kohlenbach. 2004. An arithmetical hierarchy of the law of excluded middle and related principles. In *Proceedings of the 19th Annual IEEE Symposium on Logic in Computer Science (LICS'04)*, 192–201.

Berger, Josef. 2006. The logical strength of the uniform continuity theorem. In *Logical Approaches to Computational Barriers, CiE 2006*, ed. Arnold Beckmann, Ulrich Berger, Benedikt Löwe, and John V. Tucker, 35–39, Lecture Notes in Computer Science Berlin: Springer.

Fujiwara, Makoto. 2015. Intuitionistic and uniform provability in reverse mathematics. PhD thesis, Tohoku University.

Fujiwara, Makoto. 2019. Bar induction and restricted classical logic. In *Logic, Language, Information, and Computation, WoLLIC 2019*, ed. Rosalie Iemhoff, Michael Moortgat, and Ruy de Queiroz, 236–247. Berlin, Heidelberg: Springer.

Howard, William A., and Georg Kreisel. 1966. Transfinite induction and bar induction of types zero and one, and the role of continuity in intuitionistic analysis. *Journal of Symbolic Logic* 31 (3): 325–358.

Ishihara, Hajime. 2005. Constructive reverse mathematics: compactness properties. *From sets and types to topology and analysis*, 245–267, Oxford Logic Guides Oxford: Oxford University Press.

Kawai, Tatsuji. 2019. Principles of bar induction and continuity on Baire space. *Journal of Logic and Analysis* 11 (FT3): 1–20.

Kreisel, Georg, and Anne S. Troelstra. 1970. Formal systems for some branches of intuitionistic analysis. *Annals of Mathematical Logic* 1 (3): 229–387.

Troelstra, Anne S., and Dirk van Dalen. 1988. *Constructivism in Mathematics: An Introduction. Volume I and II*. North-Holland.

Unary and Two-Variable Interval Logics

Kamal Lodaya

Abstract This paper shows that over finite word models, the interval logic of Halpern and Shoham (J Assoc Comput Mach 38(4):935–962, 1991) is expressively complete for two-variable logic with betweenness relations, introduced in Krebs et al. (Proceedings of the 31st LICS, 2016). Satisfiability of formulae can be checked in polynomial space.

1 Introduction

Soon after Halpern et al. (1983), Moszkowski (1983) came up with interval temporal logic, with its characteristic binary *chop* operation, Halpern and Shoham (1991) defined unary interval logics, using modalities based on Allen (1983) interval algebra.

In these traditional interval logics (for example, see van Benthem 1983), propositions are interpreted over intervals $[s, t]$, where s, t are positions in the word and $s \leq t$. That is, in the semantics, they represent arbitrary binary relations over the domain (the set of positions of the word $\{1, 2, \ldots\}$). This plays a key role in Halpern and Shoham showing that satisfiability of their logic *HS* was undecidable over an infinite domain. Venema (1989) showed that the *HS* modalities can be represented in a plane, which allows encoding of grid problems. This result was sharpened, for example, see Lodaya (2000), Bresolin et al. (2009b), Marcinkowski and Michaliszyn (2011). A large number of papers by Goranko, Montanari and others (see the papers Goranko et al. 2004; Bresolin et al. 2014) have classified the decidability status of the satisfiability and model checking problems for nearly all fragments of the original Halpern–Shoham logic. *HS* is one of the best understood logics.

This chapter is in its final form and it is not submitted to publication anywhere else.

K. Lodaya (✉)
The Institute of Mathematical Sciences, CIT Campus, Chennai 600113, India
e-mail: kamal@imsc.res.in

Homi Bhabha National Institute, Anushaktinagar, Mumbai 400094, India

F. Liu et al. (eds.), *Knowledge, Proof and Dynamics*,
Logic in Asia: Studia Logica Library,
https://doi.org/10.1007/978-981-15-2221-5_3

Realizing the difficulty of the two-dimensional interpretation, Halpern and Shoham suggested a different interpretation where propositions are interpreted at points, and hence their interpretation is a unary relation. Now the semantics of an interval logic formula is a first-order sentence over a linear order and hence decidability is regained. This direction led to the Duration Calculus, an extension of interval logic with measurements studied by Chaochen et al. (1991).

In this paper, we only consider finite word models (the domain is $\{1, 2, \ldots, max\}$), over a finite alphabet (set of letters) Σ. Our tools will be automata and language theoretic. The discussion can be extended to timed words. In particular, DC can be interpreted over timed words, providing a rich variety of features. The book by Chaochen and Hansen (2004) gives many details of this work. Wilke (1994), Rabinovich (2000), Lodaya and Pandya (2006) define extensions to capture monadic second-order logic, since this remains decidable over word models (Thomas 1997 is a nice survey). Surprisingly validity and model checking of many practical examples can be successfully carried out, as demonstrated by Pandya (2001) using the Mona library of Basin and Klarlund (1995) for manipulating automata. Recently, this work has been extended to synthesis for a rich variety of specifications by Wakankar et al. (2017).

Since first-order and monadic second-order logic over words have a mature algebraic theory of expressiveness (see the book of Straubing 1994), over the last 10 years, we have applied it to interval logics. By the theorem of Kamp (1968) any first-order logic sentence over words can be expressed by repeatedly using at most three variables. In a couple of papers (Lodaya et al. 2008, 2010), we characterized using interval logics two-variable logic over words $FO^2[<]$ and its extension $FO^2[<, Suc]$ with the successor relation. The characterization of the quantifier alternation hierarchy of $FO[<]$ (again see Thomas 1997) remains largely open, except for a few lower levels. We review this work in Sect. 2.

Venema (1991) showed that Halpern–Shoham logics are not expressively complete for first-order logic. Intuitively they fall within the guarded fragment of Andréka et al. (1998), specialized to words. Bresolin et al. (2009a) showed that a fragment called neighbourhood logic is expressively complete for two-variable logic. They also showed that this is not true for *HS*.

Recently, we algebraically characterized the logic $FO^2[<, bet]$, which extends two-variable logic with betweenness relations (Krebs et al. 2016, 2018), alternately with counting upto a threshold, giving a larger fragment of first-order logic than two-variable logic. In this paper, we attempt to tackle $FO^2[<, bet]$ using new fragments of Halpern–Shoham logic. In Sect. 3 we illustrate some key examples to show how they are formulated in interval logic. We are aided by a point temporal logic which we have already proved to be expressively complete in Krebs et al. (2016). In Sect. 4 we proceed in the other direction. Section 5 brings together all the results.

2 Two-Variable Logics and Interval Logics

In this paper, we confine ourselves to models which are finite words, that is, discrete linear orders with domain $\{1, \ldots, max\}$ where each position has a letter (or colour) from a finite alphabet Σ. Thus, we are in the context of formal language theory which is well-studied for decades (for example, see the book by Straubing 1994).

A language is a set of words. For example, over the alphabet $\Sigma = \{a, b, c, d\}$, those words where, between the last a and subsequent first d, there must be no b. This is usually denoted by a regular expression $\Sigma^* ac^* d\{b, c, d\}^*$. That is, any number of letters from Σ, then an a, then any number of c's, then a d, then any number of letters from b, c, d.

To avoid confusion, we do not use the word 'language' to denote the formal setup of a logic, where we prefer to use 'formulae' and 'sentences'. Our base logic will be two-variable first-order logic $FO^2[<]$ over words, with the linear order interpreting a binary predicate $<$ and letters at word positions interpreting unary predicates $a(.)$ for every $a \in \Sigma$. We will also consider $FO^2[<, Suc]$, which also has the successor partial function $y = x + 1$ interpreting a binary predicate $Suc(x, y)$. In Sect. 3 we will consider our richer logic $FO^2[<, bet]$.

Going further, one can use arbitrary numerical predicates for positions, for example, $\exists y(x = 2 \times y)$ holds when position x has an even value. The language of even-length words and the language $\{a^n b^n \mid n \geq 1\}$ are defined using this kind of predicate. Regular expressions no longer suffice to describe such languages. Circuit families were proposed for this purpose, that is, the nth circuit in the family describes the language L_n of words in a language L which are of length n. Since each L_n is a finite language, it can be defined by a boolean circuit (a directed acyclic graph with nodes, called 'gates', labelled by boolean operations) with n inputs and one output for acceptance or rejection. The implicit restriction to the alphabet $\{0, 1\}$ can be modelled by coding bigger alphabets. It turns out (Gurevich and Lewis 1984; Immerman 1987) that formulae of first-order logic on words using arbitrary numerical predicates correspond exactly to languages defined by families of circuits whose depth (length of longest path from an input to the output) is bounded by a constant.

2.1 *Unambiguous Interval Temporal Logic*

In Lodaya et al. (2008), we defined an unambiguous interval logic. For example, the language $\Sigma^* ac^* d\{b, c, d\}^*$ is defined by the formula

$$(true L_a((\neg(true F_b true)) F_d true)),$$

where L_a stands for the last a and F_b for the first b. More formally the logic has this syntax, where $a \in \Sigma$ is a letter and $P \subseteq \Sigma$ is a subset of letters:

$$\begin{aligned}\beta ::= pt \mid a \mid \lceil P \rceil \mid \lceil\lceil P \rceil \mid \lceil P \rceil\rceil \mid \lceil\lceil P \rceil\rceil \mid \beta_1 \vee \beta_2 \mid \neg\beta \mid \\ \beta_1 F_a \beta_2 \mid \beta_1 L_a \beta_2 \mid \oplus\beta \mid \ominus\beta\end{aligned}$$

Given two positions $s \leq t$ of a word w, let $[s, t]$ stand for the interval from position s to position t and $w[k]$ for the letter at position k.

$w, [s, t] \models pt$ iff $s = t$
$w, [s, t] \models a$ iff $s = t \wedge w[t] = a$
$w, [s, t] \models \lceil P \rceil$ iff for all $k : s < k < t : w[k] \in P$
$w, [s, t] \models \lceil\lceil P \rceil$ iff for all $k : s \leq k < t : w[k] \in P$
$w, [s, t] \models \lceil P \rceil\rceil$ iff for all $k : s < k \leq t : w[k] \in P$
$w, [s, t] \models \lceil\lceil P \rceil\rceil$ iff for all $k : s \leq k \leq t : w[k] \in P$
$w, [s, t] \models \beta_1 F_a \beta_2$ iff for some $k : s \leq k \leq t :$ $w[k] = a$ and
(for all $m : i \leq m < k : w[m] \neq a$) and
$w, [s, k] \models \beta_1$ and $w, [k, t] \models \beta_2$
$w, [s, t] \models \beta_1 L_a \beta_2$ iff for some $k : s \leq k \leq t :$ $w[k] = a$ and
(for all $m : k < m \leq j : w[m] \neq a$) and
$w, [s, k] \models \beta_1$ and $w, [k, t] \models \beta_2$
$w, [s, t] \models \oplus\beta$ iff $s < t$ and $w, [s + 1, t] \models \beta$
$w, [s, t] \models \ominus\beta$ iff $s < t$ and $w, [s, t - 1] \models \beta$.

Our paper proved several results about this logic. First of all, even though it does not appear so from the semantics above, the logic has the two-variable property. We showed that all formulae of the logic have a unique parsing property and a translation to specially designed automata of Schwentick et al. (2002). Using their results, the translation extends to sentences of two-variable logic $FO^2[<]$. Using the automata, we showed that the complexity of model checking is in LogDCFL, and that of satisfiability is in Pspace (and NP over a fixed alphabet). Conversely, there is an exponential construction from $FO^2[<]$ to formulae of our logic which define the same language. A more general interval logic with the same expressiveness is reported in Lodaya and Pandya (2017).

Automata have a mature algebraic theory (see the books Pin 1986; Straubing 1994). Recall that a monoid is a set with an associative operation which has an identity element (usually denoted 1). The transition monoid of an automaton is effectively constructed by looking at all the transition functions from states to states realized by different words as inputs. The associative operation is function composition and the identity element is the identity function. Thus transitions performed by letters as well as words map to monoid elements. More precisely, for a language, we use the transition monoid of the minimal automaton for the language, which is called its syntactic monoid. For the special automata of Schwentick et al. (2002), this transition monoid has nice properties, allowing an algebraic characterization of the languages in terms of an equational variety called **DA**, as also of two-variable logic from the work of Thérien and Wilke (1998), and thus of our unambiguous interval temporal logic. Tesson and Thérien (2002) is a detailed study of many properties of this variety.

We will briefly skim over this characterization. An idempotent element e in a monoid satisfies $e = ee$. Every loop in a minimal automaton (with words $l, ll, \ldots$) will ultimately map to an idempotent. M_e is the submonoid generated by factors of e, that is, generated by $\{a \mid e = paq$, for some $p, q\}$. These are the words made up of letters which go into making up the loop. A monoid is in variety **DA** if for all its idempotents e, $eM_ee = \{e\}$. The equation for **DA** is simpler, but this condition is also checkable on a monoid. As a consequence of this characterization, whether a language is definable by a sentence of two-variable logic (equivalently, of our unambiguous logic) is checkable.

To give an example, suppose the loop $(ab)^*$ maps to an idempotent e. Then M_e is the language $\{a, b\}^*$. The definition of **DA** says that in this case, all of $\{a, b\}^*$ must map to that e. This gives a non-definability result: the language $(ab)^*$ does not map to a monoid in **DA**, since it cannot be distinguished from $\{a, b\}^*$.

A similar argument shows $\Sigma^* aa \Sigma^*$ is also not two-variable definable.

2.2 Lookaround Interval Temporal Logic

In Lodaya et al. (2010), we considered an expanded version of our logic, where instead of modalities dealing with unambiguous occurrences of letters, we had modalities for unambiguous occurrences of substrings.

The formula $true F_{a\underline{a}} true$ says that there is an occurrence of the substring aa, where the underline indicates that the subintervals are positioned at the second a in the *first* occurrence of aa in the word. We saw that this language $\Sigma^* aa \Sigma^*$ is not two-variable definable. Thus this logic is more expressive. Since $(ab)^*$ is exactly those words which begin with a, end with b, and do not contain occurrences of the substrings aa and bb, it follows that this language is also definable in the lookaround logic by a longer formula.

The formula $\neg(true L_{b\underline{b}} true) F_{a\underline{a}} true$ says that the substring bb does not occur before the first occurrence of the substring aa.

Formally we have the following syntax, where $u \in \Sigma^+$ is a substring and $P \subseteq \Sigma^+$ is a finite subset of forbidden substrings of the form uav with letter $a \in \Sigma$ with prefix and suffix $u, v \in \Sigma^*$.

$$\beta ::= true \mid pt \mid u \mid \lceil \neg P \rceil \mid \lceil\!\lceil \neg P \rceil \mid \lceil \neg P \rceil\!\rceil \mid \lceil\!\lceil \neg P \rceil\!\rceil \mid$$
$$\beta_1 \vee \beta_2 \mid \neg \beta \mid \oplus \beta \mid \ominus \beta \mid \beta_1 F_{u\underline{a}v} \beta_2 \mid \beta_1 L_{u\underline{a}v} \beta_2.$$

Given two positions $s \le e$ of a word w, let $w[s \ldots t]$ stand for the substring from position s to position t. The new modalities can be defined in this way, $\lceil \neg P \rceil\!\rceil$, $\lceil\!\lceil \neg P \rceil$, $\lceil\!\lceil \neg P \rceil\!\rceil$ being analogously extended from Sect. 2.1.

$w, [s, t] \models u$ iff $w[s \ldots t] = u$

$w, [s, t] \models \lceil \neg P \rceil$ iff for all $k, m, u : s \le k < m \le t$ and $u \in P : w[k \ldots m] \ne u$

$w, [s, t] \models \beta_1 F_{u\underline{a}v} \beta_2$ iff for some $k : s + |u| \le k \le t - |v| : w[k - |u| \ldots k + |v|] = uav$
and (for all $m : s \le m < k : w[m - |u| \ldots m + |v|] \ne uav$)
and $w, [s, k] \models \beta_1$ and $w, [k, t] \models \beta_2$

$w, [s, t] \models \beta_1 L_{u\underline{a}v} \beta_2$ iff for some $k : s + |u| \le k \le t - |v| : w[k - |u| \ldots k + |v|] = uav$
and (for all $m : k < m \le t : w[m - |u| \ldots m + |v|] \ne uav$)
and $w, [s, k] \models \beta_1$ and $w, [k, t] \models \beta_2$.

Our results for the lookaround logic parallel those of the earlier paper. By following a proof analogous to our earlier one, we showed that the formulas of this logic translate to the two-variable logic $FO^2[<, Suc]$. For this purpose, we had to define and work with more general (but still specialized) automata than those defined by Schwentick et al. (2002). The model checking problem for the logic is in LogDCFL, and the satisfiability problem is in Pspace. There is an exponential construction from an $FO^2[<, Suc]$ sentence to an equivalent formula in this logic defining the same language.

As in the case of our earlier logic, the logic $FO^2[<, Suc]$ has an algebraic characterization. Thérien and Wilke (1998) showed that it corresponds to the variety **LDA** of monoids M where, for every idempotent $e \neq 1$ of M, eMe is in **DA**. Now if we suppose that the language $(ab)^*$ maps to the idempotent e, this says that the language $(ab)^* \Sigma^* (ab)^*$, which is seen to be just the language Σ^* of all words, should be definable in two-variable logic, which it is (by the sentence $true$). By our theorem, $(ab)^*$ is definable in $FO^2[<, Suc]$.

Again one can form non-definability arguments based on the algebraic characterization. Thus one shows first that the syntactic monoid for the language $c^*(ac^*bc^*)^*$ is not in **DA**. Then, since c maps to an idempotent (in fact, the identity), and $cMc = M$ is not in **DA**, the language is not in **LDA**. Hence, it is not definable by a sentence of $FO^2[<, Suc]$.

2.3 Halpern–Shoham Logics

The interval logic of Halpern and Shoham (1991) uses modalities corresponding to all the relations in the interval algebra of Allen (1983). We use some of these, B, E, D, A and $\overline{A}$, to model the logics we have seen above. All of these are subsumed by the *chop* modality of the original interval logic of Moszkowski (1983). Thus the syntax of the logic *HS* is as below, where $P \subseteq \Sigma$ is a finite subset of letters. Since we are dealing with a finite alphabet, we will freely use boolean formulas to represent P.

$$\beta ::= pt \mid a \mid \lceil P \rceil \mid \lceil\lceil P \rceil \mid \lceil P \rceil\rceil \mid \lceil\lceil P \rceil\rceil \mid \langle \mathsf{B}pt \rangle \beta \mid \langle \mathsf{E}pt \rangle \beta \mid$$
$$\beta_1 \vee \beta_2 \mid \neg\beta \mid \langle \mathsf{B} \rangle \beta \mid \langle \mathsf{E} \rangle \beta \mid \langle \mathsf{D} \rangle \beta \mid \langle \mathsf{A} \rangle \beta \mid \langle \overline{\mathsf{A}} \rangle \beta$$

One can write tricky formulas. For example, $\langle \mathsf{B}pt \rangle pt \wedge \lceil false \rceil \wedge \langle \mathsf{E}pt \rangle pt$ cannot be satisfied by a word of length more than two.

As usual in modal logic, there are derived box modalities, for example, we have $[\mathsf{D}]\beta = \neg\langle \mathsf{D} \rangle \neg\beta$. As before, the semantics is defined on intervals $[s, t]$ of word w with $s \leq t$. Our during modality is strict on the left as well as the right.

$w, [s, t] \models a$ iff $s = t$ and $w[t] = a$
$w, [s, t] \models \langle \mathsf{B}pt \rangle \beta$ iff $w, [s, s] \models \beta$
$w, [s, t] \models \langle \mathsf{E}pt \rangle \beta$ iff $w, [t, t] \models \beta$
$w, [s, t] \models \langle \mathsf{B} \rangle \beta$ iff for some $m : s \leq m < t : w, [s, m] \models \beta$
$w, [s, t] \models \langle \mathsf{E} \rangle \beta$ iff for some $m : s < m \leq t : w, [m, t] \models \beta$
$w, [s, t] \models \langle \mathsf{D} \rangle \beta$ iff for some $m : s < k < m < t : w, [k, m] \models \beta$
$w, [s, t] \models \langle \mathsf{A} \rangle \beta$ iff for some $m : t \leq m : w, [t, m] \models \beta$
$w, [s, t] \models \langle \overline{\mathsf{A}} \rangle \beta$ iff for some $m : m \leq s : w, [m, s] \models \beta$

$\langle \mathsf{A} \rangle$ and $\langle \overline{\mathsf{A}} \rangle$ move from the current interval to a neighbouring interval on the right or left, respectively, while $\langle \mathsf{B}pt \rangle$ and $\langle \mathsf{E}pt \rangle$ shrink the current interval to a point. The fragment of *HS* with just these operators is sometimes called neighbourhood logic. Their semantics can be translated into $FO^2[<]$ by rotating variables. Bresolin et al.

(2009a) showed that the converse holds and neighbourhood logic is expressively complete for two-variable logic $FO^2[<]$.

The unambiguous modalities $\beta_1 F_a \beta_2$ and $\beta_1 L_a \beta_2$ are modelled in *HS* as

$$\langle \mathsf{B}\rangle(\lceil\!\lceil \neg a\rceil \wedge \beta_1 \wedge \langle \mathsf{E}pt\rangle a \wedge \langle \mathsf{A}\rangle \beta_2), \quad \langle \mathsf{E}\rangle(\lceil \neg a\rceil\!\rceil \wedge \beta_2 \wedge \langle \mathsf{B}pt\rangle a \wedge \langle \overline{\mathsf{A}}\rangle \beta_1).$$

The first formula says that the current interval $[s, t]$ has a proper left subinterval $[s, m]$ with $m < t$ in which β_1 holds and an adjoining right subinterval after that in which β_2 holds. The second one is a mirror image. These *HS* formulae do not have identical semantics. Using the unique parsing property of our logic alluded to in Lodaya et al. (2008), the modelling does work correctly. Our results above show that we can add the unambiguous B, E modalities to A, $\overline{A}$ (and even more, the role of determinism is spelt out in Lodaya and Pandya 2017), still maintaining the expressive completeness for two-variable logic.

Some patterns are going to appear frequently. We introduce derived modalities for invariance requirements on the B, D and E operators (without any unambiguity requirement).

Definition 1 Let $[\mathsf{D}P\lceil Q\rceil R]\beta = [\mathsf{D}](\langle \mathsf{B}pt\rangle P \wedge \lceil Q\rceil \wedge \langle \mathsf{E}pt\rangle R \supset \beta)$, and for the dual modality, $\langle \mathsf{D}P\lceil Q\rceil R\rangle\beta = \langle \mathsf{D}\rangle(\langle \mathsf{B}pt\rangle P \wedge \lceil Q\rceil \wedge \langle \mathsf{E}pt\rangle R \wedge \beta)$. If the P or R requirement is trivial, we drop it. If it is the same as the Q requirement, we use the closed interval form as an abbreviation. The fragment of *HS* logic, where only these derived forms of B, D and E are used with $P, Q, R \subseteq \Sigma$, is called *InvHS*.

Note that *InvHS* defines the same languages as *HS*, since we can use Σ for the sets P, Q, R if required.

The formulae $\beta_1 F_a \beta_2$ and $\beta_1 L_a \beta_2$ above can be written in *InvHS*:

$$\langle \mathsf{B}\lceil\!\lceil \neg a\rceil a\rangle(\beta_1 \wedge \langle \mathsf{A}\rangle \beta_2), \quad \langle \mathsf{E}a\lceil \neg a\rceil\!\rceil\rangle(\beta_2 \wedge \langle \overline{\mathsf{A}}\rangle \beta_1).$$

Let $u_1 \prec u_2$ and $u_1 \succ u_2$ denote that u_1 is a proper prefix, respectively suffix, of u_2. (Note that $\prec$ and $\succ$ are not converse to each other.) The lookaround modality $\beta_1 F_{u\underline{a}v} \beta_2$ is modelled in a formula below, which is not an *InvHS* formula, and which slightly extends *HS* by matching substrings against intervals rather than just letters:

$$\langle \mathsf{B}\rangle(\lceil\!\lceil \neg uav\rceil \wedge \beta_1 \wedge \langle \mathsf{E}\rangle ua \wedge \langle \mathsf{A}\rangle(\langle \mathsf{B}\rangle av \wedge \beta_2) \wedge \bigwedge_{u \succ u', uav = u'av'} \neg(\langle \mathsf{E}\rangle u'a \wedge \langle \mathsf{A}\rangle\langle \mathsf{B}\rangle av')).$$

The conjunctions at the end rule out the possibility of an earlier match of uav which starts at the end of the beginning subinterval. (Note that $v' \prec v$ is implied.) This will not be caught by forbidding the substring uav since both occurrences of $uav = u'av'$ stretch beyond the beginning subinterval. We do not introduce derived modalities here, we will not need them in the rest of the paper.

3 Interval Logics with Invariance

In our paper (Krebs et al. 2016), we extended two-variable logic with binary predicates $a(x, y)$, $a \in \Sigma$, which say that $x < y$ and there is an occurrence of the letter a lying strictly between the two positions. The formula $\bigwedge_{a \in \Sigma} \neg a(x, y)$ says that there is no letter strictly between x and y, that is, that $y = x + 1$. Thus this logic subsumes $FO^2[<, Suc]$. There is a richer counting capability (up to a threshold) which can be modelled, we refer the reader to the paper.

In fact, the logic extends further than $FO^2[<, Suc]$, since the language $c^*(ac^*bc^*)^*$ that we saw was not definable in that logic, can be defined by a sentence in the new logic. We also started on an algebraic characterization, which was completed in a second paper (Krebs et al. 2018). Our result is that $FO^2[<, bet]$ corresponds to the variety $\mathbf{M_eDA}$ of monoids where, for every idempotent e in the monoid, the submonoid eM_ee is in $\mathbf{DA}$. Again this condition is effectively checkable on a monoid.

For example, in the syntactic monoid of the language $c^*(ac^*bc^*)^*$, c maps to an idempotent and the submonoid $cM_cc = \{c\}$ are the trivial singleton monoid, which is in $\mathbf{DA}$. This holds for all idempotents and the syntactic monoid is in $\mathbf{M_eDA}$. A more involved argument from Krebs et al. (2016) uses the characterization to show that the regular language $(a(ab)^*b)^+$ is not definable in $FO^2[<, bet]$.

3.1 *Modelling Addition*

We now attempt to design an interval logic which can describe betweenness features. Chandra et al. (1985) observed that the language $c^*(ac^*bc^*)^*$ is important in modelling integer addition. For simplicity let the base be 2 and work with the alphabet $\{0, 1\}^3$. Let ADD be the language of words representing vertically three numbers $\begin{pmatrix} m \\ n \\ m+n \end{pmatrix}$ written reversed (least significant bit first) in base 2.

For this to be correct, the ith bit of sum is computed from the ith bit of the inputs and whether or not there is a carry into bit i. If the ith bit of both inputs is 1, we map the letter to a (we think of this as a *set* operation). If the ith bit of both inputs is 0, we map to b (this is a *reset* operation). Otherwise we map to c (an *identity* operation).

Think of a monoid $(\{a, b, c\}, \cdot, c)$ representing these three operations. The carry product is $x \cdot a = a$, $x \cdot b = b$ and $x \cdot c = x$ for an element x. The carry into bit $i + 1$ of the sum is exactly when the first i monoid elements multiply to a. Now the carry computation of an addition corresponds to a word in the language $c^*(ac^*bc^*)^*$.

In *HS* logic we can say that there is a beginning subinterval ending with a and with possibly only c's preceding it, there is an ending interval beginning with b and possibly only c's following it, and there are no subintervals which begin and end with a and do not contain a b, nor those which begin and end with b and do not contain an a. That is, in *InvHS*:

$$\langle \mathsf{B}[\lceil c\rceil a\rangle true \wedge \langle \mathsf{E} b\lceil c\rceil]\rangle true \wedge \neg\langle \mathsf{D} a\lceil \neg b\rceil a\rangle true \wedge \neg\langle \mathsf{D} b\lceil \neg a\rceil b\rangle true.$$

This is also definable in $FO^2[<, bet]$. Hence the language ADD is also definable in *InvHS* and in $FO^2[<, bet]$.

3.2 Defining Circuits

In Krebs et al. (2016) we showed that our logic $FO^2[<, bet]$ can define languages arbitrarily high in the quantifier alternation hierarchy of first-order logic on words (see Thomas 1997) and the constant-depth circuit hierarchy of Sipser (1983). The entire latter hierarchy corresponds to first-order logic on words using arbitrary numerical predicates, as we said in Sect. 2. For simplicity assume that the circuits have levels consisting of gates of only one type, alternating between *or* and *and*. With alphabet $\{0, 1, o_1, a_2, o_3, \ldots, a_{2k}, o_{2k+1}, \ldots\}$ one can encode in prefix form constant-depth boolean circuits with inputs set to 0 and 1, upto any finite level. We will see examples below. In our encoding, we use $\lhd$ as a right endmarker for the input word signifying the circuit and propositions $P_0, P_1, \ldots$ for letters from subalphabets, which we will define.

Our aim is to show that the logic *InvHS* can describe circuits of arbitrary depth and hence represent regular languages at any level of the constant-depth hierarchy, using more and more such propositions. Using its semantics, it is thus a rich fragment of first-order logic $FO[<]$ on words. The formulation of Sipser (1983) is used since it is amenable for our logic.

Let $Circ_1 = o_1\{0, 1\}^+$ be the regular language encoding circuits of depth one (an *or* gate followed by inputs). Let P_1 be the subalphabet $\{o_1, a_2, o_3, \ldots, \lhd\}$. The *InvHS* formula is

$$C_1 = \langle \mathsf{B} o_1\lceil \neg o_1\rceil P_1\rangle \lceil 0 \vee 1\rceil.$$

Further, circuits which evaluate to $true$ need to have at least one 1, the language is $True_1 = o_1\{0, 1\}^*1\{0, 1\}^*$. The *InvHS* formula is

$$T_1 = \langle \mathsf{B} o_1\lceil \neg o_1\rceil P_1\rangle (\lceil 0 \vee 1\rceil \wedge \neg\lceil 0\rceil).$$

The next level is the regular language $Circ_2 = a_2(Circ_1)^+$ (an *and* gate at level two, followed by depth one circuits). Let P_2 be the subalphabet $\{a_2, o_3, \ldots, \lhd\}$. The *InvHS* formula can be written in two ways. The second formula shows how the endmarker can be used to convert invariant during formulae to invariant prefix and suffix formulae.

$$C_2 = [\mathsf{D} a_2\lceil \neg a_2\rceil P_2]C_1, \text{ or } C_2' = [\mathsf{E}\lceil \neg\lhd\rceil P_2][\mathsf{B} a_2\lceil \neg a_2\rceil P_2]C_1.$$

$True_2 = a_2(True_1)^+$ describes the circuits evaluating to $true$ at the second level. The *InvHS* logic formula forces every conjunct to have a true disjunct:

$$T_2 = [\mathsf{D}a_2\lceil\neg a_2\rceil P_2]T_1.$$

The third level is the regular language $Circ_3 = o_3(Circ_2)^+$ which is handled by defining $P_3 = \{o_3, \ldots, \lhd\}$ and

$$C_3 = [\mathsf{D}o_3\lceil\neg o_3\rceil P_3]C_2.$$

Now the circuits evaluating to $true$ are described by $True_3 = o_3(Circ_2)^* True_1$ $(Circ_2)^*$. The *InvHS* formula requires one true disjunct among all the disjuncts:

$$T_3 = C_3 \wedge (\langle \mathsf{D}o_3\lceil\neg o_3\rceil P_3\rangle T_2).$$

It should be clear how to extend this to circuits of any finite level. Thus, like $FO^2[<, bet]$, the logic *InvHS* can also define languages arbitrarily high in the quantifier alternation hierarchy.

3.3 Defining Monomials

We now show that we can define languages arbitrarily high in the nested until/since hierarchy of linear temporal logic on finite words, worked out by Thérien and Wilke (2004) (see the survey by Tesson and Thérien 2007). To do this we will show that one can use a series of between operations, briefly a *monomial* language $P_0^* a_1 P_1^* \ldots P_{n-1}^* a_n P_n^*$ for some subalphabets $P_0, P_1, \ldots, P_{n-1}, P_n$ (again interpreted by propositions), followed by a formula β. To do this in temporal logic one would require $n+1$ nested until operators.

The formula of *HS* logic is: $F_0 = \langle \mathsf{B}\rangle(F_1 \wedge \langle \mathsf{A}\rangle\beta)$, where the formulae F_m, $m \geq 1$, expand the monomial, starting with $F_1 = \langle \mathsf{B}\lceil\!\rceil P_0\rceil\rangle\langle \mathsf{A}\rangle F_2$, $F_2 = \langle \mathsf{B}a_1\lceil P_1\rceil\rangle\langle \mathsf{A}\rangle F_3$, up to $F_n = \langle \mathsf{B}a_{n-1}\lceil P_{n-1}\rceil\rangle\langle \mathsf{A}\rangle F_{n+1}$, ending with $F_{n+1} = \langle \mathsf{B}pt\rangle(a_n) \wedge \lceil P_n\rceil\!\rceil$. We can also allow, instead of a single monomial, a boolean collection of monomials before the satisfaction of β. This latter formula shows that the second level of the quantifier alternation hierarchy (see Thomas 1997) is subsumed in *HS* logic.

Another way is to run the induction down from $n+1$ to describe the languages formed as one moves left in the monomial, to obtain an *InvHS* formula scheme. The base step is $F'_{n+1} = \beta$. The induction step is $F'_{m-1} = \langle \mathsf{B}a_{m_1}\lceil P_{m-1}\rceil\rangle\langle \mathsf{A}\rangle F'_m$. The final formula is $F'_0 = \lceil\!\lceil P_0\rceil\langle \mathsf{A}\rangle F'_1$. If β was in *InvHS*, so are the formulae F'_m. It follows that the logics *InvHS* and $FO^2[<, bet]$ can define languages arbitrarily high in the until/since hierarchy of temporal logic.

Our examples have given sufficient evidence that our interval logic with invariance *InvHS* rivals the two-variable logic $FO^2[<, bet]$ in expressiveness.

Lemma 2 *InvHS subsumes* $FO^2[<, bet]$.

Proof We rely on a theorem in Krebs et al. (2016), which shows that definability in the latter logic is achieved by a point temporal logic with modalities which, for our purposes here, we can read as

$$(\neg R)\ \mathit{until}\ \alpha \text{ and } (\neg R)\ \mathit{since}\ \alpha, \text{ for } R \subseteq \Sigma.$$

But the translation from monomials to *InvHS* that we have given above is already a generalization of the first modality, and a mirrored translation will handle the second one. This gives us a direct linear translation from the point temporal logic, and hence a translation from $FO^2[<, bet]$, to *InvHS*. □

4 Staying Within Two-Variable Between Logic

As we have seen so far, the Halpern–Shoham logic with invariance *InvHS* is expressive enough to encompass our earlier two-variable logic with betweenness from Krebs et al. (2016). In this section, we see if we have gone too far and overshot this logic.

Definition 3 The fragment of *InvHS* logic, where the derived forms of the invariance requirements B and E are used only in the forms $\langle \mathsf{B} P \lceil Q \rceil R\rangle\beta$ and $\langle \mathsf{E} P \lceil Q \rceil R\rangle\underline{\beta}$, where β is a boolean combination of $\langle \mathsf{A}\rangle$ formulas or a boolean combination of $\langle \overline{\mathsf{A}}\rangle$ formulas, is called *bundled InvHS* (to borrow some jargon from Padmanabha et al. 2018). Assuming our words are provided with endmarkers, we also allow invariance D requirements.

Lemma 4 *Bundled InvHS is expressively complete for* $FO^2[<, bet]$.

Proof There is a standard linear translation from neighbourhood logic (with the $A, \overline{A}$ modalities) to two-variable logic $FO^2[<]$. From this, it follows that prefixing the neighbourhood modalities with invariance requirements to form the modalities $\langle \mathsf{B} P \lceil Q \rceil R\rangle\langle \mathsf{A}\rangle\beta$ and $\langle \mathsf{E} P \lceil Q \rceil R\rangle\langle \overline{\mathsf{A}}\rangle\beta$ linearly translates into the extended logic $FO^2[<, bet]$, the invariance requirements were designed to do so. The same holds if one takes boolean combinations of the $\langle \mathsf{A}\rangle$ or the $\langle \overline{\mathsf{A}}\rangle$ formulas. Assuming endmarkers, it follows that the invariance-guarded $\langle \mathsf{D}\rangle$ modality can be eliminated as in Sect. 3.2 and the resulting formula translated into $FO^2[<, bet]$. Combining with Lemma 2, bundled *InvHS* equals the expressive power of two-variable between logic. □

We can push this idea further. From Sect. 2.3 we know that when unambiguously guarded, the formula $\langle \mathsf{B} P \lceil Q \rceil R\rangle(\beta_1 \wedge \langle \mathsf{A}\rangle\beta_2)$ which is syntactically not in the bundled fragment, is definable in $FO^2[<]$, hence also in $FO^2[<, bet]$. What happens when we drop the unambiguity requirement? In our earlier papers (Lodaya et al. 2008, 2010) we used unique parsing properties and automaton characterizations to traverse the path from interval logic to two-variable logic. These are lacking for the logic *InvHS*. We obtain a positive result by translating *InvHS* to its bundled fragment.

Lemma 5 *InvHS is expressively complete for* $FO^2[<, bet]$.

Proof Given Lemmas 2 and 4, it is sufficient to take care of formulae which are outside the bundled fragment of *InvHS*. We take a representative type of such a formula, an invariance-guarded $\langle \mathsf{B} \rangle$ formula. The same approach works for other invariance-guarded formulae.

Consider for various cases of the subformula β_1, which we name for this proof a *loose* subformula, in an *InvHS* formula which is not inside an $\langle \mathsf{A} \rangle$ or $\langle \overline{\mathsf{A}} \rangle$ modality:

$$\langle \mathsf{B} P \lceil Q \rceil R \rangle (\beta_1 \wedge \langle \mathsf{A} \rangle \beta_2).$$

We will make use of the number of *direction alternations* in a formula: each occurrence of a $\langle \mathsf{D} \rangle$ or $\langle \mathsf{E} \rangle$ modality immediately inside a $\langle \mathsf{B} \rangle$ modality, and each occurrence of a $\langle \mathsf{D} \rangle$ or $\langle \mathsf{B} \rangle$ modality immediately inside a $\langle \mathsf{E} \rangle$ modality, and each occurrence of a $\langle \mathsf{B} \rangle$,$\langle \mathsf{E} \rangle$,$\langle \mathsf{D} \rangle$ modality immediately inside a $\langle \mathsf{D} \rangle$ modality, counts as a direction alternation.

In each case, we give rules to rewrite this formula to a formula 'closer' to the bundled fragment of *InvHS*, that is, where each invariance-guarded modality is immediately followed by an $\langle \mathsf{A} \rangle$ or $\langle \overline{\mathsf{A}} \rangle$ modality. The rules have simple ideas: requirements are pulled out to the endpoints of the interval if possible, otherwise nested modalities are converted to a sequence of modalities 'fenced' by $\langle \mathsf{A} \rangle$, $\langle \overline{\mathsf{A}} \rangle$ modalities. This reduces the modal depth of loose subformulae. This does not happen in case there is a direction alternation between adjoining modalities, the best the sequencing can do is to process them in the same direction, reducing the number of such alternations.

Let us run through the cases. If β_1 is a point formula, its point requirements can be handled in the invariance guard, with any $\langle \mathsf{A} \rangle$,$\langle \overline{\mathsf{A}} \rangle$ modal requirements being pulled out. Disjunctions can also be pulled out.

Now we have cases where β_1 is headed by a modality followed by a formula γ_1. If $\beta_1 = \langle \overline{\mathsf{A}} \rangle \gamma_1$ (or even a boolean combination of such formulas), it can be pulled out. Nested invariance-guarded B formulae first have an intersecting interval, then the remaining outer condition is fulfilled. We call this *distribution* of the invariance requirements. Nesting an invariance-guarded $[\mathsf{E}]$ formula is dealt with similarly, but a nested $\langle \mathsf{E} \rangle$ formula is linearized. Nesting an invariance-guarded $[\mathsf{D}]$ formula combines features from the treatment of nested $[\mathsf{B}]$ and $[\mathsf{E}]$ formulae, a nested $\langle \mathsf{D} \rangle$ formula combines features from the treatment of nested $\langle \mathsf{B} \rangle$ and $\langle \mathsf{E} \rangle$ formulae.

In each case either the formula mapped to is already in the bundled fragment, or the loose subformulae in the mapped formula have lesser modal depth than β_1, or the number of alternations in the mapped formula is lesser than β_1. Thus the rules can be recursively applied to obtain a reduction into the bundled fragment. Here are the rewrite rules for these cases, the left column gives the loose formula β_1 being considered. We also specify the closeness achieved in each rule.

$\langle\overline{A}\rangle\gamma_1 \wedge \neg\langle\overline{A}\rangle\gamma_2$	$\langle B P\lceil Q\rceil R\rangle(\langle\overline{A}\rangle\gamma_1 \wedge \neg\langle\overline{A}\rangle\gamma_2 \wedge \langle A\rangle\beta_2)$, *modal depth decreases* $\hookrightarrow \langle\overline{A}\rangle\gamma_1 \wedge \neg\langle\overline{A}\rangle\gamma_2 \wedge \langle B P\lceil Q\rceil R\rangle\langle A\rangle\beta_2$
$[B P'\lceil Q'\rceil R']\gamma_1$	$\langle B P\lceil Q\rceil R\rangle([B P'\lceil Q'\rceil R']\gamma_1 \wedge \langle A\rangle\beta_2)$, *modal depth decreases* $\hookrightarrow [B(P \wedge P')\lceil Q \wedge Q'\rceil R']\gamma_1 \wedge \langle E pt\rangle(R \wedge \langle A\rangle\beta_2)$
$\langle B P'\lceil Q'\rceil R'\rangle\gamma_1$	$\langle B P\lceil Q\rceil R\rangle(\langle B P'\lceil Q'\rceil R'\rangle\gamma_1 \wedge \langle A\rangle\beta_2)$, *modal depth decreases* $\hookrightarrow \langle B(P \wedge P')\lceil Q \wedge Q'\rceil(R \wedge R')\rangle(\gamma_1 \wedge \langle A\rangle\langle E R'\lceil Q\rceil R\rangle\beta_2)$
$[E P'\lceil Q'\rceil R']\gamma_1$	$\langle B P\lceil Q\rceil R\rangle([E P'\lceil Q'\rceil R']\gamma_1 \wedge \langle A\rangle\beta_2)$, *modal depth decreases* $\hookrightarrow \langle B pt\rangle P \wedge [E P'\lceil Q \wedge Q'\rceil(R \wedge R')](\gamma_1 \wedge \langle A\rangle\beta_2)$
$\langle E P'\lceil Q'\rceil R'\rangle\gamma_1$	$\langle B P\lceil Q\rceil R\rangle(\langle E P'\lceil Q'\rceil R'\rangle\gamma_1 \wedge \langle A\rangle\beta_2)$, *direction alternation decreases* $\hookrightarrow \langle B P\lceil Q\rceil P'\rangle\langle A\rangle(\langle B P'\lceil Q \wedge Q'\rceil(R \wedge R')\rangle\gamma_1 \wedge \langle A\rangle\beta_2)$
$[D P'\lceil Q'\rceil R']\gamma_1$	$\langle B P\lceil Q\rceil R\rangle([D P'\lceil Q'\rceil R']\gamma_1 \wedge \langle A\rangle\beta_2)$, *modal depth decreases* $\hookrightarrow \langle B pt\rangle P \wedge [D P'\lceil Q \wedge Q'\rceil R'](\gamma_1 \wedge \langle E pt\rangle(R \wedge \langle A\rangle\beta_2))$
$\langle D P'\lceil Q'\rceil R'\rangle\gamma_1$	$\langle B P\lceil Q\rceil R\rangle(\langle D P'\lceil Q'\rceil R'\rangle\gamma_1 \wedge \langle A\rangle\beta_2)$, *direction alternation decreases* $\hookrightarrow \langle B P\lceil Q\rceil P'\rangle\langle A\rangle(\langle B P'\lceil Q \wedge Q'\rceil(R \wedge R')\rangle(\gamma_1 \wedge \langle A\rangle\langle E P\lceil Q\rceil R\rangle\beta_2))$

Finally consider the cases where β_1 is a boolean combination of invariance-guarded modalities. We use the fact that there is a linear ordering and there are only a few order types to deal with, which we can disjunct over. Two invariance-guarded *B* formulae in the same interval can be ordered, or seen as one of two possible orderings. If there is a *B* and an *E* formula, there are three possibilities for their satisfying subintervals: either they are disjoint, or they meet, or they overlap. Each possibility leads to the distribution of the invariance requirements over upto three intervals. The *D* modality can be handled by another set of distribution disjuncts. We do not spell out the rules in detail.

As before, by recursively applying the rules we obtain formulae in bundled *InvHS*. Putting together the rules for all formulae outside the bundled fragment gives an argument that *InvHS* formulae can be equivalently put into its bundled fragment. Hence *InvHS* is expressively complete for $FO^2[<, bet]$. □

5 Drawing Some Conclusions

As we observed after Definition 1, the logic *InvHS* is semantically no different from *HS*, so Lemma 5 yields an expressiveness result for Halpern–Shoham logic. Venema (1991) showed that *HS* is weaker than full first-order logic, we can now make the weakness precise.

Theorem 6 *HS is expressively complete for* $FO^2[<, bet]$.

We put our expressiveness characterization mentioned in Sect. 3 to use, by borrowing a result from Krebs et al. (2016). Defining arbitrarily many subintervals within arbitrarily many subintervals is not in general possible in Halpern–Shoham logic:

Corollary 7 *The regular language* $(a(ab)^*b)^+$ *is not definable by an HS formula.*

To further analyse the translation provided in the above result, we measure the size of *HS* formulae using their *dag-size* (see the book of Demri et al. 2016). That is, we think of a formula as being represented by a directed acyclic graph and count only the number of distinct subformulae of the formula. Each rewrite rule in the translation of Lemma 5 changes dag-size by a constant, so the recursive translation maps an *InvHS* formula to a bundled *InvHS* formula whose dag-size only grows polynomially. Building upon this one obtains a complexity result.

Theorem 8 *The satisfiability problem for HS over finite word models is complete for polynomial space.*

Proof Since $LTL[\mathsf{F}, \mathsf{P}, \mathsf{X}, \mathsf{Y}]$ formulae can be represented in the logic, there is a polynomial space lower bound on the complexity of satisfiability, shown by Sistla and Clarke (1985).

For the upper bound, there is a trivial translation to *InvHS*. The proof of Lemma 5 gives a translation into bundled *InvHS*, which is polynomial in the dag-size of the formula. From the proof of Lemma 4 we can extract a more direct translation from bundled *InvHS*, not into two-variable between logic, but to the point temporal logic of Krebs et al. (2016) (mentioned in the proof of Lemma 2). That paper provides a further polynomial translation to standard linear temporal logic *LTL*. Finally, that the satisfiability of *LTL* formulae can be checked in space polynomial in their dag-size was shown by Sistla and Clarke (1985). □

As we mentioned in the introduction of this paper, in our interval logics, propositions are interpreted at points and not at intervals. Thus we do not encounter the high complexities of two-dimensional HS logics detailed in papers such as Bresolin et al. (2014). The expressiveness of these logics within three-variable first-order logic should be higher, but remains unexplored.

It would be interesting to extend the ideas of this paper to fragments of the Duration Calculus on timed word models. The dissertation of Shah (2012) looks at unary point temporal logics with metric interval modalities.

The work of Wakankar et al. (2017) suggests a different direction, moving from satisfiability problems to those of synthesis. This will require coming up with a suitable notion of automata.

Acknowledgements Many thanks to Fenrong Liu and Hiroakira Ono for the invitation to speak at the Asian workshop on philosophical logic in Beijing, where the beginnings of this work were presented. Junhua Yu and his team at Tsinghua University did an excellent job with the organization. I would like to thank Yanjing Wang at Peking University for all his help.

Numerous discussions with Paritosh Pandya, our earlier joint work with Simoni Shah, and our recent results with Andreas Krebs and Howard Straubing, have shaped this paper. Several questions by Johan van Benthem at the Beijing AWPL persuaded me that putting together the examples of Sect. 3 would help illustrate my ideas. Anantha Padmanabha brought to my attention a bundled fragment of his logic, and that led me to expand my logic to full *HS*. I would like to thank the anonymous referee for a careful reading of the paper.

References

Allen, James F. 1983. Maintaining knowledge about temporal intervals. *Communications of the Association for Computing Machinery* 26 (11): 832–843.

Andréka, Hajnal, Istvan Németi, and Johan van Benthem. 1998. Modal languages and bounded fragments of predicate logic. *Journal of Philosophical Logic* 27 (3): 217–274.

Basin, David A., and Nils Klarlund. 1995. Hardware verification using monadic second-order logic. In *Proceedings of the 7th CAV, Liège*, ed. Pierre Wolper, vol. 939. LNCS, 31–41.

Bresolin, Davide, Valentin Goranko, Angelo Montanari, and Guido Sciavicco. 2009a. Propositional interval neighbourhood logics: Expressiveness, decidability, and undecidable extensions. *Annals of Pure and Applied Logic* 161 (3): 289–304.

Bresolin, Davide, Dario Della Monica, Valentin Goranko, Angelo Montanari, and Guido Sciavicco. 2009b. Undecidability of interval temporal logics with the overlap modality. In *Proceedings of the 16th TIME, Bressanone-Brixen*, ed. Carsten Lutz, and Jean-François Raskin, 88–95.

Bresolin, Davide, Dario Della Monica, Angelo Montanari, Pietro Sala, and Guido Sciavicco. 2014. Interval temporal logics over strongly discrete linear orders: Expressiveness and complexity. *Theoretical Computer Science* 560 part 3: 269–291.

Chandra, Ashok, Steven Fortune, and Richard Lipton. 1985. Unbounded fan-in circuits and associative functions. *Journal of Computer and System Sciences* 30 (2): 222–234.

Chaochen, Zhou, and Michael Hansen. 2004. *Duration calculus*. Berlin: Springer.

Chaochen, Zhou, Tony Hoare, and Anders Ravn. 1991. A calculus of durations. *Information Processing Letters* 40 (5): 269–276.

Demri, Stéphane, Valentin Goranko, and Martin Lange. 2016. *Temporal logics in computer science: Finite state systems*. Cambridge: CUP.

Goranko, Valentin, Angelo Montanari, and Guido Sciavicco. 2004. A roadmap of interval temporal logics and duration calculi. *Journal of Applied Non-classical Logics* 14 (1–2): 9–54.

Gurevich, Yuri, and Harry R. Lewis. 1984. A logic for constant-depth circuits. *Information and Control* 61 (1): 65–74.

Halpern, Joseph Y., and Yoav Shoham. 1991. A propositional modal logic of time intervals. *Journal of the Association for Computing Machinery* 38 (4): 935–962.

Halpern, Joseph Y., Zohar Manna, and Ben C. Moszkowski. 1983. A hardware semantics based on temporal intervals. In *Proceedings of the 10th ICALP, Barcelona*, ed. Josep Díaz, vol. 154. LNCS, 278–291.

Immerman, Neil. 1987. Languages that capture complexity classes. *SIAM Journal on Computing* 16 (4): 760–778.

Kamp, Johan Anthony Willem. 1968. Tense logic and the theory of linear order. Ph.D. dissertation, Department of Philosophy, UCLA.

Krebs, Andreas, Kamal Lodaya, Paritosh K. Pandya, and Howard Straubing. 2016. Two-variable logic with a between relation. In *Proceedings of the 31st LICS, New York*, ed. Martin Grohe, Erik Koskinen, and Natarajan Shankar, 106–115. ACM/IEEE.

Krebs, Andreas, Kamal Lodaya, Paritosh K. Pandya, and Howard Straubing. 2018. An algebraic decision procedure for two-variable logic with a between relation. In *Proceedings of the 27th CSL, Birmingham*, ed. Dan Ghica, and Achim Jung, vol. 119. LIPIcs, 28–1. Dagstuhl, 17 pp.

Lodaya, Kamal. 2000. Sharpening the undecidability of interval logic. In *Proceedings of the 6th Asian, Penang*, ed. Jifeng He, and Masahiko Sato, vol. 1961. LNCS, 290–298.

Lodaya, Kamal, and Paritosh K. Pandya. 2006. A dose of timed logics. In *Proceedings of the 4th formats, Paris*, ed. Eugene Asarin, and Patricia Bouyer, vol. 4202. LNCS, 260–273.

Lodaya, Kamal, and Paritosh K. Pandya. 2017. Deterministic temporal logics and interval modalities. In *Proceedings of the 9th Methamphetamine Modalities*, ed. Sujata Ghosh, and R. Ramanujam, vol. 243. Eptcs, 23–40.

Lodaya, Kamal, Paritosh K. Pandya, and Simoni S. Shah. 2008. Marking the chops: An unambiguous temporal logic. In *Proceedings of the 5th IFIP TCS, Milano*, ed. Giorgio Ausiello, Juhani Karhumäki, Giancarlo Mauri, and Luke Ong, vol. 273. IFIP, 461–476.

Lodaya, Kamal, Paritosh K. Pandya, and Simoni S. Shah. 2010. Around dot depth two. In *Proceedings of the 14th DLT, London (CA)*, ed. Yuan Gao, Hanlin Lu, Shinnosuke Seki, and Sheng Yu, vol. 6224. LNCS, 303–314.

Marcinkowski, Jerzy, and Jakub Michaliszyn. 2011. The ultimate undecidability result for the Halpern-Shoham logic. In *Proceedings of the 26th LICS, Toronto*, 377–386.

Moszkowski, Ben. 1983. Reasoning about digital circuits. Ph.D. dissertation, Department of Computer Science, Stanford University.

Padmanabha, Anantha, R. Ramanujam, and Yanjing Wang. 2018. Bundled fragments of first-order modal logic: (Un)Decidability. In *Proceedings of the 38th FSTTCS, Ahmedabad*, ed. Sumit Ganguly, and Paritosh K. Pandya, vol. 122. LIPIcs, 43–1. Dagstuhl, 20 pp.

Pandya, Paritosh K. 2001. Specifying and deciding quantified discrete-time duration calculus formulae using DCVALID: An automata theoretic approach. In *Proceedings of the RTTOOLS, Aalborg*.

Pin, Jean-Éric. 1986. Varieties of formal languages. Plenum. Translated by A. Howie.

Rabinovich, Alexander. 2000. Expressive completeness of duration calculus. *Information and Computation* 156 (1/2): 320–344.

Schwentick, Thomas, Denis Thérien, and Heribert Vollmer. 2002. Partially-ordered two-way automata: A new characterization of DA. In *Proceedings of the 5th DLT, Vienna*, ed. Werner Kuich, Grzegorz Rozenberg, and Arto Salomaa, vol. 2295. LNCS, 239–250.

Shah, Simoni S. 2012. Unambiguity and timed languages: Automata, languages, expressiveness. Ph.D. dissertation, Technology and Computer Science, TIFR.

Sipser, Michael. 1983. Borel sets and circuit complexity. In *Proceedings of the 15th STOC, Boston*, 61–69. ACM.

Sistla, A. Prasad, and Edmund M. Clarke. 1985. The complexity of propositional linear temporal logics. *Journal of the Association for Computing Machinery* 32 (3): 733–749.

Straubing, Howard. 1994. *Finite automata, formal languages, and circuit complexity*. Boston: Birkhäuser.

Tesson, Pascal, and Denis Thérien. 2002. Diamonds are forever: The variety DA. In *Semigroups, algorithms, automata and languages*, ed. Gracinda Gomes, Jean-Éric Pin, and Pedro Silva, 475–499. Singapore: World Scientific.

Tesson, Pascal, and Denis Thérien. 2007. Logic meets algebra: The case of regular languages. *Logical Methods in Computer Science* 3 (1:4): 1–37.

Thérien, Denis, and Thomas Wilke. 1998. Over words, two variables are as powerful as one quantifier alternation. In *Proceedings of the 30th STOC, Dallas*, ed. Jeffrey Vitter, 234–240. ACM.

Thérien, Denis, and Thomas Wilke. 2004. Nesting until and since in temporal logic. *Theory of Computing Systems* 37 (1): 111–131.

Thomas, Wolfgang. 1997. Languages, automata and logic. In *Handbook of formal language theory III*, ed. Grzegorz Rozenberg, and Arto Salomaa. Berlin: Springer.

van Benthem, Johan. 1983. *The logic of time*. Kufstein: Reidel.

Venema, Yde. 1989. Expressiveness and completeness of an interval tense logic. *Notre Dame Journal of Formal Logic* 31 (4): 529–547.

Venema, Yde. 1991. A modal logic for chopping intervals. *Journal of Logic and Computation* 1 (4): 453–476.

Wakankar, Amol, Paritosh K. Pandya, and Raj Mohan Matteplackel. 2017. DCSYNTH: Guided reactive synthesis with soft requirements for robust controller and shield synthesis.

Wilke, Thomas. 1994. Specifying timed state sequences in powerful decidable logics and timed automata. In *Proceedings of the 3rd FTRTFT, Lübeck*, ed. Hans Langmaack, Willem-Paul de Roever, and Jan Vytopil, vol. 863. LNCS, 694–715.

Logics for Knowability Paradox with a Non-normal Possibility Operator

Youan Su and Katsuhiko Sano

Abstract This paper studies a Glivenko–Kuroda translation in logics for knowability paradox in terms of sequent calculi. Hilbert-style axiomatization and Gentzen-style sequent calculus are proposed for an intuitionistic modal logic with and without the knowability axiom. When the system does not contain the knowability axiom, it is shown to be sound and complete for an intuitionistic version of a combination of Kripke and neighborhood semantics. We define a double negation translation from our intuitionistic modal logics to the corresponding classical modal logics to establish syntactic proofs of Glivenko-style embedding results with and without the knowability axiom. In the setting without the knowability axiom, our embedding result enables us to reduce the cut-elimination theorem, Craig interpolation theorem, and decidability from the classical setting to the intuitionistic setting.

1 Introduction

The knowability paradox, also known as the Fitch–Church paradox, states that, if we claim that every truth is knowable $A \supset \Diamond KA$, then we are forced to accept its consequence that every truth is known $A \supset KA$ (Fitch 1963). This paradox is commonly recognized as a threat to Dummett's semantic anti-realism. It is because the semantic anti-realists claim the knowability principle (**KP**), i.e., every truth is knowable, but they do not accept the omniscience principle (**OP**), i.e., every truth is known.

This chapter is in its final form and it is not submitted to publication anywhere else.

Y. Su (✉)
Graduate School of Letters, Hokkaido University, Sapporo, Japan
e-mail: ariyasu613@gmail.com

K. Sano
Faculty of Humanities and Human Sciences, Hokkaido University, Sapporo, Japan
e-mail: v-sano@let.hokudai.ac.jp

F. Liu et al. (eds.), *Knowledge, Proof and Dynamics*,
Logic in Asia: Studia Logica Library,
https://doi.org/10.1007/978-981-15-2221-5_4

Since Dummett admitted he had taken some of the intuitionistic basic features as a model for the anti-realist view (Dummett 1978, p. 164), the method of switching an underlying logical system to the intuitionistic one to avoid the paradox is studied continually after its first introduction by Williamson (1982). However, this approach is criticized by Percival (1990), Kvanvig (1995) and Murzi (2010). Their points are concentrated on the unacceptability of the weak omniscience principle **(wOP)** $A \supset \neg\neg K A$, which is still an intuitionistic consequence of **(KP)**. As a reply to them,[1] DeVidi and Solomon (2001), and Dummett (2009) himself suggest that we can read double negation $\neg\neg$ as a strong possibility, then we can justify the **(wOP)** by claiming that it just expresses that it is not the case that we do not know the truth.

Artemov and Protopopescu (2010) suggest that the **(KP)** $A \supset \Diamond K A$ is intuitionistically equivalent to **(wOP)** $A \supset \neg\neg K A$, by defining $\Diamond$ as $\neg\Box\neg$ and requiring that $\Box$ satisfies T-axiom. However, we find there seems no reason why we need to define a diamond operator in terms of $\Box$ in an intuitionistic system. In our paper, we choose to treat the $\Diamond$ as primitive and we will not commit ourselves to the equivalence between **(KP)** and **(wOP)**. Then we would like to consider the following question.

- Question 1: What kind of properties should K and $\Diamond$ have in the derivation of the knowability paradox?

We show that $\Diamond$ does not need to be normal (i.e., $\Diamond$ commutes with finite disjunctions) in deriving **(OP)** and **(wOP)** in the discussions of the knowability paradox. This paper sets up a normal knowledge operator K satisfying T-axioms and a monotone $\Diamond$ in the classic and intuitionistic systems.[2]

Moreover, in order to show that the intuitionistic solution can avoid the paradox, we would like to answer the following question explicitly

- Question 2: Is (OP) $A \supset K A$ not derivable in the intuitionistic system with the knowability axiom?

We do so by giving an intuitionistic Kripke semantics and providing a counter model in Sect. 3. We also prove the soundness and the completeness of that system.

It is well-known that the double negation $\neg\neg$ plays an important role in the embedding of the classical logic into the intuitionistic logic, for example in the translation of Kolmogorov, Gödel-Gentzen, Kuroda and Glivenko (Litak et al. 2017). However, it still remains unclear about how the double negation affects a translation in the problem of knowability paradox. Since the intuitionistic approach to solve the paradox heavily relies on switching logical systems, a translation between two systems can provide a better perception. There remains a possibility that the translation of a derivation of (OP)$A \supset K A$ into the intuitionistic system may cause trouble. If so, the intuitionistic approach to solve the paradox has difficulty. We need to check if it is the case.

[1]Some goes much deeper, Hudson (2009), Artemov and Protopopescu (2016) and Usberti (2016) suggest that not only **(wOP)** but also **(OP)** is acceptable from the view of intuitionism or anti-realism.

[2]Recently, a minimal logic to derive the knowability paradox has been studied in Wang (2018).

Litak et al. (2017) give a full-view of negation translations in modal propositional logic, and specify the sufficient syntactic criteria of a successful embedding from classical modal logics to intuitionistic modal logics. However, they do not cover $\Diamond$. If we regard $\Diamond$ as primitive, the logics for the knowability paradox cannot be dealt with by their methods. When the underlying syntax contains the knowledge operator K and the diamond operator $\Diamond$ as primitive, we would like to ask the following question.

- Question 3: What is a proper Glivenko–Kuroda translation for K and $\Diamond$?

Besides, a double negation translation can provide a method to transfer the logical properties of an intuitionistic system to those of the corresponding classical system. In this paper we show that with the help of a Glivenko-style theorem we are able to transfer the cut-elimination theorem, interpolation theorem, and decidability of the intuitionistic system to that of the corresponding classical system in Sects. 4 and 5.

The method we apply is mainly on the basis of proof-theory, especially sequent calculi. Until now, a labeled structural rule free sequent calculus for the knowability paradox has been given by Maffezioli et al. (2013). They use a semantic frame condition of **(KP)** to conclude that **(OP)** does not hold in the intuitionistic system with **(KP)**. Compared to their study, our study concentrates on the translation between classical and intuitionistic systems. For this purpose, we do not employ a labeled formalism but stick to the ordinary notion of (non-labeled) formula to consider sequent calculi with structural rules. While Maffezioli et al. (2013) regard $\Diamond$ as normal, our $\Diamond$ is non-normal.

The outline of this paper is as follows: In Sect. 2, we provide two systems: **G1K**$\Diamond$**c** and **G1K**$\Diamond$**i**. Their modal operators consist of a knowledge operator K and a modal operator $\Diamond$. We give our answer to Question 1 in this section.

In Sect. 3, we provide a Kripke semantics for **G1K**$\Diamond$**i** equipped with an accessibility relation of K and a neighborhood mapping of $\Diamond$. We also give a Hilbert system **HK**$\Diamond$**i** which is equivalent to **G1K**$\Diamond$**i**, and prove the soundness and completeness for the intended semantics. Also in this section, We give a counter model to show that (OP) is not derivable in **G1K**$\Diamond$**i**, in order to answer Question 2.

Next, as an answer to Question 3, we prove the Glivenko-style theorem in Sect. 4. After proving the cut-elimination theorem of **G1K**$\Diamond$**i**, we prove that of **G1K**$\Diamond$**c** via the double negation translation. Section 5 proves the Craig interpolation theorems for both systems. Section 6 shows the decidability of **G1K**$\Diamond$**i** and **G1K**$\Diamond$**c**.

2 Syntax and Sequent Calculi for Knowability Paradox

In this section, we answer Question 1 in the introduction: what kind of properties should K and $\Diamond$ have in the intuitionistic derivation of the knowability paradox? We first define the classical and intuitionistic sequent calculi systems. Then we give the derivations of knowability paradox. We show that, for K, we follow the idea of Hintikka (2005) to define it as a normal modal operator satisfying (T)-axiom; for $\Diamond$, we define it as a non-normal but monotone modal operator.

2.1 Syntax

The set of formulas of the language $\mathcal{L}$ is defined inductively as

$$A := p \mid \bot \mid \neg A \mid A \& A \mid A \vee A \mid A \supset A \mid \Diamond A \mid K A,$$

where $p \in \mathsf{Prop}$ and Prop is a countably infinite set of propositional variables. Now let us look at an outline of a derivation of (**OP**) from (**KP**) in a classical setting, where we assume that the knowledge operator K is normal and satisfies the T-axiom scheme $KA \supset A$ and the possibility operator $\Diamond$ is also normal. In what follows in this section, "$\vdash$" means a theoremhood of bimodal logic with K and $\Diamond$ just specified. Then the knowability paradox can be formalized as follows:

(1) $\vdash \neg K(A \& \neg KA)$	T-axiom for knowledge operator
(2) $\vdash \neg \Diamond K(A \& \neg KA)$	Necessitation Law for $\Box$ and $\vdash \Box\neg\varphi \leftrightarrow \neg\Diamond\varphi$
(3) $\vdash (A \& \neg KA) \supset \Diamond K(A \& \neg KA)$	Knowability Principle
(4) $\vdash \neg(A \& \neg KA)$	(2), (3), Contraposition and Modus Ponens
(5) $\vdash \neg(A \& \neg KA) \supset (A \supset KA)$	Classical propositional Logic
(6) $\vdash A \supset KA$	(4), (5), Modus Ponens.

We note that if we block the step (5) then we can avoid concluding (**OP**). In the next section, we analyze a derivation of (**OP**) from (**KP**) in terms of sequent calculi.

2.2 The Systems of **G1K◇c** and **G1K◇i**

A *sequent*, denoted by $\Gamma \Rightarrow \Delta$, is a pair of finite multisets of formulas. The formulas Γ are called the *antecedents* of $\Gamma \Rightarrow \Delta$, instead Δ are called *succedents* of the sequent $\Gamma \Rightarrow \Delta$. When $\circ \in \{K, \Diamond, \neg\}$, we define $\circ\Gamma = \{\circ A \mid A \in \Gamma\}$. Table 1 provides four sequent calculi in this paper. The calculus **G1K◇c** captures a bimodal logic where the knowledge operator K is normal and obeys T-axiom $KA \supset A$ and $\Diamond$ is a monotone operator. The system of **G1K◇i** is obtained from **G1K◇c** by restricting our attention to the sequents whose succedents Δ contain at most one formula. Then, in the **G1K◇i** and **G1K◇c**, for K we define it as a normal modal operator satisfying (T)-axiom; for $\Diamond$, we define it as a monotone operator. If we add the axiom of knowability principle to **G1K◇c** or **G1K◇i** as an initial sequent, the extended systems are named as **G1K◇c**$^+$ or **G1K◇i**$^+$, respectively.

Definition 1 (*Context and Principal Formula*) Let $\mathcal{G}$ be one of systems of Table 1. The Γ and Δ in an inference rule of $\mathcal{G}$ except (K) and ($\Diamond$), are called the *context*. In the conclusion of each rule of $\mathcal{G}$ except (K) and ($\Diamond$), the formula(s) not in the context is called the *principal formula(s)*. For the rule of (K) and ($\Diamond$), all formulas in the conclusions of these rules are called the principal formulas.

Definition 2 (*Derivation*) A derivation $\mathcal{D}$ in $\mathcal{G}$ is a finite tree generated by rules of $\mathcal{G}$ from initial sequents of $\mathcal{G}$. We say that the *end sequent* of $\mathcal{D}$ is the sequent in the root node of $\mathcal{D}$. The *height* of a derivation is the maximum length of branches

Table 1 Sequent Calculi **G1K◊c**, **G1K◊c**$^+$, **G1K◊i** and **G1K◊i**$^+$

Sequent Calculus **G1K◊c**

Initial Sequents $A \Rightarrow A \qquad \bot \Rightarrow$

Structural Rules

$$\frac{\Gamma \Rightarrow \Delta}{A, \Gamma \Rightarrow \Delta}\,(LW) \qquad \frac{\Gamma \Rightarrow \Delta}{\Gamma \Rightarrow \Delta, A}\,(RW) \qquad \frac{A, A, \Gamma \Rightarrow \Delta}{A, \Gamma \Rightarrow \Delta}\,(LC)$$

$$\frac{\Gamma \Rightarrow \Delta, A, A}{\Gamma \Rightarrow \Delta, A}\,(RC) \qquad \frac{\Gamma \Rightarrow \Delta, A \qquad A, \Gamma' \Rightarrow \Delta'}{\Gamma, \Gamma' \Rightarrow \Delta, \Delta'}\,(Cut)$$

Logical Rules

$$\frac{\Gamma \Rightarrow \Delta, A_1 \qquad \Gamma \Rightarrow \Delta, A_2}{\Gamma \Rightarrow \Delta, A_1 \& A_2}\,(R\&) \qquad \frac{A_i, \Gamma \Rightarrow \Delta}{A_1 \& A_2, \Gamma \Rightarrow \Delta}\,(L\&)$$

$$\frac{\Gamma \Rightarrow \Delta, A_i}{\Gamma \Rightarrow \Delta, A_1 \vee A_2}\,(R\vee) \qquad \frac{A_1, \Gamma \Rightarrow \Delta \qquad A_2, \Gamma \Rightarrow \Delta}{A_1 \vee A_2, \Gamma \Rightarrow \Delta}\,(L\vee)$$

$$\frac{A, \Gamma \Rightarrow \Delta, B}{\Gamma \Rightarrow \Delta, A \supset B}\,(R\supset) \qquad \frac{\Gamma \Rightarrow \Delta, A \qquad B, \Gamma' \Rightarrow \Delta'}{A \supset B, \Gamma, \Gamma' \Rightarrow \Delta, \Delta'}\,(L\supset)$$

$$\frac{A, \Gamma \Rightarrow \Delta}{\Gamma \Rightarrow \Delta, \neg A}\,(R\neg) \qquad \frac{\Gamma \Rightarrow \Delta, A}{\neg A, \Gamma \Rightarrow \Delta}\,(L\neg)$$

$$\frac{A \Rightarrow \Delta}{\Diamond A \Rightarrow \Diamond \Delta}\,(\Diamond)\dagger \qquad \frac{\Gamma \Rightarrow A}{K\Gamma \Rightarrow KA}\,(K) \qquad \frac{A, \Gamma \Rightarrow \Delta}{KA, \Gamma \Rightarrow \Delta}\,(T)$$

†: Δ contains at most one formula.

Sequent Calculus **G1K◊i**: Δ contains at most one formula below.

Initial Sequents $A \Rightarrow A \qquad \bot \Rightarrow$

Structural Rules

$$\frac{\Gamma \Rightarrow \Delta}{A, \Gamma \Rightarrow \Delta}\,(LW) \qquad \frac{\Gamma \Rightarrow}{\Gamma \Rightarrow C}\,(RW) \qquad \frac{A, A, \Gamma \Rightarrow \Delta}{A, \Gamma \Rightarrow \Delta}\,(LC)$$

$$\frac{\Gamma \Rightarrow A \qquad A, \Gamma' \Rightarrow \Delta}{\Gamma, \Gamma' \Rightarrow \Delta}\,(Cut)$$

Logical Rules

$$\frac{\Gamma \Rightarrow A_1 \qquad \Gamma \Rightarrow A_2}{\Gamma \Rightarrow A_1 \& A_2}\,(R\&) \qquad \frac{A_i, \Gamma \Rightarrow \Delta}{A_1 \& A_2, \Gamma \Rightarrow \Delta}\,(L\&)$$

$$\frac{\Gamma \Rightarrow A_i}{\Gamma \Rightarrow A_1 \vee A_2}\,(R\vee) \qquad \frac{A_1, \Gamma \Rightarrow \Delta \qquad A_2, \Gamma \Rightarrow \Delta}{A_1 \vee A_2, \Gamma \Rightarrow \Delta}\,(L\vee)$$

$$\frac{A, \Gamma \Rightarrow B}{\Gamma \Rightarrow A \supset B}\,(R\supset) \qquad \frac{\Gamma \Rightarrow A \qquad B, \Gamma' \Rightarrow \Delta}{A \supset B, \Gamma, \Gamma' \Rightarrow \Delta}\,(L\supset)$$

$$\frac{A, \Gamma \Rightarrow}{\Gamma \Rightarrow \neg A}\,(R\neg) \qquad \frac{\Gamma \Rightarrow A}{\neg A, \Gamma \Rightarrow}\,(L\neg)$$

$$\frac{A \Rightarrow \Delta}{\Diamond A \Rightarrow \Diamond \Delta}\,(\Diamond) \qquad \frac{\Gamma \Rightarrow A}{K\Gamma \Rightarrow KA}\,(K) \qquad \frac{A, \Gamma \Rightarrow \Delta}{KA, \Gamma \Rightarrow \Delta}\,(T)$$

Additional Initial Sequent for **G1K◊c**$^+$ and **G1K◊i**$^+$

$\Rightarrow A \supset \Diamond KA$ **(KP)**

in the derivation from the end sequent to an initial sequent. A sequent $\Gamma \Rightarrow \Delta$ is *derivable* in $\mathcal{G}$ (notation: $\vdash_{\mathcal{G}} \Gamma \Rightarrow \Delta$) if it has a derivation $\mathcal{D}$ in $\mathcal{G}$ whose end sequent is $\Gamma \Rightarrow \Delta$.

In what follows in this section, we provide our proof-theoretic analysis with the knowability paradox in terms of our sequent calculi.

Proposition 1 *A sequent of the form* $A\&\neg KA \supset \Diamond K(A\&\neg KA) \Rightarrow A \supset KA$ *is derivable in* **G1K◇c**. *Therefore,* $\Rightarrow A \supset KA$ *is derivable in* **G1K◇c**$^{+}$.

Proof For the former, we provide the following derivation in **G1K◇c**

$$
\begin{prooftree}
\AxiomC{$A \Rightarrow A$}
\RightLabel{(RW)}
\UnaryInfC{$A \Rightarrow KA, A$}
\AxiomC{$KA \Rightarrow KA$}
\RightLabel{(LW)}
\UnaryInfC{$KA, A \Rightarrow KA$}
\RightLabel{$(R\neg)$}
\UnaryInfC{$A \Rightarrow KA, \neg KA$}
\RightLabel{$(R\&)$}
\BinaryInfC{$A \Rightarrow KA, A\&\neg KA$}
\AxiomC{$A \Rightarrow A$}
\RightLabel{$(L\&)$}
\UnaryInfC{$A\&\neg KA \Rightarrow A$}
\RightLabel{(K)}
\UnaryInfC{$K(A\&\neg KA) \Rightarrow KA$}
\RightLabel{$(L\neg)$}
\UnaryInfC{$\neg KA, K(A\&\neg KA) \Rightarrow$}
\RightLabel{$(L\&)$}
\UnaryInfC{$A\&\neg KA, K(A\&\neg KA) \Rightarrow$}
\RightLabel{(T)}
\UnaryInfC{$K(A\&\neg KA), K(A\&\neg KA) \Rightarrow$}
\RightLabel{(LC)}
\UnaryInfC{$K(A\&\neg KA) \Rightarrow$}
\RightLabel{$(\Diamond)$}
\UnaryInfC{$\Diamond K(A\&\neg KA) \Rightarrow$}
\RightLabel{$(L\supset)$}
\BinaryInfC{$A, A\&\neg KA \supset \Diamond K(A\&\neg KA) \Rightarrow KA$}
\RightLabel{$(R\supset)$}
\UnaryInfC{$A\&\neg KA \supset \Diamond K(A\&\neg KA) \Rightarrow A \supset KA.$}
\end{prooftree}
$$

Then we can obtain the derivability of $\Rightarrow A \supset KA$ in **G1K◇c**$^{+}$ with the help of the cut rule and an initial sequent $\Rightarrow A\&\neg KA \supset \Diamond K(A\&\neg KA)$.

We note that we have more than one formula in the succedent of sequents in some steps of the derivation above in **G1K◇c**. So, this derivation cannot be regarded as a derivation of the intuitionistic system **G1K◇i**. Instead, we have the following result.

Proposition 2 *A sequent of the form* $A\&\neg KA \supset \Diamond K(A\&\neg KA) \Rightarrow A \supset \neg\neg KA$ *is derivable in* **G1K◇i**. *Therefore,* $\Rightarrow A \supset \neg\neg KA$ *is derivable in* **G1K◇i**$^{+}$.

Proof We focus on showing the former.

$$
\begin{prooftree}
\AxiomC{$A \Rightarrow A$}
\RightLabel{(LW)}
\UnaryInfC{$\neg KA, A \Rightarrow A$}
\AxiomC{$\neg KA \Rightarrow \neg KA$}
\RightLabel{(LW)}
\UnaryInfC{$\neg KA, A \Rightarrow \neg KA$}
\RightLabel{$(R\&)$}
\BinaryInfC{$A, \neg KA \Rightarrow A\&\neg KA$}
\AxiomC{$A \Rightarrow A$}
\RightLabel{$(L\&)$}
\UnaryInfC{$A\&\neg KA \Rightarrow A$}
\RightLabel{(K)}
\UnaryInfC{$K(A\&\neg KA) \Rightarrow KA$}
\RightLabel{$(L\neg)$}
\UnaryInfC{$\neg KA, K(A\&\neg KA) \Rightarrow$}
\RightLabel{$(L\&)$}
\UnaryInfC{$A\&\neg KA, K(A\&\neg KA) \Rightarrow$}
\RightLabel{(T)}
\UnaryInfC{$K(A\&\neg KA), K(A\&\neg KA) \Rightarrow$}
\RightLabel{(LC)}
\UnaryInfC{$K(A\&\neg KA) \Rightarrow$}
\RightLabel{$(\Diamond)$}
\UnaryInfC{$\Diamond K(A\&\neg KA) \Rightarrow$}
\RightLabel{$(L\supset)$}
\BinaryInfC{$A, \neg KA, A\&\neg KA \supset \Diamond K(A\&\neg KA) \Rightarrow$}
\RightLabel{$(R\neg)$}
\UnaryInfC{$A, A\&\neg KA \supset \Diamond K(A\&\neg KA) \Rightarrow \neg\neg KA$}
\RightLabel{$(R\supset)$}
\UnaryInfC{$A\&\neg KA \supset \Diamond K(A\&\neg KA) \Rightarrow A \supset \neg\neg KA$}
\end{prooftree}
$$

The latter is similarly proved to the proof of Proposition 1. □

We remark that Dummett (2009, p. 52)'s last statement on knowability paradox claims that the intuitionistic interpretation of $A \supset \neg\neg KA$**(wOP)** is "the possibility that A will come to be known always remains open". And he suggests that the **(wOP)** "captures the relation which the justificationist believes to obtain between truth and knowledge."

3 Hilbert Systems and Their Semantics of Intuitionistic Bimodal Logic for Knowability Paradox

In this section we answer Question 2 in the introduction: is (OP) $A \supset KA$ not derivable in the intuitionistic system? We define a Kripke semantics for the **G1K**$\Diamond$**i**, then we provide a counter model to show that (OP) is not derivable.

3.1 Hilbert Systems and Intuitionistic Kripke Semantics

Let us introduce Hilbert systems called **HK**$\Diamond$**i** and **HK**$\Diamond$**i**$^+$ which are shown to be equivalent to **G1K**$\Diamond$**i** and **G1K**$\Diamond$**i**$^+$, respectively.

Hilbert system **HK**$\Diamond$**i**	
(Taut)	All instantiations of intuitionistic propositional tautologies
(K)	$K(A \supset B) \supset (KA \supset KB)$
(NC)	$\neg\Diamond\bot$
(T)	$KA \supset A$
(MP)	From A and $A \supset B$, infer B
(Nec$_K$)	From A, infer KA
(Mon$_\Diamond$)	From $A \supset B$, infer $\Diamond A \supset \Diamond B$
Hilbert system **HK**$\Diamond$**i**$^+$	
All the axioms and rules of **HK**$\Diamond$**i**	
(KP)	$A \supset \Diamond KA$

Let $\mathcal{H}$ be one of **HK**$\Diamond$**i** and **HK**$\Diamond$**i**$^+$. Let $\Gamma \cup \{A\}$ be a set of formulas, A is *derivable* in $\mathcal{H}$ from Γ (notation: $\Gamma \vdash_{\mathcal{H}} A$) if there are finite formulas $A_1, \ldots, A_n \in \Gamma$ such that $\vdash_{\mathcal{H}} (A_1 \& \cdots \& A_n) \supset A$. The following is immediate.

Theorem 1 *For any formula A, $\vdash_{\mathbf{HK}\Diamond\mathbf{i}} A$ iff $\vdash_{\mathbf{G1K}\Diamond\mathbf{i}} \Rightarrow A$, and $\vdash_{\mathbf{HK}\Diamond\mathbf{i}^+} A$ iff $\vdash_{\mathbf{G1K}\Diamond\mathbf{i}^+} \Rightarrow A$.*

Let us move to the semantics. Our semantics is based on an intuitionistic Kripke semantics equipped with an accessibility relation R for the knowledge operator. It is noted that similar definitions can be found in Proietti (2012) and Maffezioli et al. (2013). Unlike their works, our monotone diamond operator $\Diamond$ does not need to commute with the disjunction, i.e., $\Diamond(p \vee q) \supset (\Diamond p \vee \Diamond q)$ might fail (see Proposition 4).[3] We use a neighborhood mapping for our monotone diamond operator.

Definition 3 A model $M = (W, \leq, R, \tau, V)$ is a quintuple where

- W is a non-empty set of possible states,

[3]For an intuitionistic modal logic with a monotone diamond operator, the reader is referred to Kojima (2012).

- $\leq \subseteq W \times W$ is a preorder,
- $R \subseteq W \times W$ is a binary relation such that $\leq; R \subseteq R$ (where ";" means the relational composition) and $\leq \subseteq R$,
- $\tau : W \to \mathcal{PP}(W)$ is a neighborhood mapping satisfying the following two conditions: for all $w, v \in W$,

(NE) $\tau(w) \neq \emptyset$,
(DEC) $w \leq v$ implies $\tau(v) \subseteq \tau(w)$,

- V: $\mathsf{Prop} \to \mathcal{P}(W)$ is a valuation function satisfying the following *persistency* condition: if $w \leq v$ and $w \in V(p)$, then $v \in V(p)$, for all $w, v \in W$.

Remark 1 van Benthem and Pacuit (2011) developed neighborhood models for evidence and belief, in which evidence is recorded as a family of sets of worlds and a rational belief is always grounded in the evidence available to an agent. In this sense, they regarded a set $\tau(w)$ of neighborhoods around w as the set of pieces of evidence which is available to the agent at the world w.

However, the intuitionistic knowledge represents, as Brouwer suggested, the intelligent condition of an ideal mathematician who never forgets what he or she has proved or learned. Moreover, since $\tau(w) \subseteq \mathcal{P}(W)$ a neighborhood can be a set of possible worlds that may not be persistent. This fact suggests that a neighborhood may not only contain the true evidence, but also some imprecise or wrong hypothesis in our setting. We follow this idea to suggest that we can interpret $\Diamond A$ as A is consistent with any current possible hypothesis. Then the two above conditions for $\tau(w)$ can be explained as follows.

- For the (NE) non-empty condition, it requires that $\tau(w)$ is non-empty for all $w \in W$, this allows us to say that we have at least one hypothesis or conjecture to find a piece of evidence.
- For the (DEC) decreasing condition, since our neighborhood $\tau(w)$ does not consist of persistent sets alone but arbitrary (non-persistent) sets, an element of $\tau(w)$ can be regarded as a possibly imprecise or wrong hypothesis. But if we get more correct knowledge (persistent sets) along with an intuitionistic pre-ordering, such correct knowledge may exclude some imprecise hypothesis (or conjecture). This justifies $\tau(v)$ is a subset of $\tau(w)$ when w is less than or equal to v.

Definition 4 Given a model $M = (W, \leq, R, \tau, V)$, we define a satisfaction relation $M, w \models A$ as follows:

$M, w \models p$	iff	$w \in V(p)$
$M, w \not\models \bot$		
$M, w \models A \& B$	iff	$M, w \models A$ and $M, w \models B$
$M, w \models A \vee B$	iff	$M.w \models A$ or $M, w \models B$
$M, w \models A \supset B$	iff	for all $v \in W$, $w \leq v$ and $M, v \models A$ imply $M, v \models B$
$M, w \models \neg A$	iff	for all $v \in W$, $w \leq v$ implies $M, v \not\models A$
$M, w \models KA$	iff	for all $v \in W$, wRv implies $M, v \models A$
$M, w \models \Diamond A$	iff	for all $X \in \tau(w)$ there is a $x \in X$ and $M, x \models A$

For a set $\Gamma \cup \{A\}$ of formulas and a class $\mathbb{M}$ of models, $\Gamma \models_{\mathbb{M}} A$ means that for all models $M \in \mathbb{M}$ and all states w in M, for every formula $B \in \Gamma$, $M, w \models B$ implies $M, w \models A$. When Γ is an emptyset, we simply write $\models_{\mathbb{M}} A$ to say that A is *valid* in $\mathbb{M}$.

Proposition 3 *For all formulas A and all states $w, v \in W$, if $w \leq v$ and $M, w \models A$ then $M, v \models A$.*

Proof By induction on the complexity of a formula A. We note that the condition (DEC) for τ and $\leq; R \subseteq R$ are necessary here. □

Definition 5 Given any model $M = (W, \leq, R, \tau, V)$, $X \subseteq W$ is *$\leq$-closed* if $w \leq v$ and $w \in X$ jointly implies $v \in X$, for all $w, v \in W$. We use $\mathbb{M}_{\text{all}}$ to mean the class of all models and $\mathbb{KP}$ is the class of all models $M = (W, \leq, R, \tau, V)$ satisfying the following condition: for every w and $\leq$-closed subset X of W, $w \in X$ implies that every $Y \in \tau(w)$ there exists $y \in Y$ such that $\{ z \mid yRz \} \subseteq X$.

Theorem 2 *If $\vdash_{\mathbf{HK\Diamond i}} A$ then $\models_{\mathbb{M}_{\text{all}}} A$. Moreover, if $\vdash_{\mathbf{HK\Diamond i^+}} A$ then $\models_{\mathbb{KP}} A$.*

Proof By induction on the derivation of A. We focus on the cases of axioms. The validity of $A \supset (B \supset A)$ in $\mathbb{M}_{\text{all}}$ holds by Proposition 3. The axiom (K) can be proved by Proposition 3 and the condition $\leq; R \subseteq R$. For (NC), we need that the neighborhood function satisfies the condition $\tau(w) \neq \emptyset$ for an arbitrary state w. (T) can be obtained on the condition that R satisfies $\leq \subseteq R$. For the system $\mathbf{HK\Diamond i^+}$, the validity of (KP) is shown by the condition required for $\mathbb{KP}$ in the previous definition. □

By giving a counter model, we can show that our monotone diamond operator $\Diamond$ does not commute with the disjunction.

Proposition 4 *The formula $\Diamond(p \vee q) \supset (\Diamond p \vee \Diamond q)$ is not valid in $\mathbb{M}_{\text{all}}$.*

Proof Define $M = (W, \leq, R, \tau, V)$ where $W = \{1, 2\}$, $\leq = \{(1, 1), (2, 2)\}$, $R = \{(1, 1), (2, 2)\}$, $\tau(1) = \{\{1\}, \{2\}\}$, $\tau(2) = \{\{2\}\}$, $V(p) = \{1\}$, $V(q) = \{2\}$. Then we have $M, 1 \models \Diamond(p \vee q)$, and $M, 1 \not\models \Diamond p \vee \Diamond q$. □

Proposition 5 *The formula $((p \& \neg Kp) \supset \Diamond K(p \& \neg Kp)) \supset (p \supset Kp)$ is not valid in $\mathbb{M}_{\text{all}}$ and $p \supset Kp$ is not valid in $\mathbb{KP}$.*

Proof Define $M = (W, \leq, R, \tau, V)$ where $W = \{1, 2, 3\}$, $\leq = \{(1, 3)\} \cup \{ (x, x) \mid x \in W \}$, $R = \{(1, 2), (1, 3)\} \cup \{ (x, x) \mid x \in W \}$, $\tau(1) = \tau(3) = \{\{3\}\}$ and $\tau(2) = \{\{2\}\}$, and $V(p) = \{1, 3\}$. In the following picture, solid lines stand for $\leq$ and dotted lines stand for R.

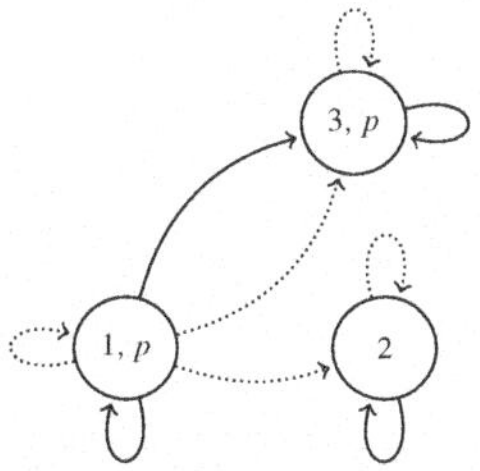

It is easy to see that M *is* a model and it also belongs to $\mathbb{KP}$. Since $M, 3 \models Kp$, we have $M, 1 \not\models p\&\neg Kp$ and $M, 3 \not\models p\&\neg Kp$, hence $M, 1 \models (p\&\neg Kp) \supset \Diamond K(p\&\neg Kp)$. Since $M, 1 \not\models p \supset Kp$, we have $M, 1 \not\models ((p\&\neg Kp) \supset \Diamond K(p\&\neg Kp)) \supset (p \supset Kp)$. This argument also shows that $p \supset Kp$ is not valid in $\mathbb{KP}$. □

By Proposition 5 and Theorems 1 and 2, we obtain the following corollary.

Corollary 1 *A sequent in the form of* $((p\&\neg Kp) \supset \Diamond K(p\&\neg Kp)) \Rightarrow (p \supset Kp)$ *is not derivable in* **G1K◊i** *and a sequent* $\Rightarrow p \supset Kp$ *is not derivable in* **G1K◊i**$^+$.

Proposition 6 *Neither* $\Diamond\top \supset \neg\neg\Diamond(p \vee \neg p)$ *nor* $\neg\neg K(p \vee \neg p)$ *is valid in* $\mathbb{M}_{\text{all}}$. *Therefore, neither* $\Diamond\top \Rightarrow \neg\neg\Diamond(p \vee \neg p)$ *nor* $\Rightarrow \neg\neg K(p \vee \neg p)$ *is derivable in* **G1K◊i**.

Proof We give one counter model against these two formulas. Let $M = (W, \leq, R, \tau, V)$ where $W = \{1, 2\}$, $\leq = \{(1, 1), (2, 2), (2, 1)\}$, $R = \{(1, 2), (2, 1), (1, 1), (2, 2)\}$, $\tau(1) = \tau(2) = \{\{2\}\}$, and $V(P) = \{1\}$. In the following picture, lines stand for $\leq$ and dotted lines stand for R.

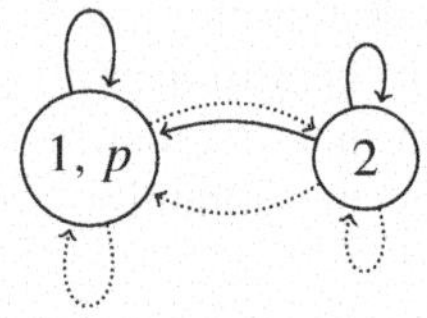

It is easy to see that M *is* a model. Then we have $M, 1 \models \Diamond\top$ since $M, 2 \models \top$, however $M, 1 \not\models \neg\neg K(p \vee \neg p)$ and $M, 1 \not\models \neg\neg\Diamond(p \vee \neg p)$. □

3.2 *Strong Completeness of* HK◊i

This section establishes a semantic completeness result of **HK◊i**, while we do not know if **HK◊i**$^+$ is strongly complete for the intended semantics.

Definition 6 Let Γ be a set of formulas, we say that Γ is a *theory* if $\Gamma \vdash_{\mathbf{HK\Diamond i}} A$ implies $A \in \Gamma$ for all formulas A. Moreover, we say that Γ is a *prime theory* if Γ is a theory and it satisfies the following condition: for all formulas A and B, if $A \vee B \in \Gamma$ then $A \in \Gamma$ or $B \in \Gamma$. Finally we say that a theory Γ is *consistent* if it is not the case that $\bot \in \Gamma$.

Lemma 1 *Let* $\Gamma \cup \{B\}$ *be a set of formulas. If* $\Gamma \nvdash_{\mathbf{HK\Diamond i}} B$ *then there exists a consistent prime theory* Γ^+ *such that* $\Gamma \subseteq \Gamma^+$ *and* $\Gamma^+ \nvdash_{\mathbf{HK\Diamond i}} B$ *and* $\bot \notin \Gamma^+$.

Definition 7 The canonical model $M^{\mathbf{HK\Diamond i}} = (W, \leq, R, \tau, V)$ is defined as follows:

- $W := \{\Gamma : \Gamma$ is a consistent prime theory of **HK◊i**$\}$.
- $\Sigma \leq \Pi$ iff $\Sigma \subseteq \Pi$.

- $\Sigma R \Pi$ iff $KA \in \Sigma$ implies $A \in \Pi$ for all formulas A.
- $\tau(\Gamma) = \{\overline{|C|} : \Diamond C \notin \Gamma\}$, where $\overline{|C|} = \{\Delta \in W : C \notin \Delta\}$.[4]
- $\Sigma \in V(p)$ iff $p \in \Sigma$.

Proposition 7 *$M^{\mathbf{HK}\Diamond\mathbf{i}} = (W, \leq, R, \tau, V)$ is a model.*

Proof It is sufficient to show that the following conditions hold in the canonical model $M^{\mathbf{HK}\Diamond\mathbf{i}}$.

1. $\leq; R \subseteq R$ and $\leq \subseteq R$.
2. $\tau(\Gamma) \neq \emptyset$ for all $\Gamma \in W$.
3. If $\Gamma \leq \Delta$ then $\tau(\Delta) \subseteq \tau(\Gamma)$, for all $\Gamma, \Delta \in W$.

(1) is easy to prove. (2) means that there exists some $X \in \tau(\Gamma)$ for an arbitrary $\Gamma \in W$. According to the definition of the consistent prime theory, we have $\Diamond\bot \notin \Gamma$, then $\overline{|\bot|} \in \tau(\Gamma)$ meets our requirement. (3) can be proved as follows: assume $\Gamma \leq \Delta$ and fix a $\overline{|C|} \in \tau(\Delta)$ for an arbitrary formula C. Then we have $\Gamma \subseteq \Delta$ and $\Diamond C \notin \Delta$, which concludes $\Diamond C \notin \Gamma$. Therefore, $\overline{|C|} \in \tau(\Gamma)$ holds. □

Lemma 2 *For an arbitrary formula A and a consistent prime theory Γ,*

$$M^{\mathbf{HK}\Diamond\mathbf{i}}, \Gamma \models A \textit{ iff } A \in \Gamma.$$

Proof We proceed by induction on the complexity of A.

First we deal with the case where A is of the form $\Diamond B$ ($\Leftarrow$) Assume $\Diamond B \in \Gamma$. Then for any formulas C, $\overline{|C|} \in \tau(\Gamma)$ such that $\Diamond C \notin \Gamma$. Then from $\Diamond B \in \Gamma$ and $\Diamond C \notin \Gamma$, we can obtain that $\Diamond B \supset \Diamond C \notin \Gamma$, which implies $\nvdash_{\mathbf{HK}\Diamond\mathbf{i}} B \supset C$. From, $\nvdash_{\mathbf{HK}\Diamond\mathbf{i}} B \supset C$, we obtain $\{B\} \nvdash_{\mathbf{HK}\Diamond\mathbf{i}} C$. By Lemma 1, we know that there exists a prime theory Δ such that, $\{B\} \subseteq \Delta$ and $\Delta \nvdash_{\mathbf{HK}\Diamond\mathbf{i}} C$, which means $C \notin \Delta$. We have $\Delta \in \overline{|C|}$ such that $B \in \Delta$. By induction hypothesis, $M^{\mathbf{HK}\Diamond\mathbf{i}}, \Delta \models B$. ($\Rightarrow$) We prove the contrapositive implication. Suppose $\Diamond B \notin \Gamma$, then $\overline{|B|} \in \tau(\Gamma)$. From the definition of $\overline{|B|}$, we know that $B \notin x$ holds for an arbitrary $x \in \overline{|B|}$. By induction hypothesis, we conclude $M^{\mathbf{HK}\Diamond\mathbf{i}}, x \not\models B$.

Secondly we deal with the case where A is of the form KB. ($\Leftarrow$) Assume $KB \in \Gamma$, we show that, for any consistent prime theory Δ, $\Gamma R \Delta$ implies $M^{\mathbf{HK}\Diamond\mathbf{i}}, \Delta \models B$. Fix any Δ such that $\Gamma R \Delta$, combining $KB \in \Gamma$, we know that $B \in \Delta$ from the definition of canonical model. Then we have $M^{\mathbf{HK}\Diamond\mathbf{i}}, \Delta \models B$ by induction hypothesis.

($\Rightarrow$) Assume $KB \notin \Gamma$, our goal is to show that there is a consistent prime theory such that $\Gamma R \Delta$ and $M^{\mathbf{HK}\Diamond\mathbf{i}}, \Delta \not\models B$. Let $\Delta^- = \{C | KC \in \Gamma\}$. Then $B \notin \Delta^-$ and $\Delta^- \nvdash B$. We can check that Δ^- is consistent. Then by Lemma 1, we can construct a consistent prime theory Δ such that $B \notin \Delta$ and $\Delta^- \subseteq \Delta$. Since for any formula C, $KC \in \Gamma$ implies $C \in \Delta$, we have $\Gamma R \Delta$. According to induction hypothesis, we can have $M^{\mathbf{HK}\Diamond\mathbf{i}}, \Delta \not\models B$ from $B \notin \Delta$. □

By Lemmas 1 and 2, we can obtain the following.

Theorem 3 (Strong Completeness of **HK◇i**) *If $\Gamma \models_{\mathbb{M}_{all}} A$ then $\Gamma \vdash_{\mathbf{HK}\Diamond\mathbf{i}} A$, for every set $\Gamma \cup \{A\}$ of formulas.*

---

[4]The definition of $\tau(w)$ is inspired by Chellas (1980, p. 257)'s definition of the canonical model.

4 Glivenko–Kuroda Embedding and Cut Elimination

In the rest of the paper, we go back to the sequent calculi and show proof-theoretic results. In this section we answer Question 3 in the introduction: what is a proper Glivenko–Kuroda translation for K and $\Diamond$? We define our appropriate translation as putting the double negation $\neg\neg$ after both K and $\Diamond$, and then show that the result of translating (OP) becomes an instance of (wOP).

4.1 *Glivenko–Kuroda Theorem: Embedding of* $\mathbf{G1K\Diamond c^+}$ *into* $\mathbf{G1K\Diamond i^+}$

As we know, the Glivenko theorem is generally known as the following equivalence: for an arbitrary propositional formula A, A is derivable in the classical propositional logic, if and only if $\neg\neg A$ is derivable in the intuitionistic propositional logic. However, this form of equivalence does not always hold between $\mathbf{G1K\Diamond c}$ and $\mathbf{G1K\Diamond i}$. For example, the sequent $\Rightarrow K(p \vee \neg p)$ is derivable in $\mathbf{G1K\Diamond c}$, but $\Rightarrow \neg\neg K(p \vee \neg p)$ is not in $\mathbf{G1K\Diamond i}$ by Proposition 6. The similar result holds also for $\Diamond\top \Rightarrow \neg\neg\Diamond(p \vee \neg p)$ by Proposition 6. Since Kuroda translation which defines the translation of the classical formula $\forall x P(x)$ as $\forall x \neg\neg P(x)$ and $\exists x P(x)$ as $\exists x P(x)$ works well in predicate logic, we follow this idea to revise the Glivenko theorem. However, we point out that the translation of $\Diamond A$ cannot be $\Diamond A$ itself.

Definition 8 The definition of the translation $(-)^\clubsuit$ is defined inductively as follows:

$$\begin{array}{llll} p^\clubsuit & := p & (A \supset B)^\clubsuit & := A^\clubsuit \supset B^\clubsuit \\ \bot^\clubsuit & := \bot & (\neg A)^\clubsuit & := \neg A^\clubsuit \\ (A \& B)^\clubsuit & := A^\clubsuit \& B^\clubsuit & (KA)^\clubsuit & := K\neg\neg A^\clubsuit \\ (A \vee B)^\clubsuit & := A^\clubsuit \vee B^\clubsuit & (\Diamond A)^\clubsuit & := \Diamond\neg\neg A^\clubsuit \end{array}$$

Moreover, we define $\Gamma^\clubsuit = \{A^\clubsuit | A \in \Gamma\}$.

Proposition 8 *1.* $\vdash_{\mathbf{G1K\Diamond c}} A^\clubsuit \Rightarrow A$ *and* $\vdash_{\mathbf{G1K\Diamond c}} A \Rightarrow A^\clubsuit$.
2. $\vdash_{\mathbf{G1K\Diamond c^+}} A^\clubsuit \Rightarrow A$ *and* $\vdash_{\mathbf{G1K\Diamond c^+}} A \Rightarrow A^\clubsuit$.

Proof By induction on the complexity of A. □

Theorem 4 *1.* $\vdash_{\mathbf{G1K\Diamond c}} \Gamma \Rightarrow \Delta$ *if and only if* $\vdash_{\mathbf{G1K\Diamond i}} \Gamma^\clubsuit, \neg\Delta^\clubsuit \Rightarrow$.
2. $\vdash_{\mathbf{G1K\Diamond c^+}} \Gamma \Rightarrow \Delta$ *if and only if* $\vdash_{\mathbf{G1K\Diamond i^+}} \Gamma^\clubsuit, \neg\Delta^\clubsuit \Rightarrow$.

Proof 1. The proof of the direction from the left to right can be shown by induction on the height of the derivation $\mathcal{D}$. The propositional cases are easy, and so we only show the case where the last rule of the derivation $\mathcal{D}$ is $(\Diamond)$.

- Suppose that there is one formula appearing in the succedent of the end sequent of $\mathcal{D}$. Then we proceed as follows:

$$\dfrac{\begin{array}{c}\mathcal{D}\\ A \Rightarrow B\end{array}}{\Diamond A \Rightarrow \Diamond B.}\,(\Diamond) \quad \rightsquigarrow \quad \dfrac{\dfrac{\dfrac{\dfrac{\dfrac{\dfrac{\begin{array}{c}\mathcal{D}'\\ A^\clubsuit, \neg B^\clubsuit \Rightarrow\end{array}}{\neg B^\clubsuit \Rightarrow \neg A^\clubsuit}\,(R\neg)}{\neg\neg A^\clubsuit, \neg B^\clubsuit \Rightarrow}\,(L\neg)}{\neg\neg A^\clubsuit \Rightarrow \neg\neg B^\clubsuit}\,(R\neg)}{\Diamond\neg\neg A^\clubsuit \Rightarrow \Diamond\neg\neg B^\clubsuit}\,(\Diamond)}{(\Diamond A)^\clubsuit \Rightarrow (\Diamond B)^\clubsuit}\,(Def)}{\neg(\Diamond B)^\clubsuit, (\Diamond A)^\clubsuit \Rightarrow,}\,(L\neg)$$

  where $\mathcal{D}'$ is obtained by induction hypothesis and we note that the latter derivation is in **G1K◊i**.
- Suppose that the succedent of the end sequent of $\mathcal{D}$ is empty. Then we proceed as follows:

$$\dfrac{\begin{array}{c}\mathcal{D}\\ A \Rightarrow\end{array}}{\Diamond A \Rightarrow .}\,(\Diamond) \quad \rightsquigarrow \quad \dfrac{\dfrac{\dfrac{\dfrac{\begin{array}{c}\mathcal{D}'\\ A^\clubsuit \Rightarrow\end{array}}{\Rightarrow \neg A^\clubsuit}\,(R\neg)}{\neg\neg A^\clubsuit \Rightarrow}\,(L\neg)}{\Diamond\neg\neg A^\clubsuit \Rightarrow}\,(\Diamond)}{(\Diamond A)^\clubsuit \Rightarrow,}\,(Def)$$

  where $\mathcal{D}'$ is obtained by induction hypothesis and so the latter derivation is in **G1K◊i**.

We have finished proving the left to right direction.
The proof of the direction from the right to left can be shown as follows: Suppose $\vdash_{\mathbf{G1K\Diamond i}} \Gamma^\clubsuit, \neg\Delta^\clubsuit \Rightarrow$, from the fact that every derivation of **G1K◊i** is also the derivation of **G1K◊c**, we have that $\vdash_{\mathbf{G1K\Diamond c}} \Gamma^\clubsuit, \neg\Delta^\clubsuit \Rightarrow$. After applying the $(R\neg)$ finite many times, we have $\vdash_{\mathbf{G1K\Diamond c}} \Gamma^\clubsuit \Rightarrow \neg\neg\Delta^\clubsuit$. Since we have $\vdash_{\mathbf{G1K\Diamond c}} \neg\neg A \Rightarrow A$ for any formula A, after applying the (Cut) finite many times we have $\vdash_{\mathbf{G1K\Diamond c}} \Gamma^\clubsuit \Rightarrow \Delta^\clubsuit$. Then after applying (Cut) with Proposition 8 (1) finite many times, we have $\vdash_{\mathbf{G1K\Diamond c}} \Gamma \Rightarrow \Delta$.

2. The proof of the direction from the left to right is based on the induction. In the case we only deal with the case of (KP). When $\vdash_{\mathbf{G1K\Diamond c^+}} \Rightarrow A \supset \Diamond K A$, we can construct the following derivation in **G1K◊i$^+$**

$$\dfrac{\dfrac{\dfrac{\dfrac{\dfrac{A^\clubsuit \Rightarrow A^\clubsuit}{\neg A^\clubsuit, A^\clubsuit \Rightarrow}\,(L\neg)}{A^\clubsuit \Rightarrow \neg\neg A^\clubsuit}\,(R\neg) \qquad \dfrac{\begin{array}{c}\mathbf{KP}\\ \neg\neg A^\clubsuit \Rightarrow \Diamond K\neg\neg A^\clubsuit\end{array} \qquad \dfrac{\dfrac{\dfrac{K\neg\neg A^\clubsuit \Rightarrow K\neg\neg A^\clubsuit}{\neg K\neg\neg A^\clubsuit, K\neg\neg A^\clubsuit \Rightarrow}\,(L\neg)}{K\neg\neg A^\clubsuit \Rightarrow \neg\neg K\neg\neg A^\clubsuit}\,(R\neg)}{\Diamond K\neg\neg A^\clubsuit \Rightarrow \Diamond\neg\neg K\neg\neg A^\clubsuit}\,(\Diamond)}{\neg\neg A^\clubsuit \Rightarrow \Diamond\neg\neg K\neg\neg A^\clubsuit}\,(Cut)}{A^\clubsuit \Rightarrow \Diamond\neg\neg K\neg\neg A^\clubsuit}\,(Cut)}{\Rightarrow A^\clubsuit \supset \Diamond\neg\neg K\neg\neg A^\clubsuit}\,(R\supset)}{\neg(A^\clubsuit \supset \Diamond\neg\neg K\neg\neg A^\clubsuit) \Rightarrow,}\,(L\neg)$$

where it is clear that $\neg\neg A^\clubsuit \Rightarrow \Diamond K\neg\neg A^\clubsuit$ can be derived from an instance of **(KP)** by the cut rule. Therefore, we have $\vdash_{\mathbf{G1K\Diamond i^+}} \neg(A \supset \Diamond K A)^\clubsuit \Rightarrow$. □

The proof of the direction from the right to left is similar to that of (1).

Then we can have the following corollary.

Corollary 2 (Glivenko–Kuroda Theorem) *For an arbitrary formula A in $\mathcal{L}$,*

1. $\vdash_{\mathbf{G1K\Diamond c}} \Rightarrow A$ *iff* $\vdash_{\mathbf{G1K\Diamond i}} \Rightarrow \neg\neg A^{\clubsuit}$.
2. $\vdash_{\mathbf{G1K\Diamond c^{+}}} \Rightarrow A$ *iff* $\vdash_{\mathbf{G1K\Diamond i^{+}}} \Rightarrow \neg\neg A^{\clubsuit}$.

Proof For (1), the direction from the left to the right is straightforward from Theorem 4 and $R(\neg)$. The direction from the right to the left is shown as follows: It can be easily proved that $\vdash_{\mathbf{G1K\Diamond i^{+}}} \neg A^{\clubsuit} \Rightarrow \neg\neg\neg A^{\clubsuit}$. Then $\neg A^{\clubsuit} \Rightarrow$ can be derived from $\Rightarrow \neg\neg A^{\clubsuit}$ by applying the $(L\neg)$ and (Cut). Then apply Theorem 4. The proof of (2) is similar. □

Now we apply our double negation translations in the study of knowability paradox. By Proposition 1, we have $\vdash_{\mathbf{G1K\Diamond c^{+}}} \Rightarrow A \supset KA$. After applying the translation from Theorem 4, we have $\vdash_{\mathbf{G1K\Diamond i^{+}}} \Rightarrow \neg\neg(A \supset KA)^{\clubsuit}$, then it concludes $\vdash_{\mathbf{G1K\Diamond i^{+}}} \Rightarrow \neg\neg A^{\clubsuit} \supset \neg\neg K\neg\neg A^{\clubsuit}$, which is an instance of **(wOP)** $A \supset \neg\neg KA$. This suggests us that, the proposal (DeVidi and Solomon 2001; Dummett 2009) to accept **(wOP)** as an intuitionistic version of **(OP)** is well reflected by our translation.

4.2 *Cut-Elimination Theorem of* G1K◇i

We show that the cut-elimination theorems hold in **G1K◇i**, and we employ our argument from Ono and Komori (1985) and Kashima (2009).

Definition 9 We define the rule $(Ecut)$ of extended cut rule as follows:

$$\frac{\Gamma \Rightarrow A \qquad A^n, \Gamma' \Rightarrow \Delta}{\Gamma, \Gamma' \Rightarrow \Delta,}\ (Ecut)$$

where A^n means n-times repetition of the formula A and $n \geq 0$. The formula A is called an *Ecut formula.*

Definition 10 A derivation $\mathcal{D}$ in **G1K◇i** is of the *Ecut-bottom form* if it has the following form

$$\frac{\dfrac{\vdots\ \mathcal{D}_1}{\Gamma \Rightarrow A}\ rule(\mathcal{D}_1) \qquad \dfrac{\vdots\ \mathcal{D}_2}{A^n, \Gamma' \Rightarrow \Delta}\ rule(\mathcal{D}_2)}{\Gamma, \Gamma' \Rightarrow \Delta,}\ (Ecut)$$

where $rule(\mathcal{D}_i)$ is the last applied rule of a given derivation $\mathcal{D}_i$ and there is no application of (*Ecut*) in $\mathcal{D}_1$ nor $\mathcal{D}_2$. Let the *weight* w of a Ecut-bottom form be the number of sequents occurring in $\mathcal{D}_1$ and $\mathcal{D}_2$. Let the *complexity* c of a Ecut-bottom form be the number of the logical symbols (&, $\vee$, $\supset$, $\neg$, $\Diamond$, and K) appearing in the *Ecut formula,* we note that $c \geq 0$ and $w \geq 2$.

Lemma 3 *Every derivation of the Ecut-bottom form can be replaced by an $(Ecut)$-free derivation with the same conclusion.*

Proof By double induction on the *complexity* and the *weight* lexicographically. Our argument is divided into four cases.

(1) $\mathcal{D}_i$ is an initial sequent.
(2) $rule(\mathcal{D}_i)$ is a structural rule.
(3) $rule(\mathcal{D}_i)$ is a logical rule where the Ecut formula is not principal.
(4) $rule(\mathcal{D}_1)$ and $rule(\mathcal{D}_2)$ are logical rules, and the Ecut formulas are principal in both rules.

Case (1) and Case (2) are straightforward. The case in which right part is the initial sequent $\bot \Rightarrow$ and $\bot$ is the cut formula, can be covered by other cases.

The rules (K) and $(\Diamond)$ cannot be included in the case (3). When rule $\mathcal{D}_i$ is the rule (T) in case (3), it is easy. As for case (4), we only show the following cases.

- Let $rule(\mathcal{D}_1)$ be $(\Diamond)$ and let $rule(\mathcal{D}_2)$ be $(\Diamond)$ as well

$$\dfrac{\dfrac{\begin{array}{c}\mathcal{D}_1\\ B \Rightarrow A\end{array}}{\Diamond B \Rightarrow \Diamond A}(\Diamond) \quad \dfrac{\begin{array}{c}\mathcal{D}_2\\ A \Rightarrow \Delta\end{array}}{\Diamond A \Rightarrow \Diamond \Delta}(\Diamond)}{\Diamond B \Rightarrow \Diamond \Delta.}(Ecut) \quad \rightsquigarrow \quad \dfrac{\dfrac{\begin{array}{c}\mathcal{D}_1\\ B \Rightarrow A\end{array} \quad \begin{array}{c}\mathcal{D}_2\\ A \Rightarrow \Delta\end{array}}{B \Rightarrow \Delta}(Ecut)}{\Diamond B \Rightarrow \Diamond \Delta.}(\Diamond)$$

  Since the complexity of the first derivation in Ecut-bottom form is less than that of the second derivation, we can use induction hypothesis to obtain the second derivation for $\Diamond B \Rightarrow \Diamond \Delta$ without applying $(Ecut)$ rule.
- Let $rule(\mathcal{D}_1)$ be (K) and Let $rule(\mathcal{D}_2)$ be (T):

$$\dfrac{\dfrac{\begin{array}{c}\mathcal{D}_1\\ \Gamma \Rightarrow A\end{array}}{K\Gamma \Rightarrow KA}(K) \quad \dfrac{\begin{array}{c}\mathcal{D}_2\\ A, (KA)^{m-1}, \Gamma' \Rightarrow \Delta\end{array}}{(KA)^m, \Gamma' \Rightarrow \Delta}(T)}{K\Gamma, \Gamma' \Rightarrow \Delta.}(Ecut) \quad \rightsquigarrow$$

$$\dfrac{\dfrac{\dfrac{\begin{array}{c}\mathcal{D}_1\\ \Gamma \Rightarrow A\end{array} \quad \dfrac{\dfrac{\begin{array}{c}\mathcal{D}_1\\ \Gamma \Rightarrow A\end{array}}{K\Gamma \Rightarrow KA}(K) \quad \begin{array}{c}\mathcal{D}_2\\ A, (KA)^{m-1}, \Gamma' \Rightarrow \Delta\end{array}}{A, K\Gamma, \Gamma' \Rightarrow \Delta}(Ecut)}{K\Gamma, \Gamma', \Gamma \Rightarrow \Delta}(Ecut)}{K\Gamma, \Gamma', K\Gamma \Rightarrow \Delta}(T^*)}{K\Gamma, \Gamma' \Rightarrow \Delta.}(LC^*)$$

  Since the weight of the first derivation in Ecut-bottom form is smaller than that of the second derivation, we can use the induction hypothesis to obtain the second derivation for $A, K\Gamma, \Gamma' \Rightarrow \Delta$ without applying $(Ecut)$ rule. □

Definition 11 The system of $\mathbf{G1K\Diamond i}^{-cut}$ is the system of $\mathbf{G1K\Diamond i}$ without (Cut) rule.

By Lemma 3 we can obtain the cut-elimination theorem of $\mathbf{G1K\Diamond i}$.

Theorem 5 *If* $\vdash_{\mathbf{G1K\Diamond i}} \Gamma \Rightarrow \Delta$ *then* $\vdash_{\mathbf{G1K\Diamond i}^{-cut}} \Gamma \Rightarrow \Delta$.

Corollary 3 (Disjunction Property of **G1K**$\Diamond$**i**) *For any formulas* A *and* B, *if* $A \vee B$ *is derivable in* **G1K**$\Diamond$**i** *then either* A *or* B *is derivable in* **G1K**$\Diamond$**i**.

Proof Suppose that $\Rightarrow A \vee B$ is derivable in **G1K**$\Diamond$**i**. Since the cut-elimination theorem has been proved, we only consider a cut-free derivation of $\Rightarrow A \vee B$. Then the last derivation can only be (RW) or $(R\vee)$. Because the empty sequent $\Rightarrow$ is not cut-free derivable in **G1K**$\Diamond$**i**, (RW) cannot be the case. Then it must be $(R\vee)$, and either $\Rightarrow A$ or $\Rightarrow B$ is derivable in **G1K**$\Diamond$**i**. □

4.3 *Cut-Elimination of* G1K$\Diamond$c

Definition 12 The system of $\mathbf{G1K\Diamond c}^{-cut}$ is the system of **G1K**$\Diamond$**c** without (Cut) rule.

By induction on derivation, we can prove the following two lemmas.

Lemma 4 (Extended Inversion Lemma of $(\neg)$ *in* $\mathbf{G1K\Diamond c}^{-cut}$). *Let* $n \geq 0$,

$$(1)\ \textit{If} \vdash_{\mathbf{G1K\Diamond c}^{-cut}} (\neg A)^n, \Gamma \Rightarrow \Delta \ \textit{then} \vdash_{\mathbf{G1K\Diamond c}^{-cut}} \Gamma \Rightarrow \Delta, A^n$$

$$(2)\ \textit{If} \vdash_{\mathbf{G1K\Diamond c}^{-cut}} \Gamma \Rightarrow \Delta, (\neg A)^n \ \textit{then} \vdash_{\mathbf{G1K\Diamond c}^{-cut}} A^n, \Gamma \Rightarrow \Delta$$

With the help of Lemma 4, we can prove the following lemma.

Lemma 5 *If* $\vdash_{\mathbf{G1K\Diamond c}^{-cut}} \Gamma^{\clubsuit} \Rightarrow \Delta^{\clubsuit}$ *then* $\vdash_{\mathbf{G1K\Diamond c}^{-cut}} \Gamma \Rightarrow \Delta$.

Theorem 6 (Cut-elimination of **G1K**$\Diamond$**c**) *If* $\vdash_{\mathbf{G1K\Diamond c}} \Gamma \Rightarrow \Delta$ *then* $\vdash_{\mathbf{G1K\Diamond c}^{-cut}} \Gamma \Rightarrow \Delta$.

Proof Suppose $\vdash_{\mathbf{G1K\Diamond c}} \Gamma \Rightarrow \Delta$. By Theorem 4, we have $\vdash_{\mathbf{G1K\Diamond i}} \Gamma^{\clubsuit}, \neg\Delta^{\clubsuit} \Rightarrow$. By cut-elimination theorem in **G1K**$\Diamond$**i**, we have $\vdash_{\mathbf{G1K\Diamond i}^{-cut}} \Gamma^{\clubsuit}, \neg\Delta^{\clubsuit} \Rightarrow$. We know that every derivation in **G1K**$\Diamond$**i** is also a derivation in **G1K**$\Diamond$**c**. So we have $\vdash_{\mathbf{G1K\Diamond c}^{-cut}} \Gamma^{\clubsuit}, \neg\Delta^{\clubsuit} \Rightarrow$. By applying Lemma 4 finitely many times, we have $\vdash_{\mathbf{G1K\Diamond c}^{-cut}} \Gamma^{\clubsuit} \Rightarrow \Delta^{\clubsuit}$. Then by Lemma 5, we conclude $\vdash_{\mathbf{G1K\Diamond c}^{-cut}} \Gamma \Rightarrow \Delta$. □

5 Craig Interpolation Theorems

In this section, we show that the Craig interpolation theorems hold in both **G1K**$\Diamond$**i** and **G1K**$\Diamond$**c**. Compared to Maffezioli et al. (2013)'s labeled approach, our non-labeled approach provides a more convenient tool for this proof-theoretic property.

5.1 Craig Interpolation Theorem of G1K$\Diamond$i

In the following part of this section, $\mathsf{V}(D)$ denotes the set of all propositional variables in a formula D. And if Γ is a finite multiset of the formulas $D_1, \ldots, D_m$, we define $\mathsf{V}(\Gamma) = \bigcup\{\mathsf{V}(D) | D \in \Gamma\}$. Especially, we have $\mathsf{V}(\bot) = \emptyset$. We say that $\langle \Gamma_1; \Gamma_2 \rangle$ is a *partition* of Γ if Γ is Γ_1, Γ_2.

Lemma 6 *Suppose that* $\Gamma \Rightarrow \Delta$ *is derivable in* **G1K$\Diamond$i**. *Then for any partition* $\langle \Gamma_1; \Gamma_2 \rangle$ *of* Γ *there exists a formula* C *(interpolant formula) such that both* $\Gamma_1 \Rightarrow C$ *and* $C, \Gamma_2 \Rightarrow \Delta$ *are also derivable in* **G1K$\Diamond$i**, *and* $\mathsf{V}(C) \subseteq \mathsf{V}(\Gamma_1) \cap \mathsf{V}(\Gamma_2, \Delta)$.

Proof If $\vdash_{\mathbf{G1K\Diamond i}} \Gamma \Rightarrow \Delta$, then by cut-elimination theorem $\vdash_{\mathbf{G1K\Diamond i}^{-cut}} \Gamma \Rightarrow \Delta$. We give the proof by induction on a cut-free derivation of $\Gamma \Rightarrow \Delta$. We only show the case where the last rule in the derivation is $(\Diamond)$

$$\frac{A \Rightarrow \Delta}{\Diamond A \Rightarrow \Diamond \Delta.}\ (\Diamond)$$

There are only two partitions of $\Diamond A$, i.e., $\langle \Diamond A; \emptyset \rangle$ and $\langle \emptyset; \Diamond A \rangle$. For the latter case, we can put an interpolant formula as $\bot$. For the former case, our goal is to show that there exists a formula C such that $\Diamond A \Rightarrow C$ and $C \Rightarrow \Diamond \Delta$ are derivable, and that $\mathsf{V}(C) \subseteq \mathsf{V}(\Diamond A) \cap \mathsf{V}(\Diamond \Delta)$. We proceed as follows: Taking a partition $\langle A; \emptyset \rangle$ of A, we use induction hypothesis to have that $A \Rightarrow C$ and $C \Rightarrow \Delta$ are derivable, and that $\mathsf{V}(C) \subseteq \mathsf{V}(A) \cap \mathsf{V}(\Delta)$ for some formula C. Then we can obtain

$$\frac{A \Rightarrow C}{\Diamond A \Rightarrow \Diamond C,}\ (\Diamond) \qquad \frac{C \Rightarrow \Delta}{\Diamond C \Rightarrow \Diamond \Delta,}\ (\Diamond)$$

and $\mathsf{V}(\Diamond C) \subseteq \mathsf{V}(\Diamond A) \cap \mathsf{V}(\Diamond \Delta)$. So $\Diamond C$ is our intended interpolant. □

Theorem 7 (Craig Interpolation theorem of **G1K$\Diamond$i**) *If* $\Rightarrow A \supset B$ *is derivable in* **G1K$\Diamond$i**, *then there exists a formula* C *such that* $\Rightarrow A \supset C$ *and* $\Rightarrow C \supset B$ *are derivable in* **G1K$\Diamond$i** *and that* $\mathsf{V}(C) \subseteq \mathsf{V}(A) \cap \mathsf{V}(B)$.

Proof Suppose that $\Rightarrow A \supset B$ is derivable in **G1K$\Diamond$i**. Then it is easy to prove that $A \Rightarrow B$ is also derivable. We consider a partition $\langle A; \emptyset \rangle$ of A, by Lemma 6, we can obtain that there exists a formula C such that $A \Rightarrow C$ and $C \Rightarrow B$ are derivable in **G1K$\Diamond$i** and $\mathsf{V}(C) \subseteq \mathsf{V}(A) \cap \mathsf{V}(B)$. After applying the $(R \supset)$, $\Rightarrow A \supset C$ and $\Rightarrow C \supset B$ are derivable. □

Now, we can prove the interpolation theorem of **G1K$\Diamond$c**.

Theorem 8 (Craig Interpolation theorem of **G1K$\Diamond$c**) *If* $\Rightarrow A \supset B$ *is derivable in* **G1K$\Diamond$c**, *then there exists a formula* C *such that* $\Rightarrow A \supset C$ *and* $\Rightarrow C \supset B$ *are also derivable in* **G1K$\Diamond$c** *and that* $\mathsf{V}(C) \subseteq \mathsf{V}(A) \cap \mathsf{V}(B)$.

Proof Suppose $\vdash_{\mathbf{G1K\Diamond c}} \Rightarrow A \supset B$, then we have $\vdash_{\mathbf{G1K\Diamond c}} A \Rightarrow B$. By Theorem 4, we have $\vdash_{\mathbf{G1K\Diamond i}} A^{\clubsuit}, \neg B^{\clubsuit} \Rightarrow$. We consider a partition $\langle A^{\clubsuit}; \neg B^{\clubsuit} \rangle$ of $A^{\clubsuit}, \neg B^{\clubsuit}$. Then by Lemma 6, there exists a formula C such that we have $\vdash_{\mathbf{G1K\Diamond i}} A^{\clubsuit} \Rightarrow C$ and $\vdash_{\mathbf{G1K\Diamond i}}$

$\neg B^{\clubsuit}, C \Rightarrow$, and that $\mathsf{V}(C) \subseteq \mathsf{V}(A^{\clubsuit}) \cap \mathsf{V}(\neg B^{\clubsuit})$. Since every derivation in **G1K◊i** is also a derivation in **G1K◊c**, we have $\vdash_{\mathbf{G1K\Diamond c}} A^{\clubsuit} \Rightarrow C$ and, $\vdash_{\mathbf{G1K\Diamond c}} \neg B^{\clubsuit}, C \Rightarrow$. Then by applying (cut) with the sequent in the form of $\Rightarrow B^{\clubsuit}, \neg B^{\clubsuit}$ (which is always derivable), we obtain $\vdash_{\mathbf{G1K\Diamond c}} C \Rightarrow B^{\clubsuit}$. Since we can establish $\vdash_{\mathbf{G1K\Diamond c}} A \Rightarrow A^{\clubsuit}$ and $\vdash_{\mathbf{G1K\Diamond c}} B^{\clubsuit} \Rightarrow B$ by Proposition 8, $\vdash_{\mathbf{G1K\Diamond c}} A \Rightarrow C$ and $\vdash_{\mathbf{G1K\Diamond c}} C \Rightarrow B$ hold after applying the (cut) rules, respectively. $\vdash_{\mathbf{G1K\Diamond c}} \Rightarrow A \supset C$ and $\vdash_{\mathbf{G1K\Diamond c}} \Rightarrow C \supset B$ hold after applying the $(R \supset)$ rules, respectively. Since the translation $(-)^{\clubsuit}$ does not change the set of variables in a formula, it follows from $\mathsf{V}(C) \subseteq \mathsf{V}(A^{\clubsuit}) \cap \mathsf{V}(\neg B^{\clubsuit})$ that $\mathsf{V}(C) \subseteq \mathsf{V}(A) \cap \mathsf{V}(B)$. □

6 Decidability of G1K◊i and G1K◊c

In this section, we show the decidability of **G1K◊i** by a method which originated from Gentzen (Cf. Ono 1998) and then we reduce the decidability of **G1K◊c** to that of **G1K◊i**.

Definition 13 We say that a derivation in **G1K◊i** satisfies the *subformula property* if every sequent appearing in this derivation only contains subformulas or formulas of the end sequent. We say that a derivation in **G1K◊i** contains *redundancies* if there is a sequent appearing at least twice in this derivation. We say that a sequent is *reduced* if every formula in the antecedent or succedent appears at most three times. We say that a sequent is *1-reduced* if every formula in the antecedent or succedent appears at most once. Given a sequent $\Gamma \Rightarrow \Delta$, we write $\Gamma^* \Rightarrow \Delta^*$ to mean 1-reduced sequent obtained from $\Gamma \Rightarrow \Delta$.

Lemma 7 *Let $\Gamma \Rightarrow \Delta$ be a sequent which is derivable in* **G1K◊i**. *Then there is a cut-free derivation $\mathcal{D}$ of 1-reduced sequent $\Gamma^* \Rightarrow \Delta^*$ of $\Gamma \Rightarrow \Delta$ such that every sequent in $\mathcal{D}$ is reduced.*

Proof Assume that $\Gamma \Rightarrow \Delta$ is derivable in **G1K◊i**. By cut-elimination theorem (Theorem 5), we can find a cut-free derivation. We show our goal by induction on the cut-free derivation. The base case is obvious. For other rules, we only show the important cases.

- Let $A \supset B, \Gamma, \Pi \Rightarrow \Delta$ be derived by the rule $(L \supset)$, i.e.,

$$\frac{\Gamma \Rightarrow A \qquad B, \Pi \Rightarrow \Delta}{A \supset B, \Gamma, \Pi \Rightarrow \Delta.}\ (L \supset)$$

  We apply the induction hypothesis to the premises and then apply $(L \supset)$. We note that when $A \supset B$ occurs in antecedents of the original premises (i.e., Γ and B, Π) then we proceed as follows:

$$\frac{A \supset B, \Gamma' \Rightarrow A \qquad B, A \supset B, \Pi' \Rightarrow \Delta}{A \supset B, A \supset B, A \supset B, \Gamma', \Pi', \Rightarrow \Delta,}\ (L \supset)$$

  where we apply (LC) finitely many times to obtain the desired 1-reduced sequent and this case also shows why we need to consider reduced sequents in a derivation.

- If $\Diamond A \Rightarrow \Diamond \Delta$ is derived by the rule $(\Diamond)$, then we apply $(\Diamond)$ to the result of the induction hypothesis.
- If $K\Gamma \Rightarrow KA$ is derived by the rule (K), then we apply (K) to the result of the induction hypothesis.
- If $KA, \Gamma \Rightarrow \Delta$ is derived by the rule (T) then we can apply (T) to the result of the induction hypothesis on its premise. If there are KA appearing in the antecedent of the premise, we can apply (LC) to the result of induction hypothesis. □

Proposition 9 *For any derivation $\mathcal{D}$ of $\Gamma \Rightarrow \Delta$ in* **G1K◊i**, *there exists a derivation $\mathcal{D}'$ of $\Gamma^* \Rightarrow \Delta^*$ such that*

1. $\mathcal{D}'$ satisfies the subformula property;
2. $\mathcal{D}'$ contains no redundancies;
3. $\mathcal{D}'$ consists only of reduced sequents.

Proof First, by cut-elimination theorem (Theorem 5), we can find a cut-free derivation of $\mathcal{D}$. Then the condition (1) can be satisfied. Then by Lemma 7, we can find a derivation of $\mathcal{D}'$ of $\Gamma^* \Rightarrow \Delta^*$ in which every sequent is reduced. It is noted that we can still keep the condition (1) in this derivation. Finally, for the second condition, a derivation containing the redundancies can always be transformed into a derivation containing no redundancies and satisfying the conditions (1) and (3). □

Theorem 9 **G1K◊i** *is decidable.*

Proof Given a sequent $\Gamma \Rightarrow \Delta$, we first consider 1-reduced sequent $\Gamma^* \Rightarrow \Delta^*$ of $\Gamma \Rightarrow \Delta$. Since the number of all reduced sequents built up from subformulas of $\Gamma^* \Rightarrow \Delta^*$ is finite, the number of all possible cut-free derivations from these reduced sequents such that they do not contain redundancies is also finite. Let us enumerate all such derivations. If we can find a derivation of $\Gamma^* \Rightarrow \Delta^*$ in the enumeration, then we can conclude that $\Gamma \Rightarrow \Delta$ is derivable in **G1K◊i** with the help of contraction rules. Otherwise, we conclude that $\Gamma \Rightarrow \Delta$ is not derivable in **G1K◊i** by Proposition 9. □

Theorem 10 **G1K◊c** *is decidable.*

Proof We use the Glivenko–Kuroda theorem to reduce the decidability of **G1K◊c** to that of **G1K◊i**. By Theorem 4, $\vdash_{\mathbf{G1K\Diamond c}} \Gamma \Rightarrow \Delta$ if and only if $\vdash_{\mathbf{G1K\Diamond i}} \Gamma^{\clubsuit}, \neg\Delta^{\clubsuit} \Rightarrow$. So it follows from the decidability of **G1K◊i** that **G1K◊c** is also decidable. □

7 Conclusion and Further Direction

Usberti (2016) criticizes that a double negation translation from the classic system to the intuitionistic system is inconsiderable in the arguments of knowability paradox, by claiming that the classical truth of a proposition A is totally different from the notion of $\neg\neg A$ (which by BHK interpretation means that, there exists an obstacle to deny the A). However, as we have shown in Sect. 4.1, the translation of **(OP)** into **G1K**$\Diamond$**i**$^+$ is an instance $\neg\neg A^\clubsuit \supset \neg\neg K\neg\neg A^\clubsuit$ of **(wOP)**. This suggests us that a translation from **G1K**$\Diamond$**c**$^+$ to **G1K**$\Diamond$**i**$^+$ does support the method to switch logical systems to solve the knowability paradox. The question of whether (**wOP**) is acceptable or not is not being answered in this paper. Our technical results tell us that if we accept (**wOP**), the Glivenko–Kuroda translation will be harmless to this approach.

However, we still have some works in the future. One is that given the fact that we did not prove the cut-elimination theorem of **G1K**$\Diamond$**c**$^+$ and **G1K**$\Diamond$**i**$^+$, we may try to convert the knowability principle into a non-logical inference rule, which allows us to prove the cut -elimination theorems; another task is the study of the completeness of **HK**$\Diamond$**i**$^+$ or **G1K**$\Diamond$**i**$^+$. These works will not only provide a better understanding of the knowability paradox but also considered as an important topic of the intuitionistic modal logic.

Acknowledgements We would like to thank Takahiro Sawasaki, Johan van Benthem, Jeremy Seligman for their helpful comments. The work of the first author was supported by the Graduate Grant Program of Graduate School of Letters, Hokkaido University and Research Assistant Program of Graduate School of Letters, Hokkaido University. The work of the second author is partially supported by JSPS KAKENHI Grant-in-Aid for Young Scientists (B) Grant Number 15K21025, JSPS KAKENHI Grant-in-Aid for Scientific Research (B) Grant Number 17H02258 and also (C) Grant Number 19K12113, and JSPS Core-to-Core Program (A. Advanced Research Networks).

Appendix: Strong Completeness of HK$\Diamond$c

Definition 14 The classical system of **HK**$\Diamond$**c** can be obtained by adding to the **HK**$\Diamond$**i** the law of excluded middle:

$$\text{(LEM)} \quad A \vee \neg A.$$

Moreover, **HK**$\Diamond$**c**$^+$ can be obtained by adding the (KP)$A \supset \Diamond KA$ to **HK**$\Diamond$**c**.

The notion of a derivation in these systems is defined similarly to those of the intuitionistic systems.

Theorem 11 *For any formula* A, $\vdash_{\mathbf{HK}\Diamond\mathbf{c}} A$ *iff* $\vdash_{\mathbf{G1}\Diamond\mathbf{c}} \Rightarrow A$. *And* $\vdash_{\mathbf{HK}\Diamond\mathbf{c}^+} A$ *iff* $\vdash_{\mathbf{G1K}\Diamond\mathbf{c}^+} \Rightarrow A$.

Definition 15 We say that a model $M = (W, \leq, R, \tau, V)$ is classical if the preorder $\leq$ is the identity relation $=$ (as a graph) on W. We use $\mathbb{C}_{\text{all}}$ to denote the class of all classical models.

It is immediate that for a classical model $M = (W, =, R, \tau, V)$ the conditions of $=; R \subseteq R$, (NE), (DEC) and the persistency condition hold trivially. Moreover, $= \subseteq R$ gives the reflexivity of R. It is obvious that the truth condition for $A \supset B$ becomes

$$M, w \models A \supset B \text{ iff } M, w \models A \text{ implies } M, w \models B.$$

Theorem 12 *If* $\vdash_{\mathbf{HK}\Diamond\mathbf{c}} A$ *then* $\models_{\mathbb{C}_{\text{all}}} A$*, for all formulas* A.

Proof It is sufficient to show that the law of excluded middle is valid in $\mathbb{C}_{\text{all}}$. The proof is straightforward. □

The notion of a consistent prime theory in **HK◇c** can be similarly defined to that of **HK◇i**.

Lemma 8 *Let* $\Gamma \cup \{B\}$ *be a set of formulas. If* $\Gamma \nvdash_{\mathbf{HK}\Diamond\mathbf{c}} B$ *then there exists a consistent prime theory* Γ^+ *such that* $\Gamma \subseteq \Gamma^+$ *and* $\Gamma^+ \nvdash_{\mathbf{HK}\Diamond\mathbf{c}} B$ *and* $\bot \notin \Gamma^+$.

By the law of the excluded middle, the following holds immediately.

Proposition 10 *Given a consistent prime theory* Γ *in* **HK◇c**, *either* $A \in \Gamma$ *or* $\neg A \in \Gamma$*, for any formula* A.

Definition 16 The canonical model $M^{\mathbf{HK}\Diamond\mathbf{c}} = (W, =, R, \tau, V)$ is defined as follows:

- $W := \{\Gamma : \Gamma$ is a consistent prime theory in $\mathbf{HK}\Diamond\mathbf{c}\}$.
- $\Sigma R \Pi$ iff $KA \in \Sigma$ implies $A \in \Pi$ for all formulas A.
- $\tau(\Gamma) = \{\overline{|C|} : \Diamond C \notin \Gamma\}$, where $\overline{|C|} = \{\Delta \in W : C \notin \Delta\}$.
- $\Sigma \in V(p)$ iff $p \in \Sigma$.

Proposition 11 *The canonical model* $M^{\mathbf{HK}\Diamond\mathbf{c}} = (W, =, R, \tau, V)$ *is a classical model, i.e.,* $M^{\mathbf{HK}\Diamond\mathbf{c}} \in \mathbb{C}_{\text{all}}$.

Proof It is sufficient to show that: $= \subseteq R$ and $\tau(\Gamma) \neq \emptyset$ for all $\Gamma \in W$. These can be easily checked.

Lemma 9 *For an arbitrary formula* A *and a consistent prime theory* Γ *in* **HK◇c**,

$$M^{\mathbf{HK}\Diamond\mathbf{c}}, \Gamma \models A \text{ iff } A \in \Gamma.$$

Proof We proceed by induction on the complexity of A. We only show the case of $B \supset C$, since the other cases are shown similarly to the proof of Lemma 2. Suppose that $B \supset C \in \Gamma$. Then we show that $M^{\mathbf{HK}\Diamond\mathbf{c}}, \Gamma \models B \supset C$. It is sufficient to show that $M, \Gamma \models B$ implies $M, \Gamma \models C$. Suppose that $M, \Gamma \models B$, we have $B \in \Gamma$ by induction hypothesis. From the supposition $B \supset C \in \Gamma$, we have $C \in \Gamma$. Then $M, \Gamma \models C$ by induction hypothesis.

Conversely, suppose that $B \supset C \notin \Gamma$. We show $M^{\mathbf{HK}\Diamond\mathbf{c}}, \Gamma \not\models B \supset C$. It is sufficient to show $M^{\mathbf{HK}\Diamond\mathbf{c}}, \Gamma \models B$ and $M^{\mathbf{HK}\Diamond\mathbf{c}}, \Gamma \not\models C$. Since $C \supset (B \supset C) \in \Gamma$, we have $C \notin \Gamma$. Since $\neg B \supset (B \supset C)$, it cannot be the case that $\neg B \in \Gamma$. By Proposition 10, we have $B \in \Gamma$. From $B \in \Gamma$ and $C \notin \Gamma$, we have $M^{\mathbf{HK}\Diamond\mathbf{c}}, \Gamma \models B$ and $M^{\mathbf{HK}\Diamond\mathbf{c}}, \Gamma \not\models C$ by induction hypothesis.

Theorem 13 (Strong Completeness of **HK◇c**) *If* $\Gamma \models_{\mathbb{C}_{\text{all}}} A$ *then* $\Gamma \vdash_{\mathbf{HK}\Diamond\mathbf{c}} A$*, for every set* $\Gamma \cup \{A\}$ *of formulas.*

References

Artemov, Sergei, and Tudor Protopopescu. 2010. Knowability from a logical point of view. In *Technical Report TR-2010008, CUNY Ph.D. Program in Computer Science*

Artemov, Sergei, and Tudor Protopopescu. 2016. Intuitionistic epistemic logic. *The Review of Symbolic Logic* 9: 266–298. https://doi.org/10.1017/S1755020315000374.

Chellas, Brian F. 1980. *Modal logic: an introduction*. Cambridge: Cambridge University Press

DeVidi, David, and Graham Solomon. 2001. Knowability and intuitionistic logic. *Philosophia* 28: 319–334. https://doi.org/10.1007/BF02379783.

Dummett, Michael. 1978. *Truth and other enigmas*, 145–165. London: Duckworth.

Dummett, Michael. 2009. Fitch's paradox of knowability. In *New essay on the knowability paradox*, ed. Joe Salerno, 51–52. Oxford: Oxford University.

Fitch, Frederick. 1963. A logical analysis of some value concepts. *Journal of Symbolic Logic* 28 (2): 135–142. https://doi.org/10.2307/2271594.

Hintikka, Jaakko. 2005. *Knowledge and belief.* College Publications

Hudson, Robert. 2009. Faint-hearted anti-realism and knowability. *Philosophia* 37: 511–523. https://doi.org/10.1007/s11406-008-9174-0.

Jing, Wang. 2018. Two solutions to knowability paradox. In *The 4th Asian workshop on philosophical logic 2018 Oct 20*, 28–41.

Kashima, Ryo. 2009. *Mathematical logic*. Asakura Publishing Co. Ltd (in Japanese)

Kojima, Kensuke. 2012. Relational and neighborhood semantics for intuitionistic modal logic. *Reports on Mathematical Logic* (47): 87

Kvanvig, Jonathan. 1995. The knowability paradox and the prospects for realism. *Noûs* 29 (4): 481–500. https://doi.org/10.2307/2216283.

Litak, Tadeusz, Miriam Polzer, and Ulrich Rabenstein. 2017. Negative translations and normal modality. In *2nd international conference on formal structures for computation and deduction (FSCD 2017)*, ed. Dale Miller, vol. 84, 27:1–27:18. Leibniz international proceedings in informatics (LIPIcs), Dagstuhl, Germany. Schloss Dagstuhl–Leibniz-Zentrum fuer Informatik. ISBN 978-3-95977-047-7, https://doi.org/10.4230/LIPIcs.FSCD.2017.27.

Maffezioli, Paolo, Alberto Naibo, and Sara Negri. 2013. The Church-Fitch knowability paradox in the light of structural proof theory. *Synthese* 190: 2677–2716. https://doi.org/10.1007/s11229-012-0061-7.

Murzi, Julien. 2010. Knowability and bivalence: intuitionistic solutions to the paradox of knowability. *Philosophical Studies* 149: 269–281. https://doi.org/10.1007/s11098-009-9349-y.

Ono, Hiroakira. 1998. Proof-theoretic methods in nonclassical logic-an introduction. *Theories of types and proofs*, vol. 2, 207–254.

Ono, Hiroakira, and Yuichi Komori. 1985. Logics without the contraction rule. *The Journal of Symbolic Logic* 50 (1): 169–201.

Percival, Philip. 1990. Fitch and intuitionistic knowability. *Analysis* 50: 182–187. https://doi.org/10.2307/3328541.

Proietti, Carlo. 2012. Intuitionistic epistemic logic, Kripke models and Fitch's paradox. *Journal of Philosophical Logic* 41: 877–900. https://doi.org/10.1007/s10992-011-9207-1.

Usberti, Gabriele. 2016. The paradox of knowability from an intuitionistic standpoint. *Advances in proof-theoretic semantics*, vol. 43, 115–137. Trends in logic (Studia Logica Library) Cham: Springer.

van Benthem, Johan, and Eric Pacuit. 2011. Dynamic logics of evidence-based beliefs. *Studia Logica* 99 (1–3): 61.

Williamson, Timothy. 1982. Intuitionism disproved? *Analysis* 42: 203–207. https://doi.org/10.2307/3327773.

Local Fact Change Logic

Declan Thompson

Abstract We investigate a modal logic of model change, introducing a new operator which allows for changing the valuation at a particular state in a model. After investigating some properties of the logic, and aspects of its expressive power, we show it to be undecidable by way of a reduction using memory logic.

1 Introduction

Extensions of the basic modal logic with model-changing operators have proven very fruitful avenues of investigation. Some such operators modify the underlying frame of evaluation, like sabotage logic (Aucher et al. 2018) and dynamic epistemic logic (van Benthem 2011). Others, like memory logic (Areces 2007; Mera 2009) and update logic (van Benthem 2011), modify the model through other means, such as changes to a valuation. In many cases, these logics are introduced in order to study graph games, like the Poison Game (Aucher et al. 2018). Another approach to studying games involves logics of propositional control in a non-modal setting, wherein players control certain variables in order to achieve their aims. Examples include logic of propositional control (Van der Hoek and Wooldridge 2005) and various Boolean Games (Harrenstein et al. 2001; Gutierrez et al. 2015).

To see where these approaches can intersect, consider a simple example. Mary wants to paint her house. Under a new civic ordinance, houses can only be painted in one of four colours, and no neighbouring houses can be painted the same colour. Mary has five neighbours. Can she paint her house? At first, we might say this is a simple graph colouring problem. The question is whether a network like that in Fig. 1 has a four-colouring. But Mary can't make her neighbours repaint; if there is a neighbouring house of each of the four colours, Mary has no option available.

This chapter is in its final form and it is not submitted to publication anywhere else.

D. Thompson (✉)
Department of Philosophy, Stanford University, Stanford, USA
e-mail: declanthompsonnz@gmail.com

F. Liu et al. (eds.), *Knowledge, Proof and Dynamics*,
Logic in Asia: Studia Logica Library,
https://doi.org/10.1007/978-981-15-2221-5_5

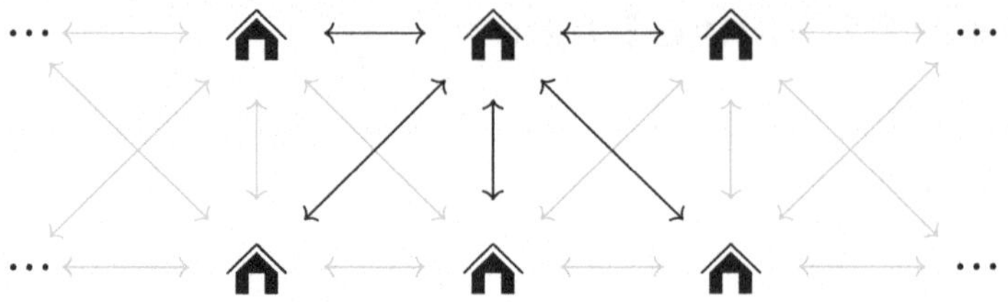

Fig. 1 Mary's neighbourhood

The question we are asking is whether Mary can obtain a certain graph property (a four-colouring) by changing only that under her control (the colour of a single house).

More generally, we're focussed on questions of *control*. Given control over one position in a network, what formulae can be made true? A key question in this area is the balance between expressive power and decidability. Highly expressive systems such as PDL+GLA (Tiomkin and Makowsky 1985, see below) often come at the cost of decidability. Pinpointing the features of a logical system that tip it to undecidability gives a better understanding of the complexity of those features. This paper (and in particular Sect. 4) can be seen as an investigation in this light.

The paper's main contribution is the introduction of a logic of local fact change (LFC), an extension of the basic modal logic containing an operator for changing the model's valuation at the point of evaluation, and nowhere else. First, we will recount some related work. Section 2 begins our investigation by looking at a very restricted logic of propositional control, containing an operator for changing the valuation outright. This allows us to set some groundwork before moving to the modal case in Sect. 3. Here, LFC is introduced, containing an operator which changes the valuation, but only at the current point in a model. Despite its apparent simplicity, Sect. 4 shows LFC to be undecidable, utilising the known undecidability of memory logic. Given how few moving parts are required on top of the minimal modal logic K, this makes LFC a candidate for a simplest undecidable modal logic. Section 4.1 shows that this undecidability result is maintained under a natural generalisation of LFC. Finally, in Sect. 5, LFC is shown not to be more expressive than memory logic.

1.1 Related Work

Fact change in a propositional context (i.e. with no modal structure) has been studied extensively in the literature on propositional control. For example, Van der Hoek and Wooldridge (2005) introduce the logic CL-PC in which coalitions of players modify partitions of a single valuation. Section 2 is essentially a simplification of this approach, and we will use results from CL-PC there. Similar work includes the Boolean Games of Harrenstein et al. (2001). Boolean Games add a goal structure to CL-PC; each player controls partition of a single valuation and has a goal formula they wish to make true. This allows for natural reasoning about strategies in situations of resource control. A connection to Boolean Games will be highlighted in Sect. 3.2.

In Public Announcement Logic (PAL, van Benthem 2011), the operator $[!p]$ represents a public announcement of the truth of p. This causes an update in an epistemic model, which makes p true at every point (i.e. makes p common knowledge). We can see this process as valuation change on a network structure. However, PAL does not model Mary's situation; the result of $[!p]$ is that p is true *everywhere*, not at a single point, and this is achieved not by changing the valuation but by removing incompatible possible worlds.

There is a system which takes a network structure and an operator for modifying the valuation at a single point: Dynamic Logic of Propositional Assignments (PA-DL, Balbiani et al. 2013). PA-DL is a variety of Propositional Dynamic Logic (PDL) where the atomic programmes are propositional assignments. The operator $\langle +p \rangle$ changes the valuation such that p becomes true. Balbiani, Herzig and Troquard give semantics in terms of a relation over the set of valuations (rather than points in a model), so networks like Mary's neighbourhood can't be represented directly. This logic can be seen as a simplification of CL-PC. Propositional Fact Change Logic (introduced in Sect. 2) and PA-DL are inter-definable. The main difference between the two systems is that PA-DL makes its assignment changes explicit, whereas Propositional Fact Change Logic keeps them implicit. Those familiar with PA-DL will also see strong connections in Sect. 3.4, where we define an operator for particularised fact change.

A great benefit of all of the systems so far mentioned is the decidability of the satisfiability problem for each. However, none precisely models Mary's situation: the result of changes to a valuation at a single point in a network. A system which can is PDL with Global and Local Assignments (PDL+GLA, Tiomkin and Makowsky 1985). This is standard PDL, but with four new model update programmes which change the valuation locally and globally. For example, $\langle p \leftarrow \mathbf{true} \rangle$ updates the model so that the valuation at the current state satisfies p (and leaves the rest of the model the same). If each state in the model represents a neighbour of Mary, this operator captures her situation: her house colour can be modelled by propositional variables and changes to it by local updates.

Tiomkin and Makowsky show that PDL+GLA is undecidable. They establish this by proving that it embeds first-order PDL. This suggests PDL+GLA is somewhat overpowered for Mary's situation. Indeed, it includes unnecessary PDL machinery. In many cases, inclusion of PDL moves a logic from decidable to undecidable. For example, where PAL is decidable, PAL with PDL operators is not (Miller et al. 2005). Hence, it is worth investigating whether the undecidability of PDL+GLA is a feature of the PDL machinery, or is present in a weaker logic which can still model Mary's situation.

A different approach can be found in Aucher et al. (2009), who provide a logic locGML containing an explicit operator for local assignment changes to specific variables, as well as global changes to variables. Thus, the logic is able to not only describe Mary's situation, but also an extension where the entire neighbourhood is repainted. Aucher et al. show their logic to be undecidable through a reduction to hybrid logic (with $\downarrow$) which relies on the global assignment change operator. For our purposes, this global operator is unnecessary. Does the undecidability result require its presence?

The results of Sect. 4 show that in both these cases (PDL+GLA and locGML), the undecidability can be traced to the local update operators. Thus, undecidability is preserved even in the simpler model presented here.

2 Propositional Fact Change

To begin, we recall some basic facts from the literature on propositional control (see, for example, Van der Hoek and Wooldridge 2005). Our setting is a simplification of standard logics of propositional control in that there are no players, and there is no partition on the propositional variables. All we are concerned with is whether propositional φ can be made true under *some* valuation. Thus, we start by modelling the situation where Mary has no neighbours.

Take a set Prop of propositional variables. The language of propositional fact change, $\mathcal{L}_{\mathsf{PFC}}$, is the standard propositional language extended with a unary $\bigcirc$ operator.

Definition 1 ($\mathcal{L}_{\mathsf{PFC}}$) $\mathcal{L}_{\mathsf{PFC}}$ is defined by

$$\varphi ::= p \mid \neg\varphi \mid (\varphi \wedge \varphi) \mid \bigcirc\varphi$$

for $p \in \mathsf{Prop}$.

Definition 2 (At) Let $\mathsf{At}(\varphi)$ be the set of propositional variables in φ.

Definition 3 (*Valuation (PFC)*) A valuation is a subset $V \subseteq \mathsf{Prop}$.

Definition 4 (*Truth under a valuation*) The truth of a formula φ under a valuation V is defined inductively as:

$$\begin{array}{lll} V \models p & \text{iff} & p \in V \\ V \models \neg\varphi & \text{iff} & V \not\models \varphi \\ V \models (\varphi \wedge \psi) & \text{iff} & V \models \varphi \text{ and } V \models \psi \\ V \models \bigcirc\varphi & \text{iff} & U \models \varphi \text{ for some } U \subseteq \mathsf{Prop}. \end{array}$$

We define the dual of $\bigcirc$ as $\bullet\varphi := \neg\bigcirc\neg\varphi$. Thus, $V \models \bullet\varphi$ iff $U \models \varphi$ for *every* $U \subseteq \mathsf{Prop}$. Other Boolean connectives are defined as usual. In the remainder of this section, we discuss a number of properties of the logic of propositional fact change (PFC).

2.1 Validities, Completeness, Reduction for PFC

For those unfamiliar with the literature on propositional control, this section is intended to provide some intuition. Those familiar may safely skip it.

First, notice that $V \models \bigcirc p \wedge \bigcirc \neg p$ for all valuations V and $p \in \mathsf{Prop}$. This is easy to establish. To show $V \models \bigcirc p$, we need some valuation under which p is true—Prop itself will do here. To show $V \models \bigcirc \neg p$, we can appeal to the valuation $\emptyset$.

We also have that $V \models \bullet \varphi$ exactly when φ is a tautology. Again, this is easy to see, since $V \models \bullet \varphi$ exactly when φ holds under every valuation. From this perspective, the new operator appears to be bringing a notion of validity into the language. We might read $\bigcirc \varphi$ as 'it is possible for φ to be true', and $\bullet \varphi$ as 'φ is necessarily true'.

This interpretation has clear parallels with the common interpretation of the modal operators $\Box$ and $\Diamond$. Indeed, if we treat each valuation V as a possible world, we can see $\bigcirc$ is a universal (and so **S5**) modality ranging over all valuations. Axiom **K**, namely, $\bullet(\varphi \rightarrow \psi) \rightarrow (\bullet \varphi \rightarrow \bullet \psi)$, holds, as well as **T** $(\bullet \varphi \rightarrow \varphi)$ and **B** $(\varphi \rightarrow \bullet \bigcirc \varphi)$. A summary of interesting validities is in Table 1.

A strong similarity can be observed between PFC and coalition logics of propositional control (CL-PC, Van der Hoek and Wooldridge 2005). These logics are used to model games of cooperation and control between a set of players W. Each player is assigned a part in a partition of Prop, and the formula $\diamond_C \varphi$, with $C \subseteq W$, indicates that under the current valuation, the players in C can change their propositional variables so as to make φ true. $\bigcirc$ is definable in this language by setting $\bigcirc \varphi := \diamond_W \varphi$.

Following the lead of van der Hoek and Wooldridge, we can give a semantics for PFC on a Kripke frame, where each valuation is a point in the model, and $\bigcirc$ is a universal accessibility relation. In fact, completeness for the case of PFC is more straightforward than that for CL-PC, since we don't need to deal with formulae of the form $\diamond_\emptyset$, for example.

Fact 1 *The formulae listed in Table 1, together with all propositional tautologies, necessitation and modus ponens, are sound and complete with respect to the semantics of PFC.*

Proof See Van der Hoek and Wooldridge (2005) for details. $\Box$

While it may be useful for modelling various situations, PFC does not have any expressive power above that of standard propositional logic. For each $U \subseteq \mathsf{Prop}$, define an operation on formulae by

$$p^U = \begin{cases} p & \text{if } p \in U \\ \neg p & \text{if } p \notin U \end{cases} \qquad (\bigcirc \varphi)^U = \bigvee_{U' \subseteq \mathsf{At}(\varphi)} \varphi^{U'}$$

$$(\neg \varphi)^U = \neg \varphi^U \qquad (\varphi \wedge \psi)^U = \varphi^U \wedge \psi^U.$$

This provides us with a translation into standard propositional logic.

Fact 2 *For every pair of valuations V, U, there is a valuation C such that for any φ,*

$$U \models \varphi \ \textit{ iff } \ V \models \varphi^{C \cap \mathsf{At}(\varphi)}.$$

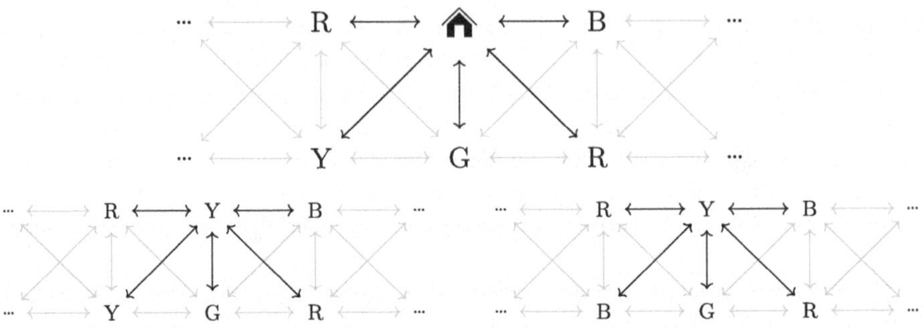

Fig. 2 Suppose the available colours are R, Y, B and G, and that Mary's neighbourhood starts as depicted in the top diagram (Mary's house is ). Then the bottom left is a move Mary could make (though it violates the civic ordinance). The bottom right is not a legal move for Mary, since it involves repainting the Y house

Table 1 Some validities of propositional $\bigcirc$. $\pm p$ and $\mp p$ are taken to be p and $\neg p$, respectively, or $\neg p$ and p, respectively. We require $p \notin \mathsf{At}(\chi)$

$(\chi \wedge \pm p) \rightarrow \bigcirc(\chi \wedge \mp p)$	$\bullet(p \vee \neg p)$	$\bullet\bullet\varphi \leftrightarrow \bullet\varphi$
$\bullet(\varphi \rightarrow \psi) \rightarrow (\bullet\varphi \rightarrow \bullet\psi)$	$\bullet\varphi \rightarrow \varphi$	$\varphi \rightarrow \bullet\bigcirc\varphi$

In particular, for every valuation V, there is a valuation C such that $V \models \varphi$ iff $V \models \varphi^{C \cap \mathsf{At}(\varphi)}$.

Proof By induction on the complexity of φ. □

3 Local Fact Change Logic

Let us review where we stand with respect to the motivating example of Mary's house painting. Suppose each of the four colours available to Mary is represented by a valuation on $\{p, q\}$. Then we can say that Mary can change house colour by writing $\bigcirc(p \wedge q)$, for example. Yet we are only able to describe changes to Mary's house in isolation. It's time to bring in the network structure from Fig. 1.

We can think of Mary's house as a point in a network. Mary can only change properties at *her* house. The rest of the neighbourhood remains fixed. This is what we're trying to capture. The valuation at the point of evaluation will be free to change, just as in the propositional case. The rest of the model will not change. Thus, we have *local* fact change. This is shown in Fig. 2.

In this section, we will first define LFC, before providing a more concrete consideration of Mary's case. Following this, we briefly consider some validities and a particularised version of LFC. Undecidability of the logic is proved in Sect. 4.

3.1 Definition of LFC

Take a standard relational frame $\mathfrak{F} = (W, R)$, where W is a set of *states* and $R \subseteq W \times W$ is an *accessibility relation* on those states. The language of local fact change $\mathcal{L}_{\mathsf{LFC}}$ is $\mathcal{L}_{\mathsf{PFC}}$ extended with the unary modality $\Diamond$.

Definition 5 ($\mathcal{L}_{\mathsf{LFC}}$) $\mathcal{L}_{\mathsf{LFC}}$ is defined by

$$\varphi ::= p \mid \neg\varphi \mid (\varphi \wedge \varphi) \mid \Diamond\varphi \mid \bigcirc\varphi.$$

Definition 6 (*Valuation (LFC)*) A valuation is a function $V : W \to 2^{\mathsf{Prop}}$. Let V_A^s be the function defined by

$$V_A^s(t) = \begin{cases} A & \text{if } t = s \\ V(t) & \text{otherwise.} \end{cases}$$

Valuations thus assign a set of propositional variables to each state. The similarity with modifications to assignment functions in first-order logic should not be understated here. On the whole, the semantics is as expected. $\bigcirc$ does just what it did in the propositional case, affecting the valuation *at the current world.*

Definition 7 (*Truth in a model*) The truth of a formula φ at a state s in $\mathfrak{F}$ under a valuation V is defined inductively as:

$$\begin{array}{lll}
\mathfrak{F}, V, s \models p & \text{iff} & p \in V(s) \\
\mathfrak{F}, V, s \models \neg\varphi & \text{iff} & \mathfrak{F}, V, s \not\models \varphi \\
\mathfrak{F}, V, s \models (\varphi \wedge \psi) & \text{iff} & \mathfrak{F}, V, s \models \varphi \text{ and } \mathfrak{F}, V, s \models \psi \\
\mathfrak{F}, V, s \models \Diamond\varphi & \text{iff} & \mathfrak{F}, V, t \models \varphi \text{ for some } t \text{ with } Rst \\
\mathfrak{F}, V, s \models \bigcirc\varphi & \text{iff} & \mathfrak{F}, V_A^s, s \models \varphi \text{ for some } A \subseteq \mathsf{Prop}
\end{array}$$

The standard Boolean connectives, $\top$ and $\bot$ are defined as usual. The dual of $\bigcirc$ is again $\bullet$.

We define $\mathfrak{F}, V \models \varphi$ iff $\mathfrak{F}, V, s \models \varphi$ for every $s \in W$. The similarities between the semantics for modal $\bigcirc$ and the standard semantics for $\exists$ should be highlighted. We can see $\bigcirc$ as treating V like an assignment function.

As mentioned in Sect. 1.1, LFC is similar to PDL+GLA, an undecidable variant of PDL which includes programmes for changing the local valuation. In fact, PDL+GLA could be seen as LFC+PDL. The natural question of whether LFC alone is decidable is answered in the negative in Sect. 4.

3.2 Application of LFC: Nash Equilibria

The ability of an agent to unilaterally change their state to achieve some goal, like in Mary's case, is studied extensively in game theory. A Nash equilibrium in a game is a strategy profile (i.e. an assignment of a strategy to each player) such that no player can do better in the game by unilaterally changing strategy. Various logics have been proposed to capture this notion (see, for example, van Benthem et al. 2009; Harrenstein et al. 2003; de Bruin et al. 2004; Bockmayr et al. 2013). However, these logics generally rely on highly abstract representations of games. In games played on a network, this often results in the network structure being implicit in the structure of payoffs. LFC provides a means to represent Nash equilibria while keeping the network structure explicit.

As an example of this type of approach, take Boolean Network Games (BNG, Seligman et al. 2015). In a BNG, each state in a frame represents a different player. The players have complete control over the valuation at their position in the network and are all assigned the same modal formula as a goal to make true. Each player's goal is evaluated at their position in the network, meaning conflict can arise. BNGs can be used to model resource sharing and scheduling problems, as well as interactions in social networks.

More formally, a BNG is a triple $\mathcal{G} = (A, R, \gamma)$, where A is a set of *agents*, R is a binary relation on A and γ is a *goal*—a formula of the basic modal language (i.e. $\mathcal{L}_{\mathsf{LFC}}$ without formulae containing $\bigcirc$). A *strategy* for $a \in A$ is a subset $V_a \subseteq \mathsf{Prop}$, and a *strategy profile* is simply a valuation on A, assigning a strategy to each player. Finally, strategy profile V is *winning for a* iff $(A, R), V, a \models \gamma$.

Example 1 Take a simplified version of Mary's neighbourhood, as shown in Fig. 3. $A = \{a, \ldots e\}$ and R is the relation indicated by arrows. The civic ordinance required that each house (agent) be painted one of four colours, and no two adjacent houses the same colour. If we take $p_1, \ldots p_4$ to be the paint colours, this gives the goal

$$\gamma = \big((p_1 \wedge \Box\neg p_1) \vee (p_2 \wedge \Box\neg p_2) \vee (p_3 \wedge \Box\neg p_3) \vee (p_4 \wedge \Box\neg p_4)\big).$$

A strategy profile V which assigns $V(a) = \{p_1\}$ and $V(i) = \{p_2\}$ for $i \neq a$ would be winning for a and losing for all other players.

A strategy profile V is a *Nash equilibrium* iff for every player a, if $(A, R), V, a \not\models \gamma$ then $(A, R), V^a_S, a \not\models \gamma$ for all $S \subseteq \mathsf{Prop}$. That is, no player can unilaterally change strategy to do better. This lends itself directly to definition using $\bigcirc$. We have

$$V \text{ is a Nash equilibrium for } \mathcal{G} = (A, R, \gamma) \quad \text{iff} \quad (A, R), V \models \gamma \vee \bullet\neg\gamma.$$

Example 1 supports a number of Nash equilibria, not all of which involve a four-colouring. If V assigns a different propositional letter to each of a, b, c and d, then $(A, R), V, e \models \bullet\neg\gamma$, since e has no remaining colour available.

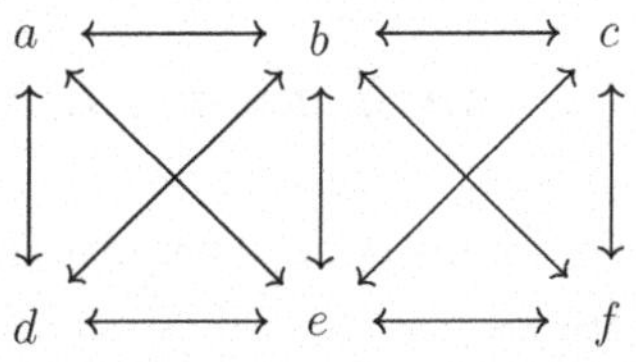

Fig. 3 A simplified version of Mary's neighbourhood

This approach can be generalised. Taking each V to represent a different possible strategy profile in a game, $\bigcirc$ provides an explicit operator for moving between strategies. This allows us to consider the logic of Nash equilibria in network games, a task left as future work.

3.3 Some Validities in LFC

The formulae listed in Table 1 remain validities under modal $\bigcirc$. This can be seen by very similar arguments to those for propositional $\bigcirc$. In addition, a number of validities arise from the interaction between $\bigcirc$ and $\Diamond$.

3.3.1 Relational Validities of LFC

One category of validities captures the fact that the structure of $\mathfrak{F}$ is not changed by $\bigcirc$. If φ contains no propositional variables but $\top$, then

$$\varphi \leftrightarrow \bullet\varphi$$

is valid. This is because in these cases φ doesn't depend on a valuation, only $\mathfrak{F}$.

For example, $\mathfrak{F}, V, s \models \Diamond\top \leftrightarrow \bullet\Diamond\top$ for all $\mathfrak{F}$, V, and s. Intuitively, this says that if a state has a neighbour, then it retains that neighbour under any local change in the valuation. More specifically, $\mathfrak{F}, V, s \models \Diamond\top$ iff there is some t such that Rst. Since the existence of this t doesn't depend on V, we have that Rst for some t iff $\mathfrak{F}, V, s \models \Diamond\top$. Using similar reasoning, we can see that if φ contains no propositional variables except $\top$ and $\bot$, then $\varphi \leftrightarrow \bigcirc\varphi$ is valid on all frames.

$\bigcirc$ can only change the valuation at one point at a time. If s has two neighbours, it cannot be that both neighbours are s. Thus

$$(\Diamond p \wedge \Diamond\neg p) \rightarrow (\bullet\Diamond p \vee \bullet\Diamond\neg p)$$

is valid. Roughly, if there is an accessible p-state and an accessible $\neg p$-state, then they cannot both be the current state. If the p-state is not the current state, then no local change in valuation will change the fact that $\Diamond p$ holds at the current state. And similarly for the case where the $\neg p$-state is not the current state. Like the previous formula, this takes advantage of the network structure of the model in order to make conclusions about the impact a local change in valuation will have.

3.3.2 Cycles and Substitution in LFC

The possibility of a state being on a cycle means we cannot be sure how the evaluation of $\Diamond\varphi$ will change under $\bigcirc$. As a simple case,

Fig. 4 A counterexample to the validity of $\Diamond\Diamond p \to \CIRCLE \Diamond\Diamond p$

$$\Diamond\Diamond p \to \CIRCLE \Diamond\Diamond p$$

is not valid. Figure 4 provides a counterexample. Suppose $V(s) = \{p\}$ so that $V, s \models \Diamond\Diamond p$. Then $V^s_\emptyset, s \not\models \Diamond\Diamond p$.

We can, however, go some way to allowing for a path around a cycle in the graph. The formula

$$\Diamond p \leftrightarrow (\bigcirc \Diamond p \wedge (p \vee \bigcirc(p \nleftrightarrow \Diamond p)))$$

attempts to capture this. $\Diamond p$ is true exactly when (a) some local valuation change allows for $\Diamond p$ and (b) either p is true (i.e. a reflexive loop) or it's possible for p and $\Diamond p$ to disagree (p true at a distinct point). This formula is valid under LFC.

This biconditional can be extended to loops of any length by stacking $\Diamond$. Hence, we have

$$\Diamond^n p \leftrightarrow (\bigcirc \Diamond^n p \wedge (\bigcirc(p \nleftrightarrow \Diamond^n p) \vee p)))$$

for every $n \in \mathbb{N}$.

It is tempting to take this as a general principle, but it is crucial that p is propositional here. The current point is guaranteed to have control over p, but the same cannot be said for arbitrary φ. This is a context in which we see substitution failing for LFC.

Fact 3 (Failure of substitution for LFC). *The validities of LFC are not closed under arbitrary substitutions.*

Proof Suppose we replace p with $\Diamond p$:

$$\Diamond\Diamond p \leftrightarrow (\bigcirc \Diamond\Diamond p \wedge (\bigcirc(\Diamond p \nleftrightarrow \Diamond\Diamond p) \vee \Diamond p))).$$

Take the frame depicted in Fig. 4 and suppose that $V(s) = \emptyset$ and $V(t) = \{p\}$. Then $V, s \models \Diamond p$ and $V, s \models \bigcirc \Diamond\Diamond p$ but $V, s \not\models \Diamond\Diamond p$. □

3.4 *Particularised Fact Change*

In many applications, it is natural to want greater control over which valuation we change to. While $\bigcirc\varphi$ changes us to *some* valuation which makes φ true, we might ask whether a *particular* change of the valuation makes φ true. The $[+p]$ operator captures something along these lines.

Definition 8 ($[+p]$) The truth of the $[+p]$ operator is defined by

$$\mathfrak{F}, V, s \models [+p]\varphi \text{ iff } \mathfrak{F}, V^s_A, s \models \varphi \text{ where } A = V(s) \cup \{p\}.$$

$[+p]\varphi$ is true if φ is true under the current valuation, except that we make p true here. For many applications, $[+p]$ seems to be a more natural operator than $\bigcirc$. For example, in the Poison Game (see Aucher et al. 2018), a demon poisons a node each turn, in an attempt to catch a traveller. $[+p]$ is an obvious choice for capturing this behaviour: set p false everywhere to begin with, and then slowly colour the graph with p. It's not immediately obvious how $\bigcirc$ can help in this situation.

The $[+p]$ operator is distinguished from similar dynamic operators in that it changes the current valuation, rather than shifting the focus to a new point of evaluation (as with the $[+\varphi]$ operator of van Benthem 2018) or removing possible points of evaluation (as with Public Announcement Logic). It most closely aligns with the $\langle +p \rangle$ operator of PA-DL (Balbiani et al. 2013, though PA-DL employs PDL modalities in a non-modal setting) and can be defined in locGML (Aucher et al. 2009) by setting $[+p]\varphi := [p + \top_{loc}]\varphi$.

Perhaps surprisingly, $[+p]$ does not add to expressive power. We can define it using $\bigcirc$.

Definition 9 (Bool) Given a finite set of propositional variables $A = \{p_0, \dots p_n\}$, let $\mathsf{Bool}(A)$ be the set of all formulae of the form $\pm p_0 \wedge \dots \wedge \pm p_n$, where $\pm$ indicates either a $\neg$ or nothing.[1] Further, let $\mathsf{Bool}(\emptyset) = \{\top\}$.

Bool takes a set of propositional variables to a set of formulae, one formula for each valuation we might have on that set of variables. The idea is that $\mathsf{Bool}(\mathsf{At}(\varphi))$ contains a distinct formula for each possible difference in valuation that could affect the truth of φ. We use this to keep relevant variables other than p constant in the valuation shift.

Definition 10 Take the following defined operator $\textcircled{p}$ for each $p \in \mathsf{Prop}$:

$$\textcircled{p}\varphi := \bigwedge_{\psi \in \mathsf{Bool}(\mathsf{At}(\varphi) \setminus \{p\})} \left(\psi \rightarrow \bigcirc(p \wedge \psi \wedge \varphi)\right).$$

If $\mathfrak{F}, V, s \models \psi_{V,s}$ for some $\psi_{V,s} \in \mathsf{Bool}(\mathsf{At}(\varphi) \setminus \{p\})$, say that $\psi_{V,s}$ *characterises* V *at* s.

Given a valuation, exactly one of the ψs will characterise V at s. So if ψ characterises V at s, we have that $\mathfrak{F}, V, s \models \textcircled{p}\varphi$ iff $\mathfrak{F}, V, s \models \bigcirc(p \wedge \psi \wedge \varphi)$. The large conjunction ensures that this ψ is held fixed under $\bigcirc$'s valuation change. For $\textcircled{p}\varphi$ to hold, it must be that making p true at s, and changing no other propositional variables which appear in φ, results in φ being true. Modulo the propositional variables needed to evaluate φ, there is only one possible valuation to change to. Using induction, we can show that $\textcircled{p}$ captures $[+p]$.

---

[1]Each $\pm$ can be independent, so we might have $p_0 \wedge \neg p_1 \wedge \dots$.

Fact 4 $V \models \textcircled{p}\varphi$ *iff* $V \models [+p]\varphi$.

Proof See appendix for details. □

Similar methods could be used to define an $[-p]$ operator which makes p false, or an operator which allows only p to vary. The focus in this paper is on $\bigcirc$, since it more closely corresponds to our motivating examples. The primary difference between the general and particularised operators is that $\bigcirc$ allows infinitely many propositional variables to be changed, whereas we can only stack finitely many $[+p]$s in a finite formula. However, since finite formulae only contain finitely many propositional variables, $\bigcirc$ can also be defined with $[+p]$ and $[-p]$. As a sketch, we could take

$$\bigcirc\varphi = \bigvee[\pm p_0][\pm p_1]\ldots[\pm p_n]\varphi,$$

where the disjunction ranges over possible values for each $\pm$, and $\{p_0, \ldots p_n\} = \mathsf{At}(\varphi)$. A more generalised operator like $\textcircled{\varphi}\psi$ (a minimal change to make φ true also makes ψ true, approaching the $[+\varphi]$ operator from van Benthem 2018) could also be defined along these lines.

4 Undecidability of Local Fact Change Logic

As discussed in Sect. 1.1, LFC is related to PDL with Global and Local Assignments (PDL+GLA, Tiomkin and Makowsky 1985). Both logics are able to model Mary's situation, but PDL+GLA is known to be undecidable. In this section, we will show that the undecidability can be traced to the presence of valuation change in a network, rather than being a product of the PDL machinery in PDL+GLA. Specifically, we show that the satisfiability problem for LFC is undecidable. The approach is inspired by that taken in Areces et al. (2017) for demonstrating the undecidability of various relation-changing modal logics: via a satisfaction preserving translation from memory logic, which is known to be undecidable (Mera 2009). Hence, we begin by introducing memory logic.

The language of memory logic $\mathcal{L}_{\mathsf{M}}$ is the basic modal language extended with a unary modality $\textcircled{r}$ and a 0-ary modality $\textcircled{k}$.

Definition 11 ($\mathcal{L}_{\mathsf{M}}$) $\mathcal{L}_{\mathsf{M}}$ is defined by

$$\varphi ::= p \mid \neg\varphi \mid (\varphi \wedge \varphi) \mid \Diamond\varphi \mid \textcircled{r}\varphi \mid \textcircled{k}.$$

We define the truth of formulae under memory logic with respect to a frame, valuation, set $C \subseteq W$ and point. The set C represents states that have been 'memorised'.

Definition 12 (*Truth in a model (memory logic)*) The truth of a formula φ at a state s in $\mathfrak{F}$ under a valuation V with a memory set C is defined inductively as:

$$
\begin{array}{lll}
\mathfrak{F}, V, C, s \models_{\mathsf{M}} p & \text{iff} & p \in V(s) \\
\mathfrak{F}, V, C, s \models_{\mathsf{M}} \neg\varphi & \text{iff} & \mathfrak{F}, V, C, s \not\models_{\mathsf{M}} \varphi \\
\mathfrak{F}, V, C, s \models_{\mathsf{M}} (\varphi \wedge \psi) & \text{iff} & \mathfrak{F}, V, C, s \models_{\mathsf{M}} \varphi \text{ and } \mathfrak{F}, V, C, s \models_{\mathsf{M}} \psi \\
\mathfrak{F}, V, C, s \models_{\mathsf{M}} \Diamond\varphi & \text{iff} & \mathfrak{F}, V, C, t \models_{\mathsf{M}} \varphi \text{ for some } t \text{ with } Rst \\
\mathfrak{F}, V, C, s \models_{\mathsf{M}} ⓡ\varphi & \text{iff} & \mathfrak{F}, V, C \cup \{s\}, s \models_{\mathsf{M}} \varphi \\
\mathfrak{F}, V, C, s \models_{\mathsf{M}} ⓚ & \text{iff} & s \in C
\end{array}
$$

The semantics for the familiar operators remains the same. The new ⓡ records the current state to memory, and the ⓚ operator tests whether the current state is in memory. The intuition is that we are able to test whether or not we are in a state that has already been visited.[2]

Now we move to the satisfaction preserving translation. To translate from $\mathcal{L}_{\mathsf{M}}$ into $\mathcal{L}_{\mathsf{LFC}}$, we first pick a propositional variable q that is not used in the formula we are translating, and ensure that q is not satisfied anywhere under the valuation in question. In effect, we choose a 'clean' q. Now using the particularised fact change operator as defined in Sect. 3.4, ⓡφ is simply $\textcircled{q}\varphi$, and ⓚ is q.

The translation comes in two parts. First, Str 'cleans' q. Second, τ replaces ⓡ with $\textcircled{q}$, and ⓚ with q. Finally, T conjoins Str and τ. Let $\mathsf{MD}(\varphi)$ be the modal $\Diamond$ depth of φ (i.e. the maximum number of nested $\Diamond$s in φ).

Definition 13 (Str) $\mathsf{Str} : \mathcal{L}_{\mathsf{M}} \times \mathsf{Prop} \to \mathcal{L}_{\mathsf{LFC}}$ is defined by

$$\mathsf{Str}(\varphi, q) = \bigwedge_{0 \leq i \leq \mathsf{MD}(\varphi)} \Box^i \neg q.$$

Definition 14 (τ) $\tau : \mathcal{L}_{\mathsf{M}} \times \mathsf{Prop} \to \mathcal{L}_{\mathsf{LFC}}$ is defined inductively by

$$
\begin{array}{ll}
\tau(p, q) = p & \tau(\varphi \wedge \psi, q) = \tau(\varphi, q) \wedge \tau(\psi, q) \\
\tau(\neg\varphi, q) = \neg\tau(\varphi, q) & \tau(\Diamond\varphi, q) = \Diamond\tau(\varphi, q) \\
\tau(ⓚ, q) = q & \tau(ⓡ\varphi, q) = \textcircled{q}\tau(\varphi, q).
\end{array}
$$

Definition 15 (T) Define a translation $\mathsf{T} : \mathcal{L}_{\mathsf{M}} \times \mathsf{Prop} \to \mathcal{L}_{\mathsf{LFC}}$ by

$$\mathsf{T}(\varphi, q) = \mathsf{Str}(\varphi, q) \wedge \tau(\varphi, q).$$

Our main task is to show that T preserves satisfiability of φ whenever $q \notin \mathsf{At}(\varphi)$. Once this is achieved, undecidability follows—a decision procedure for LFC would give us a decision procedure for ML, via T. For this, we need a way to modify a valuation so as to simulate the set C, since LFC considers no such set. The C, q-addition of V does the job.

---

[2]For this section, we will use $\models_{\mathsf{LFC}}$ to refer to the definition of truth for LFC, introduced in Definition 7.

Definition 16 (C, q-addition) Define the C, q-addition of V by

$$V_{C,q}(s) = \begin{cases} V(s) & s \notin C \\ V(s) \cup \{q\} & s \in C. \end{cases}$$

Notice that $V_{\emptyset,q} = V$ for all valuations V. First, we will show that under the assumption that q is 'clean', τ is satisfaction preserving for all possible sets C. After this, we will consider the more specific case where $C = \emptyset$, under the translation T.

Fact 5 (Adequacy of τ). *Suppose $q \notin \mathsf{At}(\varphi)$ and $q \notin V(s)$ for all $s \in W$. Then $\mathfrak{F}, V, C, s \models_{\mathsf{M}} \varphi$ iff $\mathfrak{F}, V_{C,q}, s \models_{\mathsf{LFC}} \tau(\varphi, q)$ for every $s \in W$ and all $C \subseteq W$.*

Proof By induction on the complexity of φ. The Boolean and $\Diamond$ cases are trivial.

First, consider the case where $\varphi = ⓚ$. We have $\mathfrak{F}, V, C, s \models_{\mathsf{M}} ⓚ$ iff $s \in C$. By assumption $q \notin V(s)$, and so we have that $s \in C$ iff $q \in V_{C,q}(s)$. And $q \in V_{C,q}(s)$ iff $\mathfrak{F}, V_{C,q}, s \models_{\mathsf{LFC}} q$. Hence $\mathfrak{F}, V, C, s \models_{\mathsf{M}} ⓚ$ iff $\mathfrak{F}, V_{C,q}, s \models_{\mathsf{LFC}} \tau(ⓚ, q)$.

The remaining case is where $\varphi = ⓡ\psi$ for some ψ. By the definition of truth, $\mathfrak{F}, V, C, s \models_{\mathsf{M}} ⓡ\psi$ iff $\mathfrak{F}, V, C \cup \{s\}, s \models_{\mathsf{M}} \psi$ which, by the inductive hypothesis, holds iff $\mathfrak{F}, V_{C\cup\{s\},q}, s \models_{\mathsf{LFC}} \tau(\psi, q)$. Notice that $V_{C\cup\{s\},q} = (V_{C,q})^s_A$, where $A = V_{C,q}(s) \cup \{q\}$. It follows from Fact 4 that $\mathfrak{F}, V, C, s \models_{\mathsf{M}} ⓡ\psi$ iff $\mathfrak{F}, V_{C,q}, s \models_{\mathsf{LFC}}$ ⓠ$\tau(\psi, q)$. □

Fact 6 (Adequacy of T). *For all frames $\mathfrak{F}$, valuations V, points s and formulae $\varphi \in \mathcal{L}_{\mathsf{M}}$, when $q \notin \mathsf{At}(\varphi)$:*

1. *$\mathfrak{F}, V, \emptyset, s \models_{\mathsf{M}} \varphi$ iff $\mathfrak{F}, V', s \models_{\mathsf{LFC}} \mathsf{T}(\varphi, q)$, where $V'(t) = V(t) \setminus \{q\}$ for every $t \in W$; and*
2. *If $\mathfrak{F}, V, s \models_{\mathsf{LFC}} \mathsf{T}(\varphi, q)$ then $\mathfrak{F}, V', s \models_{\mathsf{LFC}} \mathsf{T}(\varphi, q)$, where $V'(t) = V(t) \setminus \{q\}$ for every $t \in W$.*

Proof First, part 1. Take $q \notin \mathsf{At}(\varphi)$. We have that $q \notin V'(t)$ for all $t \in W$, and so $(*)$ $\mathfrak{F}, V', s \models_{\mathsf{LFC}} \mathsf{Str}(\varphi, q)$. Since V and V' differ only with respect to q, and $q \notin \mathsf{At}(\varphi)$, we have $\mathfrak{F}, V, \emptyset, s \models_{\mathsf{M}} \varphi$ iff $\mathfrak{F}, V', \emptyset, s \models_{\mathsf{M}} \varphi$. From Fact 5, we know that $\mathfrak{F}, V', \emptyset, s \models_{\mathsf{M}} \varphi$ iff $\mathfrak{F}, V'_{\emptyset,q}, s \models_{\mathsf{LFC}} \tau(\varphi, q)$. Since $V'_{\emptyset,q} = V'$, we then have $\mathfrak{F}, V, \emptyset, s \models_{\mathsf{M}} \varphi$ iff $\mathfrak{F}, V', s \models_{\mathsf{LFC}} \tau(\varphi, q)$. Finally, we use $(*)$ to reach the required result.

For part 2, it is sufficient to note that $\mathfrak{F}, V, s \models_{\mathsf{LFC}} \mathsf{T}(\varphi, q)$ requires that q is false at every point reachable from s in at most $\mathsf{MD}(\varphi)$ steps (since $\mathsf{Str}(\varphi, q)$ is satisfied). Since no other points are considered in the evaluation of $\mathsf{T}(\varphi, q)$, we know that every point could have q false. Hence, $\mathfrak{F}, V', s \models_{\mathsf{LFC}} \mathsf{T}(\varphi, q)$. □

At last, we reach the main result, using T to show LFC is undecidable.

Theorem 1 *The satisfiability problem for LFC is undecidable.*

Proof Fact 6 shows that there is a satisfiability preserving reduction from memory logic into LFC. Hence, if the satisfiability problem for LFC were decidable, we would have a decision procedure for memory logic's satisfiability problem. But, as shown in Mera (2009), memory logic is undecidable. □

Given the conceptual simplicity of LFC, this result is somewhat surprising. Even without the PDL modalities of PDL+GLA, or the global assignment changes of locGML, we still have undecidability. The undecidability seems to stem from the potential for cycles of arbitrary length in a graph. If frames are required to be acyclic, decidability follows. But if cycles are allowed, difficulties like those discussed in Sect. 3.3 begin to arise.

4.1 Generalised Assignment Models

As noted above, the definition of $\bigcirc$ bears a strong similarity to the standard truth definition for the first-order quantifier $\exists$. This becomes stronger under approaches where $\exists x\varphi$ is interpreted in a multimodal model as a modality moving through a state space of assignment functions along a binary relation R_x. This approach is described in van Benthem (1997). Under the standard first-order semantics, R_x is the relation $=_x$ (i.e. identity up to x-value) and the state space contains all possible assignment functions. However, if the assumption of a complete assignment state space is lifted, the resulting logic is decidable. We might take from this that some part of the undecidability of first-order logic is due to the assumption that all assignment functions are available.

Can a parallel approach provide decidability for a similarly generalised LFC? It does not, but the reason for failure is not immediately clear. *Generalised LFC models* are pairs $(\mathfrak{F}, \mathfrak{V})$, where $\mathfrak{F}$ is a relational frame, and $\mathfrak{V}$ a set of valuations on $\mathfrak{F}$. The generalised truth definition follows Definition 7, with a new clause substituted for $\bigcirc$:

$$(\mathfrak{F}, \mathfrak{V}), V, s \models_g \bigcirc\varphi \text{ iff } (\mathfrak{F}, \mathfrak{V}), V^s_A, s \models_g \varphi \text{ for some } V^s_A \in \mathfrak{V}$$

This semantics provides a problem for the adequacy of T. The formula $\textcircled{r}\top \to \bot$ is unsatisfiable in memory logic, yet its translation

$$\mathsf{T}(\textcircled{r}\top \to \bot, q) \quad = \quad \neg q \wedge (\textcircled{q}\top \to \bot)$$

is satisfiable on any generalised LFC model where no valuation in $\mathfrak{V}$ allows q to be true.

T fails to preserve satisfiability because it relies on the assumption that a q-assignment is available, and so that $\textcircled{q}$ will always succeed. T also relies on the assumption that q begins false at all relevant points in the model. We already guarantee this latter assumption with Str, and slight modification can guarantee the former also.

Definition 17 (Str') $\mathsf{Str}' : \mathcal{L}_{\mathsf{M}} \times \mathsf{Prop} \to \mathcal{L}_{\mathsf{LFC}}$ is defined by

$$\mathsf{Str}'(\varphi, q) = \bigwedge_{0 \leq i \leq \mathsf{MD}(\varphi)} \left(\Box^i \neg q \wedge \bigwedge_{\star_j \in \{ⓠ\Box, \Box\}} \star_1 \cdots \star_i ⓠ\top \right).$$

There are exponentially many new conjuncts, each consisting of a sequence of ⓠ$\Box$ and $\Box$, followed by ⓠ$\top$. These ensure that for every path of length $\mathsf{MD}(\varphi)$ from the point of evaluation, any subset of points along that path can be made to satisfy q.

Definition 18 (T') Define a translation $\mathsf{T}' : \mathcal{L}_{\mathsf{M}} \times \mathsf{Prop} \to \mathcal{L}_{\mathsf{LFC}}$ by

$$\mathsf{T}'(\varphi, q) = \mathsf{Str}'(\varphi, q) \wedge \tau(\varphi, q).$$

For $\mathsf{T}'(\varphi, q)$ to be satisfied, we need a generalised LFC model where q is false at relevant points *and* q *can be made true at all those points*, as well as $\tau(\varphi, q)$ being satisfied. The problem is resolved. We have that

$$\mathsf{T}'(ⓡ\top \to \bot, q) \quad = \quad \neg q \wedge ⓠ\top \wedge (ⓠ\top \to \bot),$$

which is equivalent to $\bot$, and so not satisfiable.

Theorem 2 *The satisfiability problem for generalised LFC is undecidable.*

Proof Fact 5 holds once we add the assumption that $V_{C,q} \in \mathfrak{V}$ for all $C \subseteq W$. A similar modification is needed for Fact 6, ensuring that all required valuations are available in $\mathfrak{V}$. For part 1, we choose $\mathfrak{V}$ to be the set of all valuations. For part 2, the satisfaction of $\mathsf{Str}'(\varphi, q)$ guarantees that $\mathfrak{V}$ contains the required valuations. The rest of the proof follows that in Theorem 1. $\Box$

The new translation works because of the meagre amount of change to valuations memory logic requires. We need change only one specified propositional variable, and the changes happen only at a restricted number of points. A similar approach does not work to embed memory logic into the generalised first-order semantics, since $\exists$ is not the counterpart to ◯, but rather to $\Diamond$, moving between points in the model. ◯ more closely resembles a second-order quantifier, changing the referent of unary predicates.

5 Expressive Power

The undecidability of LFC raises the question of just how powerful it is. Memory logic (ML) is a fairly weak undecidable logic, so the relationship between LFC and ML is of interest. In this section, we investigate the expressive power of LFC through Ehrenfeucht–Fraïssé games and show that LFC is not more expressive than ML. An appropriate definition of bisimulation can be found in the Appendix.

To demonstrate that LFC is not more expressive than ML, we will construct a pair of pointed models which ML can distinguish but LFC cannot. For this, we need a notion of *model equivalence*. We will define such a notion based on Ehrenfeucht–Fraïssé games. Our definition follows the one presented for ML in (Areces et al. 2011).

Definition 19 (*Ehrenfeucht–Fraïssé Games*) Let $\mathfrak{F}_1, \mathfrak{F}_2$ be frames, with V_1 a valuation on $\mathfrak{F}_1$, V_2 a valuation on $\mathfrak{F}_2$, and $s_1 \in W_1$, $s_2 \in W_2$ states in the respective frames.

An Ehrenfeucht–Fraïssé*game* $EF(\mathfrak{F}_1, \mathfrak{F}_2, V_1, V_2, s_1, s_2)$ is defined as follows. There are two players, Spoiler and Duplicator. The rounds of the game are played recursively as follows:

If $V_1(s_1) \neq V_2(s_2)$ then Spoiler wins.
Else if both s_1 **and** s_2 **have no neighbours** then Duplicator wins.
Else Spoiler chooses one of the following two moves:

1. Spoiler modifies *both* V_1 and V_2, sending s_1 and s_2 to a new set A of propositional variables. The players play the game $EF(\mathfrak{F}_1, \mathfrak{F}_2, V_{1A}^{s_1}, V_{2A}^{s_2}, s_1, s_2)$.
2. The following occur in order:
   (a) For some $i \in \{1, 2\}$, Spoiler chooses $t_i \in W_i$ such that $R_i s_i t_i$. Let $j \neq i$ be the other element of $\{1, 2\}$.
   (b) If s_j has no neighbour t_j such that $R_j s_j t_j$ and $V_i(t_i) = V_j(t_j)$, Spoiler wins. Otherwise, Duplicator picks such a t_j. The players play the game $EF(\mathfrak{F}_1, \mathfrak{F}_2, V_1, V_2, t_1, t_2)$.

In an infinite game, Duplicator wins.

If Duplicator has a winning strategy in $EF(\mathfrak{F}_1, \mathfrak{F}_2, V_1, V_2, s_1, s_2)$, we write $(\mathfrak{F}_1, V_1, s_1) \equiv_{EF} (\mathfrak{F}_2, V_2, s_2)$.

Informally, Spoiler is trying to demonstrate that the two pointed models are different by making moves in one model that cannot be copied in the other. Duplicator has the task of trying to copy all the moves spoiler makes, in the other model. If Spoiler wins, then there is some formula of LFC which corresponds to the strategy spoiler has chosen. That formula distinguishes the two models. If Duplicator wins, then no such distinguishing formula exists; the two models are equivalent from the perspective of LFC.

Fact 7 (Correctness of Ehrenfeucht–Fraïssé Games). *For all formulae* $\varphi \in \mathcal{L}_{\mathsf{LFC}}$, *if* $(\mathfrak{F}_1, V_1, s_1) \equiv_{EF} (\mathfrak{F}_2, V_2, s_2)$ *then* $\mathfrak{F}_1, V_1, s_1 \models \varphi$ *iff* $\mathfrak{F}_2, V_2, s_2 \models \varphi$.

Proof By induction on φ. For the base case, notice that if $(\mathfrak{F}_1, V_1, s_1) \equiv_{EF} (\mathfrak{F}_2, V_2, s_2)$ then $V_1(s_1) = V_2(s_2)$, otherwise Spoiler wins. The Boolean cases are straightforward, and the $\Diamond$ case standard. We prove only the case for $\bigcirc$.

Suppose $(\mathfrak{F}_1, V_1, s_1) \equiv_{EF} (\mathfrak{F}_2, V_2, s_2)$. $\mathfrak{F}_1, V_1, s_1 \models \bigcirc\psi$ iff $\mathfrak{F}_1, V_{1A}^{s_1}, s_1 \models \psi$ for some $A \subseteq \mathsf{Prop}$. Since in $EF(\mathfrak{F}_1, \mathfrak{F}_2, V_1, V_2, s_1, s_2)$ Duplicator has a winning strategy, they have a winning response when Spoiler opts for move 1, giving the game

$EF(\mathfrak{F}_1, \mathfrak{F}_2, V_1{}^{s_1}_A, V_2{}^{s_2}_A, s_1, s_2)$. Hence, by the inductive hypothesis, $\mathfrak{F}_1, V_1{}^{s_1}_A, s_1 \models \psi$ iff $\mathfrak{F}_2, V_2{}^{s_2}_A, s_2 \models \psi$. But this holds iff $\mathfrak{F}_2, V_2, s_2 \models \bigcirc\psi$. □

Having established that our definition of Ehrenfeucht–Fraïssé games is correct, we move to constructing two pointed models that memory logic can distinguish, which LFC cannot. In what follows, we make no assumptions about the cardinality of Prop (in particular, it may be finite).

Let $\mathfrak{G} = (W_\mathfrak{G}, R_\mathfrak{G})$, where $W_\mathfrak{G} = 2^{\mathsf{Prop}} \times \omega$ and $R_\mathfrak{G} = W_\mathfrak{G} \times W_\mathfrak{G}$. That is, $\mathfrak{G}$ is a frame with ω copies of each subset of Prop for states, and R is the complete relation on these states. Let $V_\mathfrak{G}$ be the function taking (A, α) to A. That is, the valuation on each state is simply the subset of Prop that state corresponds to. This is somewhat similar to a canonical model construction.

Now consider two extensions of $\mathfrak{G}$. In both extensions, we add a *spy point* s. First,

$$\mathfrak{G}_1 = (W_\mathfrak{G} \cup \{s\}, R_\mathfrak{G} \cup (\{s\} \times W_\mathfrak{G})).$$

In $\mathfrak{G}_1$, s is related to every point in $W_\mathfrak{G}$. Second,

$$\mathfrak{G}_2 = (W_\mathfrak{G} \cup \{s\}, R_\mathfrak{G} \cup (\{s\} \times W_\mathfrak{G}) \cup \{(s, s)\}).$$

In $\mathfrak{G}_2$, s is also related to itself. Now extend $V_\mathfrak{G}$ such that $V_\mathfrak{G}(s) = \emptyset$. These models are depicted in Fig. 5.

Claim 1 $(\mathfrak{G}_1, V_\mathfrak{G}, s) \equiv_{EF} (\mathfrak{G}_2, V_\mathfrak{G}, s)$.

Proof We must show Duplicator has a winning strategy. Both $\mathfrak{G}_1$ and $\mathfrak{G}_2$ are serial, so Duplicator will never win due to a lack of neighbours. Hence, Duplicator must win due to an infinite game. We will show Duplicator has such a strategy by induction on the number of the rounds: Spoiler does not win at round 1, and if Spoiler does not win in round n, then Duplicator can prevent Spoiler winning at round $n + 1$.

Base case $V_\mathfrak{G}(s) = V_\mathfrak{G}(s)$, so the first Spoiler win condition fails. If Spoiler chooses move 1, Spoiler does not win in this round. Suppose Spoiler chooses move 2.

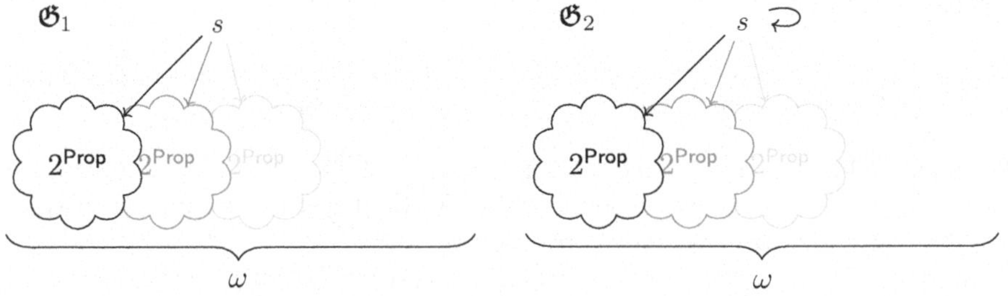

Fig. 5 The models $\mathfrak{G}_1$ and $\mathfrak{G}_2$

If Spoiler chooses to move in $\mathfrak{G}_1$, then they must move to an element (A, α) of $2^{\mathsf{Prop}} \times \omega$ (there are no other neighbours of s in $\mathfrak{G}_1$). Since (A, α) is also in $\mathfrak{G}_2$, Duplicator simply chooses (A, α) also. Since the two models have the same valuation, (A, α) satisfies all the requirements of 2b.

If Spoiler chooses to move in $\mathfrak{G}_2$ and chooses an element (A, α) of $2^{\mathsf{Prop}} \times \omega$, the same argument follows. If Spoiler chooses s, then Duplicator instead chooses $(\emptyset, \emptyset)$. Since $V_{\mathfrak{G}}(s) = \emptyset = V_{\mathfrak{G}}((\emptyset, \emptyset))$, and $Rs(\emptyset, \emptyset)$, this choice satisfies the requirements of 2b.

In all cases, the round ends with players playing a game of the form $EF(\mathfrak{G}_1, \mathfrak{G}_2, V^1_{\mathfrak{G}}, V^2_{\mathfrak{G}}, u_1, u_2)$.

Inductive step Suppose round n concluded with the players moving to play the game $EF(\mathfrak{G}_1, \mathfrak{G}_2, V^1_{\mathfrak{G}}, V^2_{\mathfrak{G}}, u_1, u_2)$. There are two cases:

1. In round n, Spoiler chose move 1. So $V^1_{\mathfrak{G}} = V^{x\,u_1}_{\mathfrak{G}\,A}$ and $V^2_{\mathfrak{G}} = V^{y\,u_2}_{\mathfrak{G}\,A}$ for some $A \subseteq \mathsf{Prop}$.[3] In particular, we know that $V^1_{\mathfrak{G}}(u_1) = V^2_{\mathfrak{G}}(u_2)$, so Spoiler doesn't immediately win round $n + 1$.
2. In round n, Spoiler chose move 2. So by clause 2b, we know $V^1_{\mathfrak{G}}(u_1) = V^2_{\mathfrak{G}}(u_2)$, and Spoiler doesn't immediately win round $n + 1$.

Hence, in neither case does Spoiler immediately win round $n + 1$.

As noted above, Duplicator does not immediately win round $n + 1$ either, since both $\mathfrak{G}_1$ and $\mathfrak{G}_2$ are serial.

If Spoiler chooses move 1, the round will conclude with the players moving to play the game $EF(\mathfrak{G}_1, \mathfrak{G}_2, V^{1\,u_1}_{\mathfrak{G}\,B}, V^{2\,u_2}_{\mathfrak{G}\,B}, u_1, u_2)$ for some $B \subseteq \mathsf{Prop}$.

Suppose Spoiler chooses move 2 and picks $i \in \{1, 2\}$. Spoiler chooses t_i in $\mathfrak{G}_i$ such that $R_i u_i t_i$. Let j be the other element of $\{1, 2\}$. Duplicator chooses $t_j = (V^i_{\mathfrak{G}}(t_i), \beta)$, where β is chosen such that t_j has never been visited before. An available β is guaranteed, since there are always ω states with the valuation $V^i_{\mathfrak{G}}(t_i)$. Since every state is related to t_j, $R_j u_j t_j$. Since t_j has not been visited before, Spoiler has not had a chance to modify the valuation there. Hence, we know that $V^i_{\mathfrak{G}}(t_i) = V^j_{\mathfrak{G}}(t_j)$. Thus, t_j is a legal choice for Duplicator. The round will conclude with the players moving to play the game $EF(\mathfrak{G}_1, \mathfrak{G}_2, V^1_{\mathfrak{G}}, V^2_{\mathfrak{G}}, t_1, t_2)$.

Whichever move Spoiler chooses, the round ends with the players playing a game of the required form.

Hence, there is no round at which Spoiler wins. Thus, Duplicator wins and $(\mathfrak{G}_1, V_{\mathfrak{G}}, s) \equiv_{EF} (\mathfrak{G}_2, V_{\mathfrak{G}}, s)$. □

Claim 2 $\mathfrak{G}_1, V_{\mathfrak{G}}, \emptyset, s \not\models_{\mathsf{M}}$ ⓡ$\Diamond$ⓚ but $\mathfrak{G}_2, V_{\mathfrak{G}}, \emptyset, s \models_{\mathsf{M}}$ ⓡ$\Diamond$ⓚ.

Proof Once we notice the reflexive loop on s in $\mathfrak{G}_2$, trivial. □

[3]The notations $V^x_{\mathfrak{G}}$ and $V^y_{\mathfrak{G}}$ are intended to indicate possibly different modifications to $V_{\mathfrak{G}}$.

Thus, we have shown a pair of pointed models that ML can distinguish which LFC cannot. It is worth noting that the same pair of models demonstrates that LFC is not as expressive as locGML.

If Prop is countably infinite, a minor modification of Ehrenfeucht–Fraïssé games allows the proof go to through with simplified versions of $\mathfrak{G}_1$ and $\mathfrak{G}_2$, each containing only one copy of 2^{Prop}. The question remains whether LFC is strictly less expressive than ML, or incomparably expressive. This is left for future work.

6 Conclusion and Further Work

LFC provides a new account of the interaction between graph games, logics of control and network games. From one perspective, it models incremental updates to a graph. From another perspective, it naturally extends logics of propositional control to the modal case. From a third perspective, it introduces an operator which asks whether a particular formula is satisfiable at a point in a network. That such a small addition to the minimal modal logic takes us to undecidability is surprising, demonstrating that decidability can be difficult to obtain even with basic model-changing operations.

One direction for future work is in learning more about LFC itself. As mentioned in Sect. 3, the $\bigcirc$ operator's semantics has strong parallels with the standard semantics for $\exists$. In fact, a better parallel is with hybrid logic (Blackburn and Seligman 1995). In future work, we will show that LFC is equi-satisfiable with the hybrid logic containing $\downarrow$, though not as expressive. Finding a suitable axiomatisation of LFC is likely to pose a challenge, due to the failure of substitutability. This is in line with difficulties in axiomatising logics like sabotage logic, and an investigation into the source of these difficulties would prove worthwhile.

A second direction is to consider broader applications of the logic. Here, we focussed on modelling network games like BNGs, but a wider range of graph games might be well modelled also. Games like the Poison Game or the Sabotage Game are natural first steps. There may also be surprising applications to even broader domains such as epistemic logic and provability logic.

Finally, weakenings of the logic to find decidable fragments would frame further investigations as to which properties cause undecidability.

Acknowledgements I would like to thank Carlos Areces for very helpful input, especially in connection with memory logics. Ben Sparkes encouraged investigating similarities with hybrid logics, which led to many of the results in this paper. I also received very helpful feedback from Johan van Benthem, the members of the Logic Workshop at Stanford, and the participants of the Fourth Asian Workshop in Philosophical Logic. Finally, the anonymous reviewers provided very useful feedback.

Appendix

Proof of Fact 4

Fact 8 *Let* $A \subseteq \mathsf{Prop}$. *If* $\psi \in \mathsf{Bool}(\mathsf{At}(\varphi))$ *and* $\mathfrak{F}, V, s \models \psi$ *and* $\mathfrak{F}, V^s_A, s \models \psi$ *then*

$$\mathfrak{F}, V, s \models \varphi \quad \textit{iff} \quad \mathfrak{F}, V^s_A, s \models \varphi.$$

Proof Recall that ψ is of the form $\pm p_0 \wedge \cdots \pm p_n$, where $p_0, \ldots p_n \in \mathsf{At}(\varphi)$. Since ψ has no modal operators and its evaluation only depends on the values of propositional variables in $\mathsf{At}(\varphi)$, we know $V(s) \cap \mathsf{At}(\varphi) = V^s_A(s) \cap \mathsf{At}(\varphi)$. From this, we conclude $V(t) \cap \mathsf{At}(\varphi) = V^s_A(t) \cap \mathsf{At}(\varphi)$ for all t. Observing that φ also only depends on the values of propositional variables in $\mathsf{At}(\varphi)$ gives us the result. □

Fact 9 $\mathfrak{F}, V, s \models \textcircled{p}\varphi$ *iff* $\mathfrak{F}, V, s \models \neg \textcircled{p} \neg\varphi$.

Proof For simplicity, we will omit $\mathfrak{F}$ below. We have the following equivalences:

$$
\begin{aligned}
V, s \models \neg\textcircled{p}\neg\varphi \quad &\text{iff} \quad V, s \models \neg \bigwedge_{\psi \in \mathsf{Bool}(\mathsf{At}(\neg\varphi)) \setminus \{p\})} (\psi \to \bigcirc(p \wedge \psi \wedge \neg\varphi)) && (1)\\
&\text{iff} \quad V, s \models \bigvee_{\psi \in \mathsf{Bool}(\mathsf{At}(\neg\varphi)) \setminus \{p\})} (\psi \wedge \neg\bigcirc(p \wedge \psi \wedge \neg\varphi)) && (2)\\
&\text{iff} \quad V, s \models \psi_{V,s} \wedge \neg\bigcirc(p \wedge \psi_{V,s} \wedge \neg\varphi) && (3)\\
&\text{iff} \quad V, s \models \bullet(\neg p \vee \neg\psi_{V,s} \vee \varphi) && (4)\\
&\text{iff} \quad V^s_B, s \models \neg p \vee \neg\psi_{V,s} \vee \varphi \text{ for all } B && (5)\\
&\text{iff} \quad \text{for all } B, \text{ either } V^s_B, s \not\models p \text{ or } V^s_B, s \not\models \psi_{V,s} \text{ or } V^s_B, s \models \varphi. && (6)
\end{aligned}
$$

Equation (1) follows from the definition of $\textcircled{p}$, and (2) via basic logic. Taking $\varphi_{V,s}$ to characterise V at s, equivalence (3) holds, and since $V \models \psi_{V,s}$ by the definition of $\psi_{V,s}$, we get (4). Equations (5) and (6) then follow from the truth definitions. In the remainder of the proof, we will establish that (6) holds iff $V, s \models \textcircled{p}\varphi$.

($\Rightarrow$) Suppose $V^s_B, s \not\models p$ or $V^s_B, s \not\models \psi_{V,s}$ or $V^s_B, s \models \varphi$ for all B. Let $A = V(s) \cup \{p\}$. We have that $V^s_A, s \models p$ trivially, and $V^s_A, s \models \psi_{V,s}$ since $V, s \models \psi_{V,}$ and $p \notin \mathsf{At}(\psi_{V,s})$, so we must have $V^s_A, s \models \varphi$. Hence $V^s_A, s \models p \wedge \psi_{V,s} \wedge \varphi$ and so $V, s \models \bigcirc(p \wedge \psi_{V,s} \wedge \varphi)$. From this it quickly follows that $V, s \models \textcircled{p}\varphi$ from the definition of $\textcircled{p}$.

($\Leftarrow$) Suppose $V, s \models \textcircled{p}\varphi$. By similar reasoning to the move from (2) to (4), we know that $V, s \models \bigcirc(p \wedge \psi_{V,s} \wedge \varphi)$. Thus, $V^s_C, s \models p \wedge \psi_{V,s} \wedge \varphi$ for some C. Take arbitrary B and suppose $V^s_B, s \models p$ and $V^s_B, s \models \psi_{V,s}$. Since V^s_B and V^s_C agree on p and $\psi_{V,s}$ at s, we know that $V^s_B, s \models \varphi$ iff $V^s_C, s \models \varphi$, by Fact 8. Since $V^s_C, s \models \varphi$, we have $V^s_B, s \models \varphi$. Since B was arbitrary, we can conclude that for all B, either $V^s_B, s \not\models p$ or $V^s_B, s \not\models \psi_{V,s}$ or $V^s_B, s \models \varphi$. □

Fact 4 *$\mathfrak{F}, V, s \models \textcircled{p}\varphi$ iff $\mathfrak{F}, V, s \models [+p]\varphi$, for every formula φ and all frames $\mathfrak{F}$, valuations V and points s.*

Proof By induction on the complexity of φ. For simplicity, we will omit reference to $\mathfrak{F}$ in what follows. Let $A = V(s) \cup \{p\}$.

For φ propositional there are two cases. If $\varphi = p$, then $V, s \models \textcircled{p}\varphi$ iff $V, s \models \top \rightarrow \bigcirc(p \wedge \top \wedge p)$, which is the case since $V, s \models \bigcirc p$. It's trivial to see that $V^s_A, s \models p$. The second case is when $\varphi = q \neq p$. In this case, we conclude that $V, s \models \textcircled{p}q$ iff $V, s \models q \rightarrow \bigcirc(p \wedge q \wedge q)$ and $V, s \models \neg q \rightarrow \bigcirc(p \wedge \neg q \wedge q)$. Since we know $V, s \models \bigcirc(p \wedge q)$ and $V, s \not\models \bigcirc\bot$, we conclude that $V, s \models \textcircled{p}q$ iff $V, s \models q$. By the definition of truth, $V, s \models q$ iff $q \in V(s)$ iff $q \in V(s) \cup \{p\}$ (since $q \neq p$). This holds iff $V^s_A, s \models q$ iff $V, s \models [+p]q$.

Next consider the case where $\varphi = \neg\psi$ for some ψ. Using Fact 9, $V, s \models \textcircled{p}\neg\psi$ iff $V, s \models \neg\textcircled{p}\psi$. The required result follows quickly using the inductive hypothesis. Similar arguments to those given in Fact 9 can be used to establish the case for $\varphi = \psi \wedge \chi$ for some ψ and χ.

Now we take $\varphi = \bigcirc\chi$ for some χ. Using a similar argument to Fact 9, we know that $V, s \models \textcircled{p}\bigcirc\chi$ iff $V, s \models \bigcirc(p \wedge \psi_{V,s} \wedge \bigcirc\chi)$, where $\psi_{V,s} \in \mathsf{Bool}(\mathsf{At}(\chi) \setminus \{p\})$ characterises V at s. By definition, this holds iff $V^s_B, s \models p \wedge \psi_{V,s} \wedge \bigcirc\chi$ for some B. This immediately establishes that $V^s_A \models \bigcirc\chi$ since we can move to V^s_B. Hence, $V, s \models [+p]\bigcirc\chi$. For the other direction, notice that $V^s_A, s \models p$ and $V^s_A, s \models \psi_{V,s}$ since $V, s \models \psi_{V,s}$ and $p \notin \mathsf{At}(\psi_{V,s})$.

Finally,[4] suppose $\varphi = \Diamond\chi$. Unpacking definitions, we know that $V, s \models \textcircled{p}\Diamond\chi$ iff $V^s_B, s \models p \wedge \psi_{V,s} \wedge \Diamond\chi$ for some B, where $\psi_{V,s} \in \mathsf{Bool}(\mathsf{At}(\chi) \setminus \{p\})$ characterises V at s. So $V^s_B, s \models p \wedge \psi_{V,s}$ and by the definition of V^s_A and $\psi_{V,s}$ we know $V^s_A, s \models p \wedge \psi_{V,s}$. It follows that V^s_B and V^s_A agree on all propositional variables in $\mathsf{At}(\Diamond\chi)$, and so by Fact 8 we can conclude that $V^s_B, s \models \Diamond\chi$ iff $V^s_A, s \models \Diamond\chi$.

Initially, we found that $V, s \models \textcircled{p}\Diamond\chi$ iff $V^s_B, s \models p \wedge \psi_{V,s} \wedge \Diamond\chi$, and now we have shown that $V^s_B, s \models \Diamond\chi$ iff $V^s_A, s \models \Diamond\chi$. From these, we establish that $\mathfrak{F}, V, s \models \textcircled{p}\Diamond\chi$ iff $\mathfrak{F}, V, s \models [+p]\Diamond\chi$. □

Bisimulation

We can define a natural notion of bisimulation for LFC. Like the Ehrenfeucht–Fraïssé games, this is similar to the approach for ML in Areces et al. (2011).

Definition 20 Let $\mathfrak{F}_1 = (W_1, R_1)$ and $\mathfrak{F}_2 = (W_2, R_2)$.

A relation Z between $W_1 \times \mathsf{Prop}^{W_1}$ and $W_2 \times \mathsf{Prop}^{W_2}$ is a *bisimulation* if the following clauses hold:

Non-empty $Z \neq \emptyset$.
Agree If $(s_1, V_1)Z(s_2, V_2)$ then $V_1(s_1) = V_2(s_2)$.

[4] Since $\textcircled{q}$ is defined, we need not show this case.

Zig If $(s_1, V_1)Z(s_2, V_2)$ and $R_1 s_1 t_1$ then there is $t_2 \in A_2$ with $R_2 s_2 t_2$ and $(t_1, V_1)Z(s_t, V_2)$.

Zag If $(s_1, V_1)Z(s_2, V_2)$ and $R_2 s_2 t_2$ then there is $t_1 \in A_1$ with $R_1 s_1 t_1$ and $(t_1, V_1)Z(s_t, V_2)$.

Change If $(s_1, V_1)Z(s_2, V_2)$ then $(s_1, {V_1}^{s_1}_{A})Z(s_2, {V_2}^{s_2}_{A})$ for every $A \subseteq \mathsf{Prop}$.

This is the standard definition of bisimulation, but with valuations included. This is because we must keep track not only of which state we're in, but what the current valuation is. The last clause simply ensures that a change of valuation at bisimilar points preserves bisimilarity (assuming the valuation is changed uniformly).

Fact 5 (Correctness of bisimulation). *Let $\mathfrak{F}_1 = (W_1, R_1)$ and $\mathfrak{F}_2 = (W_2, R_2)$. For every formula $\varphi \in \mathcal{L}_{\mathsf{LFC}}$, $s_1 \in W_1$ and $s_2 \in W_2$, if there is a bisimulation Z with $(s_1, V_1)Z(s_2, V_2)$ then $\mathfrak{F}_1, V_1, s_1 \models \varphi$ iff $\mathfrak{F}_2, V_2, s_2 \models \varphi$.*

Proof By induction on φ. The Boolean and $\Diamond$ cases are standard.

The final case is where $\varphi = \bigcirc\psi$. $\mathfrak{F}_1, V_1, s_1 \models \bigcirc\psi$ iff $\mathfrak{F}_1, {V_1}^{s_1}_{A}, s_1 \models \psi$ for some $A \subseteq \mathsf{Prop}$. Since $(s_1, V_1)Z(s_2, V_2)$, by **Change** we know $(s_1, {V_1}^{s_1}_{A})Z(s_2, {V_2}^{s_2}_{A})$. Applying the inductive hypothesis, $\mathfrak{F}_1, {V_1}^{s_1}_{A}, s_1 \models \psi$ iff $\mathfrak{F}_2, {V_2}^{s_2}_{A}, s_2 \models \psi$. But this holds exactly when $\mathfrak{F}_2, V_2, s_2 \models \bigcirc\psi$, as required. □

References

Areces, Carlos, Raul Fervari, Guillaume Hoffmann, and Mauricio Martel. 2017. Undecidability of Relation-Changing Modal Logics. In *Dynamic Logic. New Trends and Applications*, vol. 23, 1–16. Lecture Notes in Computer Science. Cham: Springer. ISBN: 978-3-319-73578-8 978-3-319-73579-5.

Areces, Carlos. 2007. Hybrid Logics: The Old and The New. In *Proceedings of LogKCA-07*, 15–29.

Areces, Carlos, Diego Figueira, Santiago Figueira, and Sergio Mera. 2011. The Expressive Power of Memory Logics. *The Review of Symbolic Logic* 4 (02): 290–318.

Aucher, Guillaume, Philippe Balbiani, Luis Fariñas del Cerro, and Andreas Herzig. 2009. Global and Local Graph Modifiers. *Electronic Notes in Theoretical Computer Science* 231(25): 293–307; In *Proceedings of the 5th Workshop on Methods for Modalities (M4M5 2007)*.

Aucher, Guillaume, Johan van Benthem, and Davide Grossi. 2018. Modal Logics of Sabotage Revisited. *Journal of Logic and Computation* 28 (2): 269–303.

Balbiani, Philippe, Andreas Herzig, and Nicolas Troquard. 2013. Dynamic Logic of Propositional Assignments: A Well-Behaved Variant of PDL. In *Proceedings of the 2013 28th Annual ACM/IEEE Symposium on Logic in Computer Science*, 143–152. LICS '13. Washington: IEEE Computer Society. ISBN: 978-0-7695-5020-6.

Benthem, van, and Johan. 1997. Modal Foundations for Predicate Logic. *Logic Journal of the IGPL* 5: 259–286.

Benthem, van, Patrick Girard Johan, and Olivier Roy. 2009. Everything Else Being Equal: A Modal Logic for Ceteris Paribus Preferences. *Journal of Philosophical Logic* 38 (1): 83–125.

Blackburn, Patrick, and Jerry Seligman. 1995. Hybrid Languages. *Journal of Logic, Language and Information* 4 (3): 251–272.

Bockmayr, Alexander, and Heike Siebert. 2013. Bio-Logics: Logical Analysis of Bioregulatory Networks. In *Programming Logics*, vol. 7797, eds. Andrei Voronkov and Christoph Weidenbach, 19-34. Berlin: Springer. ISBN: 978-3-642-37650-4 978-3-642-37651-1.

de Bruin, B.P. 2004. *Explaining Games: On the Logic of Game Theoretic Explanations*. Amsterdam: Institute for Logic, Language and Computation. ISBN: 978-90-5776-129-4.
der Hoek, Van, and Wiebe, and Michael Wooldridge. 2005. On the Logic of Cooperation and Propositional Control. *Artificial Intelligence* 164(1–2): 81–119.
Gutierrez, Julian, Paul Harrenstein, and Michael Wooldridge. 2015. Iterated Boolean Games. *Information and Computation* 242: 53–79.
Harrenstein, Paul, Wiebe van der Hoek, John-Jules Meyer, and Cees Witteveen. 2001. Boolean Games. In *Proceedings of the 8th Conference on Theoretical Aspects of Rationality and Knowledge*, 287–298. Burlington: Morgan Kaufmann Publishers Inc.
Harrenstein, Paul, Wiebe Van der Hoek, John-Jules Meyer, and Cees Witteveen. 2003. A Modal Characterization of Nash Equilibrium. *Fundamenta Informaticae* 57(2-4): 281–321.
Mera, Sergio Fernando. 2009. Modal Memory Logics. PhD Thesis, Universidad de Buenos Aires.
Miller, Joseph S., and Lawrence S. Moss. 2005. The Undecidability of Iterated Modal Relativization. *Studia Logica* 79, no. 3 (1): 373–407.
Seligman, Jeremy, and Declan Thompson. 2015. Boolean Network Games and Iterated Boolean Games. In *Logic, Rationality, and Interaction*, eds. Wiebe van der Hoek, Wesley H. Holliday, and Wen-fang Wang, 353–365. Lecture Notes in Computer Science. Berlin: Springer. ISBN: 978-3-662-48561-3.
Tiomkin, M.L., and J.A. Makowsky. 1985. Propositional Dynamic Logic with Local Assignment. *Theoretical Computer Science* 36 (1): 71–87.
van Benthem, Johan. 2011. *Logical Dynamics of Information and Interaction*. Cambridge: Cambridge University Press. ISBN: 978-0-521-76579-4.
van Benthem, Johan. 2018. Constructive Agents. *Indagationes Mathematicae* 29(1): 23–35.

Solution Complexity of Local Variants of Sabotage Game

Tianwei Zhang

Abstract The study of graph games serves as a way to analyze an existing logic, as well as an inspiration for designing new logics. Given the fact that game-theoretic analysis is reviving in AI study, a new stress in the study of graph games should be the performance of standard algorithmic tasks that are conducted on graphs. In this paper, we carry out a case study on the respective graph game for three main local variants of sabotage modal logic, which have a broad range of applications in various other fields. We analyze the solution complexity for each game and show the implications these results have on their corresponding logic. This work is a first attempt to understand why similar-looking variants of a graph game and their corresponding logics can have drastically different computational complexities, with the goal to bring up a more general topic that requires further studies, namely to identify the parameters of games and logic that crucially affect complexity.

1 Introduction

Sabotage modal logic was proposed in 2003 as a format for analyzing games that modify graphs they are played on. Since its first introduction, it has inspired a line of systems in a dynamic-epistemic spirit (van Benthem 2011). Along with different variants that stems from the minimal language such as 'graph modifier logic' (Aucher et al. 2009), 'swap logic' (Areces et al. 2014), 'arrow update logic' (Areces et al. 2012; Kooi and Renne 2011) and so on, various graphs games of interactive settings are proposed to capture the essences of such logics. Meanwhile examples of how

This chapter is in its final form and it is not submitted to publication anywhere else.

T. Zhang (✉)
Tsinghua University-University of Amsterdam Joint Research Center for Logic, ILLC, University of Amsterdam, Amsterdam 1098 XG, The Netherlands
e-mail: thss15_zhangtw@163.com

F. Liu et al. (eds.), *Knowledge, Proof and Dynamics*,
Logic in Asia: Studia Logica Library,
https://doi.org/10.1007/978-981-15-2221-5_6

real life scenarios can be depicted by graph games can be found in a broad collection of areas including theoretical computer science (Grüner et al. 2013; Radmacher and Thomas 2008), learning theory (Gierasimczuk et al. 2009), logics for social networks (Liu et al. 2014; Seligman et al. 2013), argumentation (Grossi 2010a, b), and many more.

In light of the avid interest that has recently arisen in the game-theoretic analysis of AI study, a new stress in the study of graph games is the performance of standard algorithmic tasks conducted on graphs. This paper is a case study. We focus on localized sabotage games, whose global versions can have either link deletion or node-deletion. We take three of the existing variants of localized sabotage games, namely local link-cutting sabotage (adjacent sabotage game in Rohde 2005), local node-deleting sabotage (van Benthem and Klein 2018), as well as poison game (Blando et al. 2018), and focus on their solution complexity. We show that local link-cutting has a low complexity, namely PTIME, while the corresponding logic is undecidable. Meanwhile, local node-deletion and poison game do not have this effect: model checking and game solution in both cases are PSPACE-complete.

2 Preliminary

In this preliminary section, we provide formal definitions of the games and the corresponding solution problems in a brief way.

Definition 2.1 A graph or an arena $\mathcal{A}$ is a pair (V, η) where V is a non-empty, finite set of vertices, and $\eta : V \times V \to \{true, false\}$ is an edge function. For $u, v \in V$, the value $\eta(u, v)$ indicates whether there exists an edge from u to v in the arena. It is worthwhile to mention that we can use a similar formulation for depiction of multi-graphs, simply by using a function $\eta : V \times V \to \mathbb{N}$ instead of $\eta : V \times V \to \{true, false\}$.

Definition 2.2 A *sabotage game* is given by $\mathcal{G} = (\mathcal{A}, v_{in})$, where $\mathcal{A} = (V, \eta_{in})$ is an arena as defined above, and $v_{in} \in V$ is an initial vertex. The game is played by the two players Traveler and Demon. A *position* of the game is a triple (τ, v, η), where $\tau \in \{0, 1\}$ signifies whose turn it is, where we adopt the convention that 0 stands for Traveler and 1 for Demon; $v \in V$ is the current vertex and $\eta : V \times V \to \{true, false\}$ gives the connectivity function of the currently available edges and vertices. If it is clear from the context whose turn it is and which edges are available, we use vertex v to refer to the current position of the game.

In both local link-cutting and node-deleting variants, the game proceeds with the two players alternating their turns. In the local link-cutting version, Traveler chooses a local vertex to move to at each of her steps. Then Demon acts in his turn by removing one local edge from the current position. In the local node-deleting sabotage game, Traveler chooses one of the local vertexes to move to as well. The only difference is that Demon acts by removing one local *vertex* from the current position. Both games have *reachability* as their game goal. Given a set $F \subseteq V$ of goal points, Traveler wins if and only if she can reach some vertex in F. In other words, given a finite sequence of positions $\pi_0 \ldots \pi_n$ showing the progression of a game, Traveler wins at the n-th position if and only if $\pi_n = (\tau, v, \eta)$ where $v \in F$.

In contrast, the poison game has a slightly different setting (Blando et al. 2018). During the game, the two players, Traveler and Demon (sometimes called Poisoner), alternate their turns and, at each step, choose a vertex that is a successor of the vertex previously chosen by the other player. Traveler makes the first move by choosing a vertex $s \in G$. Demon responds by choosing a successor s' of s, which he poisons with a poison that affects only the Traveler. This means that the Demon's move renders vertex s' inaccessible to Traveler but not to Demon himself. Then, Traveler has to choose a non-poisoned successor of s', and so on. The winning conditions are as follows: Traveler wins the poison game if either (i) she manages to keep choosing non-poisoned successors no matter what vertices Demon selects, or (ii) she begins her turn at a vertex with no successors, or (iii) Demon begins his turn at a vertex with no successors. If at least one of (i)–(iii) is satisfied, we say that Traveler *survives* the game. Demon, on the other hand, wins the poison game if he manages to poison a vertex that (a) has at least one successor and (b) all of whose successors have already been poisoned. The poison game is close to the sabotage modal logic in spirit, since we can think of $\langle t \rangle$ as the modality associated with a second graph relation, one that corresponds to Traveler's 'safe accessibility' relation and shrinks over time, as more and more states get poisoned. Under this interpretation, the poison modality $\langle p \rangle$ behaves analogously to the sabotage modality, in that poisoning moves result in the deletion of links from the basic graph relation.

In this paper, we will focus on the solution complexity of each of the games mentioned above. Traditionally, in game theory, the solution of a game is the set of all the equilibria. In two-player zero-sum finite games like the ones we are going to discuss, the search of such equilibria can be simplified, since (i) in such games an action profile is an equilibrium if and only if it is the combination of one player's winning strategy and any strategy of the other player and (ii) at any position of the game, one and only one player has a winning strategy, finding all the equilibria is reduced to finding all the winning strategies of each player. Although one player can have more than one winning strategy from the same initial position, we will again simplify our task by only caring about the set of vertices starting from which Traveler has a winning strategy (then the set of vertices from which Demon has a winning strategy will simply be the complement).

Definition 2.3 The solution problem of a sabotage game $\mathcal{G}$ with the arena $\mathcal{A}$ is to find all the vertices in $\mathcal{A}$ starting from which Traveler has a winning strategy against Demon.

3 Local Link-Cutting Variant

In this section, we show that the solution problem of the local link-cutting sabotage game is in the complexity class of PTIME. For this aim, we introduce an algorithm that solves the problem in PTIME and prove the correctness of the algorithm.

3.1 A Solution Algorithm

Here we present a polynomial-time algorithm for solving the local link-cutting sabotage in pseudo-code. V is the set of all the vertices in the given graph. F is the set of all the goal points of the graph (a fortiori a subset of V). η is a function from $V \times V$ to $\{true, false\}$ used to indicate whether two nodes in the graph are connected. Here we assume that there are no multiple links between nodes, but with some modifications, this algorithm can also be used to calculate all the winning positions in a multi-linked graph with a new η from $V \times V$ to the set of natural numbers.

Algorithm 1 Algorithm to solve local link-cutting sabotage

```
procedure LocalLinkCutting(V, F, η)
    input: arena(V, η); F ⊆ V final vertices
    output: L ⊆ V winning positions
    L = F
    count = |F|
    while count > 0 do
        T = the set of v in V − L with two or more neighbors in L
        count = |T|
        L = L ∪ T
    end while
end procedure
```

In practice, if provided with $O(|V|^2)$ space to store the pointer list of neighbors for each node and $O(|V|)$ space for a queue for scheduling, the algorithm can solve the problem in $O(|V|^2)$ time.

Intuitively, the algorithm can be understood as the propagation of a signal. At the very beginning, the only sources of the signal are the goal points. They emit the signal to each of their immediate neighbors. The other nodes in the graph turn themselves into a source and, in turn, emit the signal to their neighbors as soon as they have received two signals themselves. This way the algorithm goes iteratively until it reaches an equilibrium.

Also note that in the multi-linked version of the algorithm, signals transmit through links, so multiple links should be counted multiple times, that is, if a non-source node has only one neighbor emitting signals but has multiple links to this neighbor, then

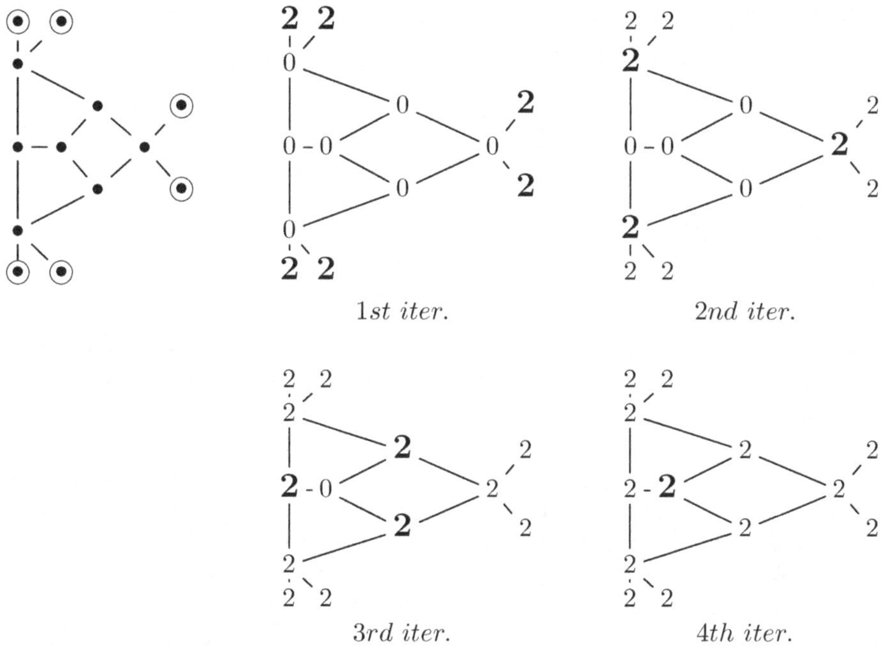

Fig. 1 Example for the algorithm

this node will also become a source and start emitting signals itself. The reason for this is easy to understand since the Demon can only cut one link at each turn.

An example of how the signal propagates in each iteration is shown in Fig. 1, with the circled nodes being goal points. In each iteration, the number on each node denotes how many links it has to sources in total. The goal points begin with 2 for the consistency of notation. In the beginning, the only sources are the goal points. In the second iteration, a, b, and c also become sources since they each have two links that connect to a source. In the third iteration, d, e, and f are also converted into sources because of their connection to a, b, and c. Eventually, in the fourth iteration, g is converted into a source because of its connection to d, e, and f. At this point, there are no more nodes to be converted, and thus the situation reaches an equilibrium and the algorithm ends. As a result, all the nodes in the example are winning positions for Traveler since they all are sources in the end.

3.2 Correctness Proof of the Algorithm

To prove the correctness of the algorithm, we first prove the following useful lemma.

Definition 3.1 Define the degree $\xi(v)$ of a node v in L as follows: $\forall v \in L\ \xi(v) = n$ if and only if v is added to L in the n-th iteration. We stipulate that for all $v \in F$, $\xi(v) = 0$.

Lemma 3.2 *For all $v \in L$, if there is a winning strategy in $((V, \eta), F, v)$ for Traveler, then there is a winning strategy for Traveler where $v_1, v_2, \ldots v_n$ are the nodes visited by Traveler in a sequential order, and $\xi(v_1) > \xi(v_2) > \cdots > \xi(v_n)$. For convenience, we will call such a strategy a 'decreasing strategy' in the following paragraphs.*

Proof We prove the lemma by induction on $\xi(v)$.

If $\xi(v) = 0$, Traveler wins at the very beginning, so the lemma is trivially true.

Suppose $\forall v \in L$ with $0 \leq \xi(v) \leq n$ the statement is true. Suppose Demon cuts l and changes the game to $((V, \eta'), F, v)$. By the algorithm, v has two or more neighbors with a degree less than $n + 1$, so Traveler can then opt to land on a node u of degree less than $n + 1$. By the assumption, Traveler has a decreasing strategy in $((V, \eta), F, u)$. It is easy to see that the strategy only involves links between nodes in $\{v \in L \mid \xi(v) \leq n\}$, and, therefore, does not involve l. This shows that this strategy is also a decreasing strategy for $((V, \eta'), F, u)$.

Now we can create a decreasing strategy for $((V, \eta), F, v)$. In the first round, after Demon cuts a link and changes the game to $((V, \eta'), F, v)$, move to any node u of degree less than $n + 1$. In all the following rounds, move according to the decreasing strategy for $((V, \eta'), F, u)$.

Thus, for all $v \in L$ Traveler has a decreasing strategy in $((V, \eta), F, v)$. □

Theorem 3.3 *A node $v \in V$ in a local link-cutting sabotage game $((V, \eta), F)$ is a winning position for Traveler if and only if $v \in L = LocalLinkCutting(V, F, \eta)$.*

Proof $\rightarrow$: Suppose Traveler is currently at $v \notin L$. Then by the description of the algorithm, at most one of v's neighbors is in L. In light of that, Demon can make sure that Traveler next lands on a node $v' \notin L$ by cutting the only (if any) link to L. This way, if the Traveler starts from a node in $V - L$, then there's a strategy for Demon that guarantees that Traveler always stays in $V - L$. Eventually, Traveler finds herself at a node with no successor and the Demon wins.

$\leftarrow$: We can prove this direction of the proposition by applying Lemma 3.2 directly since a decreasing strategy is by nature a winning strategy, the $\leftarrow$ direction is trivially proven. □

4 Local Node-Deleting Variant

In this section, we establish the PSPACE-Completeness of the solution problem of the local node-deleting variant of sabotage games. For this aim, we provide an alternating algorithm for solving the decision problem of the variant that runs in polynomial time, along with a polynomial-time reduction from the problem of Quantified Boolean Formula, which is known to be PSPACE-complete.

4.1 PSPACE Upper Bound by an Alternating Algorithm

In this and the next section, we use the existence of an alternating algorithm to show the upper bounds of the problems. *Alternating Turing machines*, ATM for short, are a generalization of non-deterministic Turing machines. Their state sets are divided into two components: existential states and universal states. An existential state is acceptable if and only if some transition leads to an accepting state; while a universal state is acceptable if and only if every transition leads to an accepting state. For more information on alternating turing machines and alternating algorithms, please see Balcázar et al. (2012). We use APTIME to denote the class of problems that can be solved by Alternating Turing machines in polynomial time. According to Balcázar et al. (2012), we have the following result:

Theorem 4.1 *APTIME = PSPACE.*

In this and next section, we use pseudocode to specify alternating algorithms. In codes, we use instruction 'guess' to signify non-deterministic choices of successor corresponding to the existential states of an ATM, and instruction 'choose all' to signify the branching of universal states of an ATM. Apart from inheriting notations V, F, η, we also introduce new ones: v for denoting the starting point of Traveler and L for an initially empty set for keeping track of the visited nodes at the current state.

Algorithm 2 Alternating algorithm to solve local node-deleting sabotage

procedure LOCALNODEDELETING(V, F, η, v, L)
  input: arena (V, η); $F \subseteq V$ final vertices; $v \in V$; $L \subseteq V$
  if $v \in F$ **then**
    accept
  else if $v \in L$ or $\forall v' \in V, \eta(v, v') = false$ **then**
    reject
  else
    guess $v' \in V$ with $\eta(v, v') = true$
    universally choose $u \in V$ with $\eta(v', u) = true$
    let $V' := V - \{u\}$
    let $F' = F - \{u\}$
    let $L' := L \cup \{v\}$
    let $\eta' = \eta_{|V' \times V'}$
    call LocalNodeDeleting(V', F', η', v', L')
  end if
end procedure

We start the algorithm with the initial call

$$\text{LocalNodeDeleting}(V_{in}, F_{in}, \eta_{in}, v_{in}, \emptyset).$$

Lemma 4.2 *Let $\mathcal{G} = ((V_{in}, \eta_{in}), F_{in}, v_{in})$ be a local node-deleting sabotage game. The alternating algorithm LocalNodeDeleting accepts its initial input if and only if*

Traveler has a winning strategy in $\mathcal{G}$. The running time is polynomial with respect to $|V|$.

Proof First, we prove the correctness of the algorithm.

Let $c_0 c_1 \ldots$ be some computation branch of LocalNodeDeleting and assume that $(V_i, F_i, \eta_i, v_i, L_i)$ is the parameter of call c_i. Let

$$\pi := (0, V_0, F_0, v_0, \eta_0)(1, V_1, F_1, v_1, \eta_1)(0, V_2, F_2, v_2, \eta_2)(1, V_3, F_3, v_3, \eta_3)\ldots$$

By the update of parameters in Line 8–13, π forms a legal play of the sabotage game. In fact, by the universal choice in Line 9, there is a one-to-one correlation between computation trees and game trees of Traveler. Further, by Line 12, we have $L_i := \{v_j | j < i\}$, that is, L_i gives the set of already visited nodes during the prefix of the play π from v_0 to v_i.

If Traveler has a winning strategy from v_{in} in the game, then she can ensure that she reaches some final vertex starting from $v_i n$. By a lemma yet to be proven, Lemma 4.11, she can do so without visiting any vertex twice. The algorithm can guess the successor v_{i+1} of v_i in Line 8 according to the winning strategy. Since it is winning, there is some natural number n such that $v_n \in F$. Further, no node v_i for $i < n$ is a dead end at the moment when it is Traveler's turn to continue the play from v_i, i.e. at position $(0, v_i, \eta_i)$. Thus, the algorithm does not reject its input in Line 6 during some call c_i with $i < n$ before it finally terminates in an accepting state in Line 4 during call c_n. Therefore, every computation branch terminates in an accepting state and the algorithm accepts its input.

Conversely, if Demon has a winning strategy, then he can delete nodes in such a way that every play leads either to some already visited node or to some dead end. Thus, a winning strategy of Demon ensures that for each guess of successors in Line 8, the corresponding computation tree contains at least one non-accepting branch. Hence, the algorithm rejects the input.

Because parameter L in each embedded call is increasing, it restricts the depth of recursion up to $|V| + 1$. It is also clear that in each layer of recursion, only polynomial space is required. Finally, checking the conditions and updating the parameters take a time that is polynomial with respect to $|V|$. By the correspondence of alternating polynomial time and deterministic polynomial space, the last part of the proposition is proven. □

Lemma 4.2 and Theorem 4.1 automatically give us the following theorem

Theorem 4.3 *Solving the local node-deleting sabotage game is at most PSPACE.*

4.2 Reduction from QBF

Definition 4.4 (*QBF*) Let ϕ be a quantifier-free Boolean formula in conjunctive normal form over a set of Boolean variables $x_1 \ldots x_n$ for some $n \in \mathbb{N}$. An instance of the problem Quantified Boolean Formula, or QBF for short, is deciding whether the following formula is satisfiable:

$$\exists x_1 \forall x_2 \exists x_3 \ldots Qx_n : \varphi$$

here Q is equal to $\exists$ for n odd and to $\forall$ otherwise. Note that our definition of QBF requires the formula to begin with an existential quantification, but this is no loss of generality.

In 1973, Stockmeyer and Meyer (1973) determined the complexity of this problem:

Theorem 4.5 *The problem QBF is PSPACE-complete.*

In the remainder of this section, we present a polynomial time reduction from QBF to the solution problem of a local node-deleting sabotage game. This method of reduction is inspired by the proof of PSPACE-Completeness of the global link-cutting variance of the sabotage game by Rohde (2005).

Let φ be an instance of QBF in conjunctive normal form with Boolean variables $x_1 \ldots x_n$. We construct an arena $\mathcal{A}_{\simeq}$ for a local node-deleting sabotage game $\mathcal{G}_{\simeq}$ such that

Proposition 4.6 *Let φ be as in Definition 4.4. Traveler has a winning strategy in $\mathcal{G}_{\simeq}$ if and only if there exists an assignment for x_1 such that for all assignments for $x_2 \ldots \varphi$ is satisfied.*

4.2.1 Construction of the Arena

The arena consists of $n + 2$ components, one for each variable plus two verification components. Generally, the traversal of the arena starts with an existential component, which, by design, is modified to be slightly different from the standard existential component. Following it is a universal component, and then again an existential component, and so on. At the bottom of the arena, we find two verification components. In a verification component, C_is are called clause vertices. Connected to them below are literal vertices L_{ijk}s, each representing one literal in the clause. The curve line stemming from each of the literal vertices represents a link that connects the literal vertex to a duo (definition see Definition 4.7). A curve line leads to duo X_i if x_i is positive in l, and to $\neg X_i$ if x_i is negative in l. The complete lines are omitted for the sake of neatness.

Definition 4.7 (*a duo*) Let $\mathcal{A} = (V, \eta)$ be the arena of $\mathcal{G}$. A duo $D \subseteq V$ consists of two vertices x and y such that $\forall z \in V, \eta(x, z) = \eta(y, z)$ and $\eta(x, y) = false$. Duos are represented by X_i and $\neg X_i$ in figures.

Intuitively, a duo is two vertices that are not connected and have the exact same connections to the other nodes. The introduction of duos serves the purpose of not

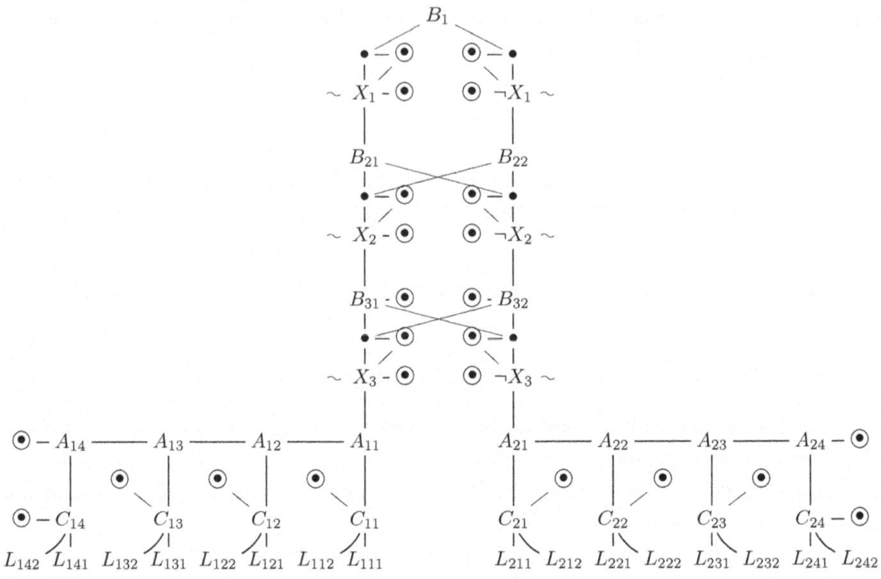

Fig. 2 Local node-deleting: arena for Example 4.8

only simplifying the arena in the display but also ensuring Proposition 4.6. Note that an arena with a duo is bisimilar to an arena with single vertices in place of duos. To be specific, the two vertices of a duo is bisimilar to its corresponding single vertex, and the rest of the vertices are bisimilar to themselves. During the game, this bisimulation will be valid as long as we do not delete vertices belonging to a duo, which is the case most of the time as we will see later in this subsection. In light of that, for the rest of the section, without further explanation, we will address the duos as if they were single vertices. In 3.2.4, we will encounter situations where we have to consider the deletion of one of the vertices belonging to a duo, and we will explain why we need to introduce duos in the first place.

Example 4.8 Let $\varphi = c_1 \wedge c_2 \wedge c_3 \wedge c_4$ be a Boolean formula in conjunctive normal form with Boolean variables x_1, x_2, x_3. We assume that each clause consists of exactly two literals, i.e., for $k \in [1, 4]$, we have $c_k = l_{k1} \vee l_{k2}$ with $i_{kj} \in \{x_1, \neg x_1, x_2, \neg x_2, x_3, \neg x_3\}$. By definition, the satisfiable problem of the following formula is an instance of QBF:

$$\exists x_1 \forall x_2 \exists x_3 : \varphi$$

The arena for Example 4.8 is depicted in Fig. 2.

4.2.2 Properties of the Game

Before we delve into the details of each component, we first prove some useful properties of the game. These properties are also discussed in van Benthem and Liu (2018). Along with several others, the discovery of these properties might help to understand the disparity between the computational complexites in similarly-appearing graph games, which we will discuss in greater details in Sect. 5.

Definition 4.9 Given two local node-deleting sabotage games $\mathcal{G}_1 = (\mathcal{A}_1, F_1, x_1)$ and $\mathcal{G}_2 = (\mathcal{A}_2, F_2, x_2)$ where $\mathcal{A}_1 = (V_1, \eta_1)$, $\mathcal{A}_2 = (V_2, \eta_2)$, we say $\mathcal{G}_1 \leq \mathcal{G}_2$ if and only if there exists an injective map φ from V_1 to V_2 such that
(i) $\varphi(x_1) = x_2$
(ii) $\forall v \in V_1, \varphi(v) \in F_2$ if and only if $v \in F_1$
(iii) $\forall y_1, y_2 \in V_1, \eta_1(y_1, y_2) \leq \eta_2(\varphi(y_1), \varphi(y_2))$

Lemma 4.10 (Monotonicity) *Given two local node-deleting sabotage game $\mathcal{G}_1$ and $\mathcal{G}_2$ as in Definition 4.9, if Traveler has a winning strategy in $\mathcal{G}_1$, and $\mathcal{G}_1 \leq \mathcal{G}_2$, then Traveler has a winning strategy in $\mathcal{G}_2$.*

Proof The key point of the proof is to use the map φ in Definition 4.9 to transform the winning strategy in $\mathcal{G}_1$ to a winning strategy in $\mathcal{G}_2$. We first apply φ to the strategy for $\mathcal{G}_1$. This gives us a fragment of the strategy for $\mathcal{G}_2$ in which we only consider the situations where Demon deletes nodes in $\varphi(V_1)$. Then we augment the fragment in the following way: when Traveler is at $u \in \varphi(V_1) - F_2$, if Demon deletes a node $v \notin \varphi(V_1)$, Traveler then (randomly) picks a node $v' \in \varphi(V_1)$ such that $\eta_2(v', u) = true$ (the existence of such v' is guaranteed by the fact that Traveler has a winning strategy in $\mathcal{G}_1$) and follows the strategy as if Demon had deleted v'. Note that in this way Traveler will always stay in $\varphi(V_1)$. It's easy to see that the augmented fragment is a winning strategy for $\mathcal{G}_2$ since it provides Traveler with the reaction for all the possible situations and guarantees that Traveler always wins. □

Lemma 4.11 (Acyclicity) *If Traveler has a winning strategy in $\mathcal{G}_{\simeq}$, she can win without visiting any node twice.*

Proof Suppose Traveler has a winning strategy that involves visiting a node v twice. Suppose when she first visits v, the game is $\mathcal{G}_1 = (\mathcal{A}_1, F_1, v)$, and when she visits v the second time, the game is $\mathcal{G}_2 = (\mathcal{A}_2, F_2, v)$. It is obvious that $\mathcal{G}_2 \leq \mathcal{G}_1$ with φ being the inclusion map. Since Traveler has a winning strategy in $\mathcal{G}_2$, we can use φ to transform this strategy into a winning strategy in $\mathcal{G}_1$. In the transformed strategy, the number of circles is reduced by (at least) 1 by avoiding the journey in the original strategy from $\mathcal{G}_1$ to $\mathcal{G}_2$. Since any winning strategy in a local node-deleting sabotage game is finite in terms of steps, it consequently has a finite number of circles. By removing circles for a finite number of times, we can obtain a winning strategy that does not involve visiting a vertex twice. □

Given Lemma 4.11, we will assume that Traveler never visits any vertex twice in the whole game and will not repeat this assumption unless necessary. Note that by our discussion before, the two vertices of a duo should be considered as one single vertex, and thus visiting one vertex of a duo and then later visiting the other should count as visiting a vertex twice as well.

4.2.3 Components and Their Functions

In the rest of this section, we describe the details of each component (existential, universal and verification component), show a few consequences of the structure and how these consequences lead to Proposition 4.6.

The structures of each component (existential, universal and verification) are shown in Fig. 3 on the left, on the right, and in Fig. 4, respectively.

Lemma 4.12 *When traversing an existential component, Traveler has the choice not to visit either* X_i *or* $\neg X_i$*.*

Proof We prove the statement by analyzing how the two players interact in the competitive setting. Depending on the choice of Demon in the previous component, Traveler enters an existential component from B_{i1} or B_{i2}. We will see that this does not affect her strategy. Without loss of generality, suppose Traveler starts from B_{i1} and wants to avoid visiting X_i. Given the way components are assembled, we assume that Demon makes the first move. At B_{i1}, Demon has to delete the immediate goal point or the Traveler can move straight to it and win. Then the Traveler can move to d_2. At d_2, Demon again has to remove the immediate goal point for the same reason. Then Traveler moves to $\neg X_i$. At $\neg X_i$, similarly, Demon has to delete the immediate goal point. Now Traveler can go down and exit the component, leaving X_i not visited. □

The previous lemma shows that an existential component is a place where the Traveler gets to choose. Later, as we will see, when Traveler comes back to verify the QBF, the unvisited duo will provide Traveler with access to goal points. This way Traveler chooses the valuation of the variable of an existential quantifier by choosing which duo to not visit.

Lemma 4.13 *When traversing a universal component, Demon can force Traveler to visit either* X_i *or* $\neg X_i$ *and prevent her from reaching any goal point inside the component.*

Proof Similarly, we prove the statement by analyzing the process of traversal. Depending on the choice of herself in the previous component, Traveler enters a universal component from B_{i1} or B_{i2}. We will see that this does not affect the Demon's strategy. Without loss of generality, suppose Traveler starts from B_{i1} and Demon wants to make sure that she visits X_i. Given the way components are assembled, we assume that Demon makes the first move. At B_{i1}, Demon deletes d_2. Then Traveler can only move to d_1. At d_1, Demon has to remove the immediate goal point. Then Traveler moves to X_i. At X_i, similarly, Demon has to remove the immediate goal point. Now Traveler can only exit the component. □

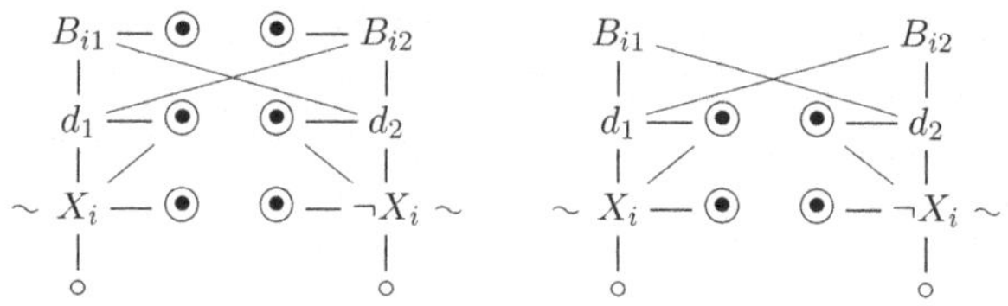

Fig. 3 Local node-deleting: existential and universal component

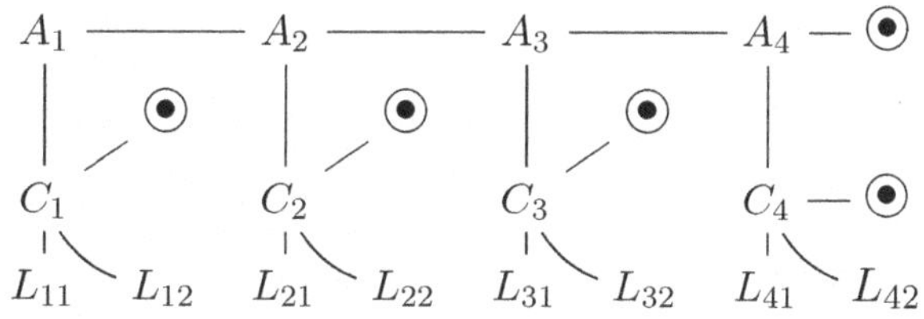

Fig. 4 Local node-deleting: verification component

Here the idea is similar. Demon gets to choose the valuation of the variable of the universal quantifier and he does that by forcing Traveler to visit the opposite duo. This way the duo Traveler does not visit remains access to goal points and in turn corresponds to the literal being true.

Lemma 4.14 *When traversing a verification component, Demon can force Traveler to visit one of* C_1, C_2,...,C_m *with* m *being the number of clauses in the QBF, and prevent her from reaching any goal point inside the component.*

Proof Again, we prove the statement by showing the reasoning of both players. Given the way the components are put together, we assume that Demon acts first. Suppose Demon wants Traveler to visit C_i. Traveler starts from A_1. For all $1 \leq j < i$, when Traveler arrives at A_j, Demon deletes C_j. Then Traveler has to move to A_{j+1}. If $i < m$, when Traveler arrives at A_i, Demon deletes A_{i+1}, else Demon deletes the immediate goal point. When Traveler arrives at C_i, Demon deletes the immediate goal point. □

Lemma 4.15 *At* C_i*, Traveler can choose any one of the literal nodes directly linked to* C_i *to move to.*

Proof This is obvious since when Traveler arrives at C_i, Demon has to delete the immediate goal point and this leaves Traveler to choose freely which literal nodes to move to. □

The previous two corollaries show that we are able to mimic a verification game in a verification component. Demon, the falsifier, chooses the clause in which to falsify the formula by choosing the C_i for Traveler to visit. Traveler, the verifier, in turn, chooses the literal that is true (if any) in the clause that Demon has just chosen.

Next we prove Proposition 4.6.

4.2.4 Correctness of the Reduction

With Lemma 4.12–4.15, we can prove an equivalent version of Proposition 4.6:

Proposition 4.16 *Let φ be as in Definition 4.4.*

(1) If there exists an assignment for x_1 such that for all assignments for x_2 ..., φ is satisfied, then Traveler has a winning strategy in $\mathcal{G}_{\simeq}$.

(2) If for all assignments for x_1 there exists an assignment for x_2 such that ..., $\neg\varphi$ is satisfied, then Demon has a winning strategy in $\mathcal{G}_{\simeq}$.

Proof (1) If there exists an assignment for x_1 such that for all assignments for $x_2 \ldots \varphi$ is satisfied, then a winning strategy for Traveler can be constructed as follows: Traverse the existential and universal components by the order that they are put together (if at any X_i or $\neg X_i$ Demon delete the vertex directly below, move to the immediate goal point and win), and when at each existential component, avoid X_i if x_i is true in the assignment and avoid $\neg X_i$ if false. When at C_i, move to the literal node that corresponds to the literal that makes the clause true according to the assignment. It is clear that the duo connected to the literal node has not been visited.

Demon now may delete one node of the duo, but Traveler can move to the remaining vertex of the duo. Note that this is the only chance for Demon to delete a vertex that belongs to a duo without allowing Traveler to move to a goal point in the next turn. This is also why we need such structures as 'duos' in the first place: if there were single vertices in place of the duos in the arena, then at this point of the game Demon could simply delete the unvisited X_i or $\neg X_i$ to prevent Traveler from winning.

When Traveler moves to the unvisited X_i or $\neg X_i$ (now may consist of only one node instead of two), there are two immediate goal points. Demon may delete one of them, but Traveler can still move to the remaining one and win.

(2) We first show the following fact: If Traveler moves through a curve link from X_i or $\neg X_i$ to a literal node, since Demon can then delete the immediate clause node, Traveler will have no choice but go back through the curve link. Therefore, suppose Traveler does not visit any vertex twice, she will always exit a universal or an existential component by going down through the straight link, that is, Traveler will traverse the existential and universal components by the order that they are put together.

If for all assignments for x_1 there exists an assignment for x_2 such that $\ldots \neg\varphi$ is satisfied, then a winning strategy for Demon can be as follows: When Traveler traverses the existential and universal components, at each universal component Demon forces Traveler to visit the duo corresponding to the assignment that satisfies $\neg\varphi$. Then when Traveler arrives at a verification component, there will be at least one clause node that is only connected to literal nodes whose immediate duo has already been visited. Demon then forces Traveler to that clause node, and then Traveler has to visit a duo for the second time, which indicates that she loses. □

Theorem 4.17 *The solution problem of local node-deleting sabotage games is PSPACE-complete.*

Proof Since the reduction only involves generating verification components that have $O(l)$ clause nodes and literal nodes and assembling all the components, the time complexity of the reduction is $O(l)$, where l is the length of the QBF. Thus, we have a polynomial-time reduction from the problem of true QBF to the solution problem of local node-deleting sabotage games. This establishes the PSAPCE-hardness of the problem. With Lemma 4.2, we conclude that the problem is PSPACE-complete. □

5 Poison Game

In this section, we establish the PSPACE-completeness of the solution problem of the poison game. To this end, we provide an alternating algorithm for solving the problem that runs in polynomial time, along with a polynomial-time reduction from the problem of Quantified Boolean Formula.

5.1 *PSPACE Upper Bound by an Alternating Algorithm*

Here we present an alternating algorithm for solving poison games. Besides the familiar notations, we also utilize an integral variable *count* to keep record of the number of steps for which Demon hasn't poisoned an unpoisoned vertex. As we will see later, the usage of this extra variable ensures that the algorithm terminates.

Algorithm 3 Alternating algorithm to solve poison game

```
procedure SolvePoison(V, η, v, L, count)
    input: arena (V, η); v ∈ V; L ⊆ V
    if count = |L| then
        accept
    else
        universally choose u ∈ V with η(u, v) = true
        guess v' ∈ V with η(v, v') = true
        if v' ∈ L then
            reject
        end if
        let L' := L ∪ {u}
        if u ∈ L then
            call SolvePoison(V, η, v', L', count + 1)
        else
            call SolvePoison(V, η, v', L', 0)
        end if
    end if
end procedure
```

We start the algorithm with the initial call

$$\text{SolvePoison}(V, \eta, v_{in}, \emptyset, 0).$$

Lemma 5.1 *Let $\mathcal{G} = ((V, \eta_{in}), v_{in})$ be a poison game. The alternating algorithm SolvePoison accepts its initial input if and only if Traveler has a winning strategy in $\mathcal{G}$. The running time is polynomial with respect to $|V|$. Consequently the problem of solving poison games belongs to PSPACE.*

Proof First, we prove the correctness of the algorithm.

Note that Traveler wins if she can play infinitely, and since the arena itself is finite, some sort of repetition is inevitable. By the pigeonhole principle, when Demon hasn't poisoned any unpoisoned vertices in the most recent $|L| + 1$ steps, he must have visited/poisoned a vertex for at least twice. Since no vertices were added between the repeated visits, we can prune the branches that have such repetition.

Let $c_0 c_1 \dots$ be some computation branch of SolvePoison and assume that $(V, \eta, v_i, L_i, count_i)$ is the parameter of call c_i. Let

$$\pi := (0, v_0, L_0, count_0)(1, v_1, L_1, count_1)(0, v_2, L_2, count_2)(1, v_3, L_3, count_3)$$

By the update of parameters in Line 6–16, π forms a legal play of the poison game. In fact, by the universal choice in Line 6, there is one-to-one correlation between the computation tree and the game tree of Traveler (pruned in the way mentioned above). Further, by Line 11, L_i gives the set of vertices already poisoned by Demon during the prefix of the play π from v_0 to v_i.

If Traveler has a winning strategy from v_i in the game, then she can survive for an infinite number of steps. The algorithm can guess the successor v_{i+1} of v_i in Line 6–7 according to the winning strategy. Since it is winning, for any natural number i, $v_{i+1} \notin L_i$, and since after pruning the algorithm always terminates after finite steps, it follows that every computation branch terminates in an accepting state and the algorithm accepts its input.

Conversely, if the Demon has a winning strategy then he can poison vertices in such a way that there is some natural number n such that $v_{n+1} \in L_n$. Thus, a winning strategy of Demon ensures that for each guess of successors in Line 7, the corresponding computation tree contains at least one non-accepting branch. Hence, the algorithm rejects the input.

Complexity-wise speaking, since $|L| \leq |V|$ and every time Demon poisons a new vertex $count$ is reset, the depth of recursion is at most $|V|^2$, which is polynomial with respect to $|V|$. As is in the proof for Lemma 4.2, in each layer of recursion only polynomial space is required, and checking the conditions and updating the parameters take a time that is polynomial with respect to $|V|$. By the correspondence of alternating polynomial time and deterministic polynomial space, the last part of the lemma is fulfilled. □

5.2 Reduction from QBF

In this part, our aim is to construct a poison game $\mathcal{G}_{\simeq}$ for any quantified boolean formula ϕ such that Traveler has a winning strategy in $\mathcal{G}_{\simeq}$ if and only if there is an assignment for x_1 such that for all assignments for x_2 there exist ... and φ is satisfied by the overall assignment, in which case φ belongs to QBF.

5.2.1 Construction of the Arena

The arena also consists of $n + 1$ components of three kinds (existential, universal, and verification component). Starting from the existential component, the arena alternates between existential and universal components, until the verification component is reached. For the sake of convenience, we continue to use Example 4.8 for illustration in this section.

The arena for Example 4.8 is depicted in Fig. 5. Each of the wavy arrows in the verification component represents multiple outward arrows, the number of which equals the number of literals in each clause. Each of these arrows points to the variable vertex X_i if x_i is positive in l, and in $\neg X_i$ if x_i is negative in l. The complete arrows are omitted for the sake of neatness.

5.2.2 Components and Their Function

In this subsection, we go into details and describe the structure and function of each component.

Proposition 5.2 *When Demon poisons a vertex with no successors, Traveler wins.*

Proof It follows directly from the winning conditions for Traveler. □

Existential Component. Shown in Fig. 6. To show how this component fits in the whole arena, the hollow bullet at the bottom represents the starting vertex in the next component. The same representation is also used in the universal component. Traveler enters an existential component from s_i. Given Proposition 5.2, Demon avoids esc and poisons a. Now, Traveler can choose the assignment of x_i simply by choosing which node between b_1 and b_2 to travel to. She moves towards b_1 if she wants x_i to be false and towards b_2 if she wants x_i to be true. Whichever branch she chooses, the variable vertex on that branch will be poisoned by Demon in the next step, which is his only option. Traveler then moves onto the hollow bullet and exits the component.

Universal Component. Shown in Fig. 7. Similarly, Traveler enters such a component from s_i, and for the same reason, Demon does not poison esc but instead chooses the assignment for x_i by poisoning either of the two variable vertices. He poisons b_1 if he wants x_i to be false and b_2 if he wants x_i to be true. Then Traveler moves to the hollow bullet and exists the component.

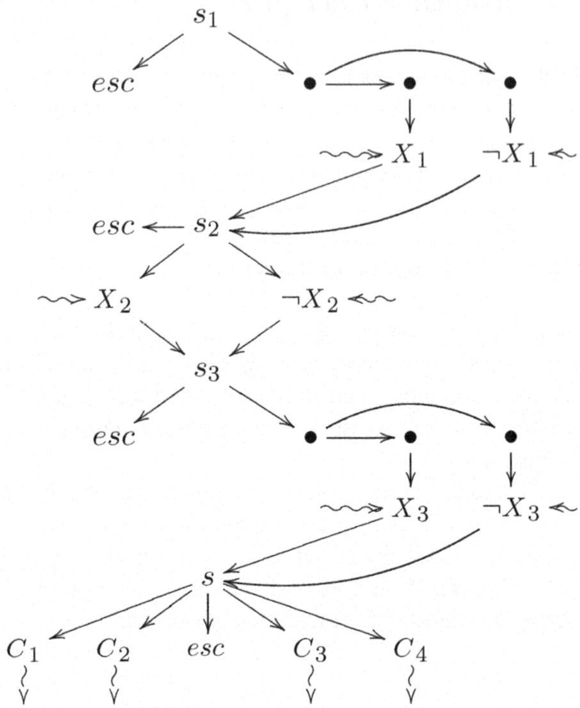

Fig. 5 Poison game: arena for Example 4.8

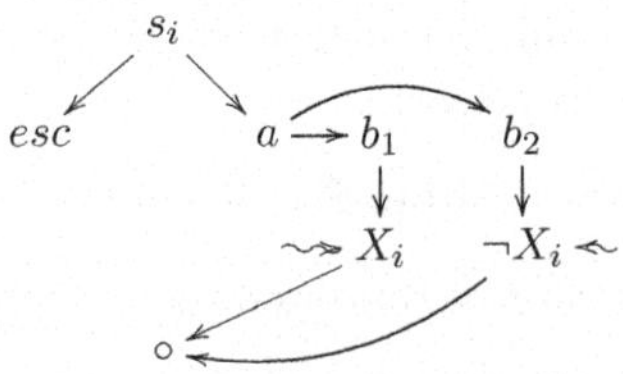

Fig. 6 Poison game: existential component

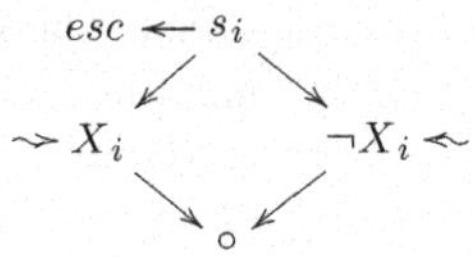

Fig. 7 Poison game: universal component

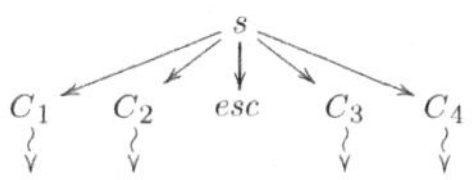

Fig. 8 Poison game: verification component

Verification Component. Shown in Fig. 8. Traveler enters the verification component from s. Given Proposition 5.2, Demon only considers poisoning one of the clause vertices (marked as C_i in the figures). Exchange the order of these two sentences. Also in the first sentence it should be Example 4.8. Demon chooses in which clause to falsify the QBF by poisoning the corresponding clause vertex. Suppose Demon chooses to poison C_i, given the way clause vertices are linked to variable vertices, Traveler chooses among the literals in c_i by moving towards the corresponding variable vertex. If the variable vertex of choice has already been poisoned when the players traversed the other components, the Traveler loses. Otherwise, the game continues and Demon has no choice but poison s_i/s of the next component, as there is only one outward arrow at a variable vertex. Then, from that s_i/s, Traveler can move towards esc (short for escape), and win the game.

5.2.3 Correctness of the Reduction

Lemma 5.3 *For any QBF φ, Traveler has a winning strategy in $\mathcal{G}_{\simeq}$ if and only if φ is a true QBF.*

Proof If φ is a true QBF, then there is an assignment for x_1 such that for all assignments for x_2 there exist $\ldots$ φ is true, i.e., there is at least one true literal in each clause. If Traveler follows the truth value of the variable in such an assignment in each existential component, then it is guaranteed that each C_i will leads to at least one literal vertex that has not been poisoned. Traveler can move to that literal vertex and then win by moving to esc.

If φ is a false QBF, then for all assignments for x_1 there is an assignment for x_2 such that $\ldots\varphi$ is false, i.e., there is at least one clause of QBF in which all the literals are false. If Demon follows the truth value of the variable in such an assignment in each universal component, then it is guaranteed that in the verification component, Traveler will have to choose to move from a point whose successors (one or more) have all been poisoned, and thus loses the game. □

Theorem 5.4 *The solution problem of poison games is PSPACE-complete.*

Proof Since the reduction only involves generating verification components that have $O(l)$ clause nodes and assembling all the components, the time complexity of the reduction is $O(l)$, where l is the length of the QBF. Thus, we have a polynomial-time reduction from the problem of true QBF to the solution problem of poison games. This establishes the PSPACE-hardness of the problem. Together with Lemma 5.1, we conclude that the problem is PSPACE-complete. □

6 Logical Aspect

There is a close connection between a modal language of games and its corresponding evaluation game, or so called 'logic game' (van Benthem 2014; van Benthem and Klein 2018). All the games mentioned above are no exceptions. For the modal language for local link-cutting sabotage game, see *loc*SML in Aucher et al. (2017). For a modification of *loc*SML that considers deleting links of certain property called *definable sabotage logic* and its corresponding logic game, see Li (2018). For the modal language that depicts local node-deleting sabotage game and its modification considering deleting nodes with certain property please refer to Chen (2018). For poison game, see Stockmeyer and Meyer (1973) for its original formalization and Aucher et al. (2017), Grossi and Rey (2019), Blando et al. (2018) for later analyses.

Note that in all three logics, the winning positions of a graph game can be defined using the modal μ-calculus, see van Benthem and Liu (2018), Areces and van Benthem (2019), Blando et al. (2018), for definition respectively.

Given a specific language, the winning positions for Traveler in a local node-deleting sabotage game and a poison game can be expressed as a formula in the given language combined with a μ-calculus. For example, we adopt the least expressive language PSL in Blando et al. (2018) for poison game

Definition 6.1 (*Syntax of PSL*) Let *Prop* be a countable set of propositional atomic formulas. We define the language *PSL* by defining formulas φ as follows:

$$\varphi := q \mid \neg\varphi \mid (\varphi \wedge \psi) \mid \langle t\rangle\varphi \mid \langle p\rangle\varphi$$

where $q \in Prop$. For $\langle t\rangle$ and $\langle p\rangle$, their duals are defined as $[t] := \neg\langle t\rangle\neg$ and $[p] := \neg\langle p\rangle\neg$, respectively.

Definition 6.2 (*Semantics of PSL*) A model for *PSL* is a tuple $\mathcal{M} = (W, R, V, M)$, where (W, R, V) is a standard relational structure, and $M \subseteq W$ is the memory of the model. The semantics of the new operators is then given by

$\mathcal{M}, w \models \langle p\rangle\varphi$ iff there exists w' s.t. wRw' and $\mathcal{M}[w'], w' \models \varphi$

$\mathcal{M}, w \models \langle t\rangle\varphi$ iff there exists w' s.t. wRw', $w' \notin M$ and $\mathcal{M}, w' \models \varphi$

where $\mathcal{M}[w'] = (W, R, V, M \cup \{w'\})$.

From the analysis in the section before we know that given a finite arena, there's always a $n \in \mathbb{N}$ s.t. Traveler can survive forever if she can survive for n steps. Then the winning positions of Traveler on such finite graph can be captured by the validity of the following sentence in the language:

$$([t]\bot \vee [p]\bot \vee \langle t\rangle[p])^{n-1}(\langle t\rangle\top)$$

Similarly, for the local node-deleting sabotage game, we have Chen (2018)'s formulation.

Definition 6.3 (*Syntax of SML_s*) Let *Prop* be a countable set of propositional atomic formulas. We define the language *PSL* by defining formulas φ as follows:

$$\varphi := p \mid \neg\varphi \mid (\varphi \wedge \psi) \mid \Diamond\varphi \mid \langle -\psi \rangle \varphi$$

where $p \in Prop$. For $\langle -\psi \rangle$, its dual is defined as $[-\psi] := \neg\langle -\psi \rangle \neg$.

Definition 6.4 (*Semantics of SML_s*) A model for SML_s is a standard relational structure $\mathcal{M} = (W, R, V)$. The semantics of the new operator is then given by

$$\mathcal{M}, w \models \langle -\psi \rangle \varphi \text{ iff there exists } w' \text{ s.t. } wRw',\ w \neq w',\ \mathcal{M}, w' \models \psi \text{ and } \mathcal{M}^{w'}, w \models \varphi$$

where $\mathcal{M}^{w'} = (W - \{w'\}, R, V)$.

For any local node-deleting sabotage game with a finite arena, the game ends with in a n steps for some natural number n. Thus, the following sentence captures the characteristics of the winning positions for Traveler

$$f^n(\bot)$$

where $f(x) = \gamma \vee (\Diamond\top \wedge \langle -\top \rangle \Diamond x)$, and γ is the goal points that we start with.

In case of local node-deleting sabotage and poison game, whose upper bound for model checking is PSPACE (Areces et al. (2009), Blando et al. (2018), Chen (2018)[1] respectively). Since the complexity for model checking should be no less than the complexity for checking any specific formula, we have the follow theorems

Theorem 6.5 *Model checking of PSL, PML and $ML_\emptyset$ is PSPACE-complete.*

Theorem 6.6 *Model checking of SML_s is PSPACE-complete.*

Let us take a look at an alternative formulation of winning positions with μ-calculus. For a local link-cutting sabotage game the winning positions of Traveler is captured by

$$\mu p \cdot (\gamma \vee (\Diamond\top \wedge \blacksquare\Diamond p))$$

where γ is the proposition representing the set of goal points. Given the polynomial-time solution algorithm in Sect. 3, we find that locality brings about a significant change in the modal fixed-point definition for Traveler's winning positions in the sabotage game. Instead of using the dynamic link deletion modality, we can use the following formula instead:

[1]In Areces et al. (2009), it is been proven that the model checking for $ML(Ⓚ, ⓡ, ⓔ)$ is PSPACE-complete and since $ML(Ⓚ, ⓡ)$ (which is, in turn, the same language as $ML_\emptyset$ in Blando et al. (2018)) is a fragment of $ML(Ⓚ, ⓡ, ⓔ)$ and thus the upper bound for the model checking for $ML_\emptyset$ is PSPACE, and by the inclusion of expressive power of the three modal languages for poison game and the clearly polynomial-time translation between the languages, all three of the languages has PSPACE as the upper bound for model checking.

$$\mu p \cdot (\gamma \vee \langle \geq 2 \rangle p)$$

Here $\langle \geq 2 \rangle$ is a static graded modality stating the existence of at least two successors of the current point satisfying p. The equivalence between these two modal fixed-point formulas is shown in Sect. 3, and but the statement fails to hold for the global sabotage game, since the solution complexity for the latter is PSPACE-complete (Löding and Rohde 2003). Interested readers can find more about graded modalities in Lassiter (2017).

Some might suspect that the algorithm in Sect. 3 can somehow be expended to produce a polynomial-time algorithm for the purpose of model checking. However, this idea is not viable given the fact that model checking for local link-cutting sabotage game is PSPACE-complete (Rohde 2005).

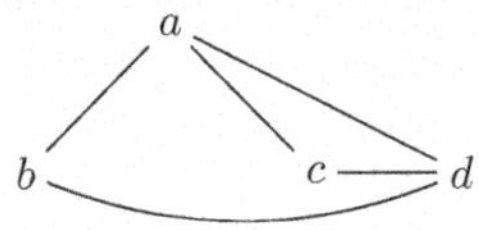

The failure of a similar recursive algorithm for model checking of the local link-cutting sabotage can be illustrated in the following example: Suppose we want to use a similar recursive algorithm to determine the set of vertices in the arena ($V(p) = \{a\}$) above that satisfy

$$\blacksquare\Diamond\Diamond\Diamond p$$

First, in the spirit of the analysis above, we replace $\blacksquare\Diamond$ with $\langle \geq 2 \rangle$ and get

$$\langle \geq 2 \rangle\Diamond\Diamond p$$

Then in a recursive manner we can calculate

$V(p) = \{a\}$, $V(\Diamond p) = \{b, c, d\}$, $V(\Diamond\Diamond p) = \{a, b, c, d\}$, $V(\langle \geq 2 \rangle\Diamond\Diamond p) = \{a, b, c, d\}$.

At the same time, starting from a, if Demon cuts the link between a and d, a will no longer satisfy $\Diamond\Diamond p$. Hence, $a \notin V(\blacksquare\Diamond\Diamond\Diamond p)$, which shows that $V(\blacksquare\Diamond\Diamond\Diamond p) \neq V(\langle \geq 2 \rangle\Diamond\Diamond p)$.

This inequality results from the fact that there's a bidirectional dependency between a and d such that if we cut the link between them, for both of them there is a sub-clause of $\blacksquare\Diamond\Diamond\Diamond p$ changes from true to false, namely $\Diamond\Diamond\Diamond p$ for a and $\Diamond p$ for d. This indicates that we can no longer establish a tree-like dependence relation among the vertices as we did in Sect. 3 since, for example, the link between a and d bears a bidirectional dependency. The property of having a tree-like hierarchy of dependence among vertices might be crucial to the existence of a polynomial algorithm for the solution problem, and after proper abstraction and formalization, it is a promising candidate parameter to account for the discrepancy of complexities between superficially similar games.

7 Related Work

This paper focuses on the complexity analysis of several localized variants of sabotage games and its implication. Rohde (2005) provides yet another proof of the solution problem of local link-cutting sabotage being PTIME using a double tractor. The work also provides insights for the design of arenas in Sects. 4 and 5.

For a thorough introduction to SML, Aucher et al. (2017) formulates the latest results in the minimal sabotage modal logic of arbitrary edge deletion. Stemming from this minimal logic, a number of authors have studied other graph games using their corresponding modal logics. The logic of Li (2018) characterizes the scenario where Demon all the local links that lead to nodes of a certain property. Chen (2018) provides a logic for analyzing a similar game in which Demon deletes all the local nodes of a certain property instead of links. For the poison game, Blando et al. (2018) provides three languages of different expressive power for characterization, all of which took from memory logics (Mera 2009). Thompson (2018) proposes a scenario called Boolean network games, combined with a logic of local fact change that characterizes Nash equilibria. All of these studies provide new ways of looking at the interaction between graph games, network games, and logics of control.

For more logical systems inspired by the dynamic-epistemic tradition, see 'graph modifier logic' (Aucher et al. 2009), 'swap logic' (Areces et al. 2014), and 'arrow update logic' (Areces et al. 2012; Kooi and Renne 2011). To see how the study of sabotage modal logic interacts with other areas of study, for theoretical computer science see Grüner et al. (2013), Radmacher and Thomas (2008), for learning theory see Gierasimczuk et al. (2009), for logics of social networks see Liu et al. (2014), Seligman et al. (2013) and for argumentation see Grossi (2010a, b). For a much broader inspection of analysis and design for graph games in tandem with the corresponding modal logics, see van Benthem and Liu (2018), which proposes various meaningful new games and identifies general questions between logic and games.

8 Conclusion

In this paper, we took out a case study of three of the local variants of sabotage games. For the local link-cutting variant, we propose a polynomial-time algorithm for calculating the winning positions for Traveler in the corresponding game. The static nature of the algorithm inspired the substitution of the dynamic operator with a static graded one in the μ-calculus formula in *loc*SML expressing the proposition that Traveler has a winning strategy. We provide alternating algorithms simulating the process of the games that run in polynomial time for both the local node-deleting sabotage game and the poison game to show that the complexity of the game solution is within PSPACE. We also describe ways to reduce instances of Quantified Boolean Formula to instances of both of the games. Since it is well established that QBF is a PSPACE-complete problem, it shows that the solution problems of both games are PSPACE-hard. This combined with the upper bound we mentioned establishes the fact that both the local node-deleting sabotage and the poison game are PSPACE-complete in terms of game solution. Using existing results showing that the complexity of model checking for the corresponding logics of both languages is up to PSPACE, it is easy to deduce that model checking for these corresponding logics is PSPACE-complete.

In a broader perspective, what these results show more generally is that slight modifications to a graph game can have huge effects on its complexity properties. In light of this fact, further research is needed to understand this phenomenon in a more systematic way and to identify which parameters of games and logics that crucially affect complexity.

Acknowledgements This research is supported by Tsinghua University Initiative Scientific Research Program (2017THZWYX08). I wish to thank Johan van Benthem for his invaluable advice throughout the development of this project. I also wish to thank my parents and several teachers and friends, whose company has given me the courage to discover and create, and has made this journey meaningful.

References

Areces, C., R. Fervari, and G. Hoffmann. 2012. Moving arrows and four model checking results. *Logic, Language, Information and Computation*.

Areces, C., R. Fervari, and G. Hoffmann. 2014. Swap logic. *Logic Journal of Igpl* 22 (2): 309–332.

Areces, C., D. Figueira, D. Gorín, and S. Mera. 2009. Tableaux and model checking for memory logics. In *International conference on automated reasoning with analytic tableaux and related methods*.

Areces, C. and J. van Benthem. 2019. The logic of stepwise removal.

Aucher, G., P. Balbiani, L.F.D. Cerro and A. Herzig. 2009. Global and local graph modifiers. *Electronic Notes in Theoretical Computer Science* 231 (none): 293–307.

Aucher, G., J.V. Benthem and D. Grossi. 2017. Modal logics of sabotage revisited. *Journal of Logic and Computation* 28 (2): 269–303.

Balcázar, J.L., J. Díaz and J. Gabarró. 2012. *Structural complexity II*, vol. 22. Springer Science & Business Media.

Blando, M., Areces, Franchesca Zaffora, K., and Carlos. 2018. The modal logics of the poison game.

Chen, Y. 2018. Modal logics of definable point deletion.

Gierasimczuk, N., L. Kurzen and F.R Velázquezquesada. 2009. Learning and teaching as a game: A sabotage approach. In *International Conference on Logic*.

Grossi, D. 2010a. Argumentation in the view of modal logic. In *International conference on argumentation in multi-agent systems*.

Grossi, D. 2010b. On the logic of argumentation theory. In *International conference on autonomous agents and multiagent systems*.

Grossi, D. and S. Rey. 2019. Credulous acceptability, poison games and modal logic. arXiv:1901.09180.

Grüner, S., F.G. Radmacher, and W. Thomas. 2013. Connectivity games over dynamic networks. *Theoretical Computer Science* 493: 46–65.

Kooi, B., and B. Renne. 2011. Arrow update logic. *Review of Symbolic Logic* 4 (4): 536–559.

Lassiter, D. 2017. *Graded modality: Qualitative and quantitative perspectives*. Oxford: Oxford University Press.
Li, D. 2018. Losing connection: The modal logic of definable link deletion.
Liu, F., J. Seligman, and P. Girard. 2014. Logical dynamics of belief change in the community. *Synthese* 191 (11): 2403–2431.
Löding, C. and P. Rohde. 2003. Model checking and satisfiability for sabotage modal logic. In *International conference on foundations of software technology and theoretical computer science*, pp. 302–313. Springer.
Mera, S. 2009. *Modal memory logics*. PhD thesis, Citeseer.
Radmacher, F.G., and W. Thomas. 2008. A game theoretic approach to the analysis of dynamic networks 1 abstract. *Electronic Notes in Theoretical Computer Science* 200 (2): 21–37.
Rohde, P. 2005. On games and logics over dynamically changing structures.
Seligman, J., F. Liu, and P. Girard. 2013. Facebook and the epistemic logic of friendship. *Computer Science*.
Stockmeyer, L.J. and A.R. Meyer. 1973. Word problems requiring exponential time (preliminary report). In *Proceedings of the fifth annual ACM symposium on theory of computing*, pp. 1–9. ACM.
Thompson, D. 2018. Local fact change logic. *Manuscript*.
van Benthem, J. 2011. *Logical dynamics of information and interaction*. Cambridge: Cambridge University Press.
van Benthem, J. (2014). *Logic in games*. Cambridge: MIT Press.
van Benthem, J. and D. Klein. 2018. Interfaces of logic and games.
van Benthem, J. and F. Liu. 2018. Graph games and logic design.

Invited Papers

Graph Games and Logic Design

Johan van Benthem and Fenrong Liu

Abstract Graph games are interactive scenarios with a wide range of applications. This position paper discusses old and new graph games in tandem with matching logics and identifies general questions behind this match. Throughout, we pursue two strands: logic as a way of analyzing existing graph games, and logic as an inspiration for designing new graph games. Our aim is modest: we propose a perspective that complements existing game-theoretic and computational ones, we raise questions, make observations, and suggest research directions—technical results are left to future work. But frankly, our main aim with this survey paper is to show that graph games are concrete, fun, easy to grasp, and yet challenging to study.

1 Two Graph Games

Graph games are played on graphs, directed or undirected, with one or many relations, perhaps with annotations at vertices. This setting may seem specialized—but the scope of these games is broad, with applications to computation, argumentation, communication, social networks, warfare, and other scenarios. To introduce our perspective, we start with two concrete examples that illustrate the issues leading up to our theme. The first 'travel game', played on a fixed graph, is a showcase of the close match between a standard graph game modeling a computational search task and logics describing game solution—in this case, modal logic. The second 'sabotage game' is a travel game with an obstructing player changing the initial graph, where a standard reachability task now has to be performed under adverse circumstances. This game is a pilot for studying basic features of realistic contemporary scenarios of, say,

This chapter is in its final form and it is not submitted to publication anywhere else.

J. van Benthem
Stanford University, Stanford, CA, USA

J. van Benthem · F. Liu (✉)
Tsinghua University, Beijing, China
e-mail: fenrong@tsinghua.edu.cn

F. Liu et al. (eds.), *Knowledge, Proof and Dynamics*,
Logic in Asia: Studia Logica Library,
https://doi.org/10.1007/978-981-15-2221-5_7

AI systems designed for cooperative purposes, but now being used by players with adversarial goals. In this case, a significant threshold is crossed. The computational complexity of solving sabotage games, as opposed to travel games, jumps from Ptime to Pspace-complete, and the match with logic design for the game becomes much more complex, as we shall see. Such transitions in game design and logic design are the trigger for this position paper. For more general background on logic and games presupposed in what follows, see van Benthem (2014, 2019).

1.1 Traveling Along a Graph

Travel games Travel games take place on a graph **G** with a starting point s. Two players A, E take turns in going along available edges, thus moving to new points—A starts. A player loses at any point if it is her turn to move, but there is no available edge to travel. Infinite runs are considered a win for player E.

Example A travel game.

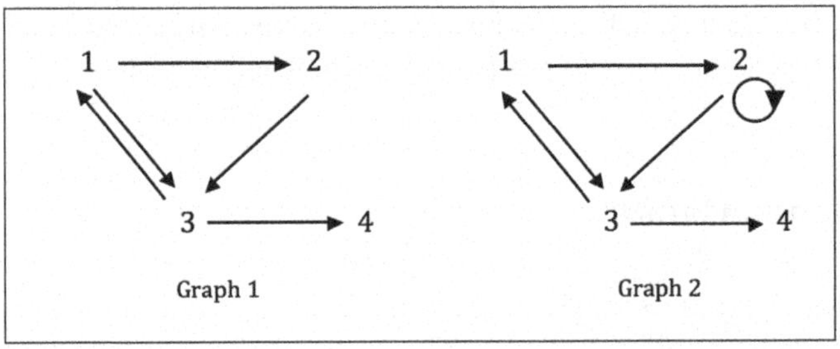

First consider Graph 1. If player A starts at point 1, and moves to 3, then E can move to the endpoint 4, and A loses. Note that E also has another strategy in this eventuality, which is to stay in the two-cycle $\{1, 3\}$. However, overall, A has a winning strategy: first, go to 2, then player E must go to 3, and A can go to 4. Other positions can be analyzed in a similar manner.

Next, consider Graph 2 with a slight change in accessibility. This time, if A starts at 1, it is E who has a winning strategy. If A moves to 3, E wins as before. And if A moves to 2, E chooses the reflexive loop and stays there. If A wants to prevent E from winning an infinite history staying at 2, then he must move to 3—but then E wins again by moving to 4.

There are also variations of the travel game, with other winning conventions, or with separate relations for the two players—and we will see others below.

Under suitable choices of vertices and edges, travel games have a broad range of applications: from analyzing the complexity of evaluating logical formulas in models, or of finding models, to solving general search problems (Grädel et al. 2002).

Theoretically, by the Gale–Stewart Theorem, many travel games are 'determined', thanks to the simplicity of their setup and their winning conditions: in any starting position (**G**, s), either player A or E has a winning strategy.

Modal logic There is an illuminating logical side to travel games. Graphs are natural models for a modal logic, and the moves are described by modalities. In particular, a winning response pattern for player E typically involves an iterated universal-existential modality of the form $\Box\Diamond$: E has a responding move to whatever A chooses. In a more elaborate modal formalism, the winning positions for E are defined by the following formula of the modal μ-calculus

$$\nu p \cdot \Box\Diamond p$$

This fixed-point formula defines the largest subset P of the graph where player E can always keep moving inside P, no matter which move is chosen by A.

Thus, an elementary theory of travel games is contained in a well-known system of modal logic. As a benefit, model checking for modal properties in graphs can be done fast, often in polynomial time. Moreover, the modal language is invariant for bisimulation between graphs on which the game is played, and this leads to standard model-theoretic results on preservation under graph change and other topics. Finally, the valid modal laws of travel games are decidable.

Coda: logic games Interestingly, the travel game itself is not only described by modal formulas. It may also be seen as an evaluation game for a modal language of games, applied to the above special formula $\Box\Diamond p$ describing the existence of winning strategies, suitably iterated to match the size of the graph.[1] Thus, the travel game returns higher up as a 'logic game' for its own modal logic. This intriguing entanglement will return in what follows.

1.2 Travel and Damage

The sabotage game The following graph game has been proposed to model what happens to standard computational tasks, such as path search, in a perturbed environment (van Benthem 2014). Starting from an initial point s, a 'Traveler' T moves through a graph **G** with possibly multiple edges, one link at a time. However, in each round of the game, a 'Demon' D first cuts one link, anywhere in the graph. Traveler wins when she reaches her goal point, or more generally, any point in her perhaps larger goal region in the graph. Also, players lose when they cannot move: this would happen to Demon when there are no more links, or to Traveler if she stands at an isolated point. This time, unlike in the Travel Game, the graph itself changes as the game proceeds.

Many further interpretations exist for sabotage games, including blocking access in information networks or learning viewed as a process of guiding players into certain subregions of a graph.

Again the game is determined. The conditions of Zermelo's Theorem or the Gale–Stewart Theorem are met, so Traveler or Demon has a winning strategy.

[1] The travel game matches the usual evaluation game for the μ-calculus formula $\nu p \cdot \Box\Diamond p$.

Example A sabotage game.
Let Traveler start in point 1 of the graph depicted, trying to reach point 4.

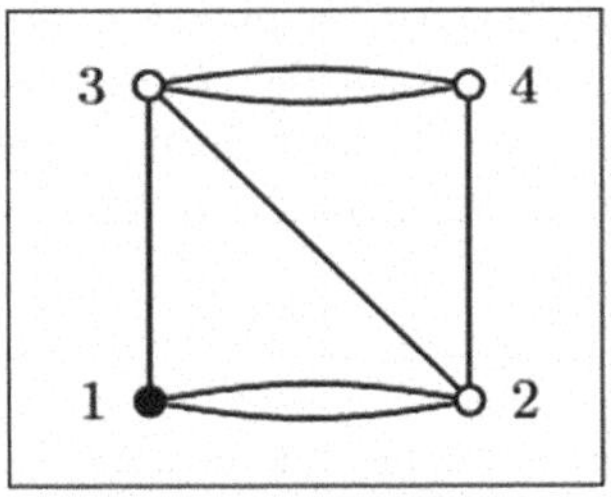

It is the Demon who has the winning strategy, by first cutting one upper link between points 3 and 4, and then responding appropriately to whatever Traveler does. Note, however, that Demon has no winning strategy if he is restricted to 'local sabotage', cutting links at the current position of Traveler. Other positions and other goals can be analyzed in a similar fashion. Also, the example is easily modified to work with directed instead of undirected graphs.

Sabotage modal logic Again there is a logical background here. Just as the travel game, the sabotage game suggests a modal language with standard modalities $\Diamond$ describing moves by Traveler, but it has to add new less standard modalities $\blacksquare$ for Demon's graph change, describing what is true at the current point after one edge has been deleted from the current accessibility relation. Winning strategies for Traveler are then associated with modal patterns

$$\blacksquare\Diamond\blacksquare\Diamond \ldots$$

More general winning positions for players can be defined in the 'sabotage μ-calculus' which adds the $\blacksquare$ modality to the usual syntax of the modal μ-calculus. An example is the formula for winning positions of Traveler with a goal region defined as all points satisfying the formula γ (Aucher et al. 2017):

$$\nu p \cdot (\gamma \vee \Box\bot \vee \blacksquare\Diamond p)$$

Unlike with travel games, even the modal logical theory of sabotage games is not so smooth and simple, and it contains several surprises. Model checking for the basic modal sabotage logic takes Pspace instead of Ptime, so a complexity jump occurs in the logic of gamified simple algorithmic tasks involving perturbation. Moreover, while this modal logic is still effectively translatable into first-order logic, satisfiability is undecidable—and despite its being axiomatizable in principle, for instance using suitably adapted semantic tableaux, no perspicuous axiomatization is known.[2] And while there is a model theory based on a natural notion of sabotage bisimulation, open problems abound.

[2]However, van Benthem, Mierzewski and Zaffora (2019) has new axiomatization techniques combining hybrid logic and dynamic-epistemic logic.

These issues extend to the sabotage μ-calculus, whose semantics differs from what one might expect prima facie, due to the fact that evaluation of approximations for fixed-points may require jumping across different graphs (Aucher et al. 2017).

The other side of this additional complexity is, of course, that sabotage games are more interesting to play than travel games.

Digression: concrete cases To explore how this general perspective works in practice, one can consider solvability of the games on special classes of graphs.[3]

Sabotage in trees Consider finite intransitive trees, with arrows pointing from parent to child nodes. In this case, the sabotage game simplifies considerably. First, there is no loss of power for Demon if only local cutting is allowed, that is, just on immediate successors of Traveler's position. For, such moves block the whole subsequent subtree for Traveler—so, if a winning strategy for Demon involved cutting further down at some stage, he could just as well have cut at the top of the subtree where Traveler is standing. Given this, it can be proved that Traveler has a winning strategy standing at a node s if there is a 'rich subtree' T starting at s with out-degree 2 (there are at least two successors in T for every parent node in T) all of whose end nodes are in the goal region.[4]

Sabotage in grids In trees, travel decisions are irrevocable. Choosing a daughter node makes all sibling nodes and their subtrees inaccessible. An almost opposite structure is that of a grid, where any two paths from a current node can still converge to a common point by suitable further moves. For a concrete example, consider a geometric 8-point *Cube*, with a goal region of two points.[5] These points must lie on some face of the cube. Now the game acquires a concrete geometrical meaning. Clearly, Traveler has a winning strategy in some vertices of this face, depending on where the goal points lie, but Demon has a winning strategy when Traveler stands at the opposite four vertices. When the goal region contains three points, Traveler may

[3]For a general study of graph change encompassing sabotage, but also other destructive mechanisms, see the systematic approach in Areces et al. (2015), which studies more general hybrid logics of graph change.

[4]If there is such a rich subtree, Traveler can obviously stay inside it, no matter where Demon cuts, and eventually, as the game moves down the tree, Traveler ends up in an endpoint in the goal region. If there is no rich subtree, then each point t in the subtree has at most one successor u whose subtree is rich. For, if there were two such successors, then there is a rich subtree at s after all. But then, Demon can cut the connection to this single rich subtree (if one exists at all) and force Traveler to a situation without a rich subtree. In the end, this forces Traveler into an endpoint outside of the goal region.—Note that, if Traveler can move back and forth in the tree, this outcome changes. Suppose s is not a winning node for Traveler, but there are three sibling nodes that are winning. Then, no matter how Demon cuts, Traveler can move back one step, and then still reach a winning node. For a more precise version of this argument, see the fixed-point analysis in Sect. 2.2. The existence of a rich subtree is not definable in basic modal logic since one can always create rich subtrees modulo bisimulation by duplicating parts of trees. It is not even first-order definable, Valentin Goranko (p.c.).

[5]This classical geometric setting is also a good antidote to a hidden bias in the literature toward using only simple two-dimensional 'planar graphs' in sabotage scenarios.

or may not have a winning strategy everywhere, depending on the distribution of the goal points.

Logic games once more As with the travel game, one can also view the sabotage game as a semantic evaluation game for special formulas of sabotage modal logic with alternating travel and deletion modalities, or in infinite versions, for similar formulas of the sabotage μ-calculus.

Our examples show how design of graph games and design of matching logics go hand in hand, both with fixed and with changing graphs. Of course, one can also study the above graph games on their own, using techniques from computer science or game theory, but the special focus in this paper will be on this game–logic connection: how it works, and what light it sheds.[6]

2 Modified Scenarios

Next, we broaden the analysis beyond the specific games considered so far. We look at variations on travel and sabotage that model further realistic scenarios to see how the match with logics persists, but in a fine balance that raises interesting further questions. The examples to follow here will not be discussed in as much detail as the above games, given our more methodological purpose.

2.1 Modified Travel Games

Two graph travelers This time, both players are localized, so the game starts in a position $(\mathbf{G}, s, t)$ with a starting point s for A and t for E. We can assume that A starts, but we could also let the two players move simultaneously.[7] Now we need to stipulate a winning condition, or rather a goal. Let us say that, for both players, the goal is to keep going forever. This is not a zero-sum game: infinite histories for both agents are a draw—though we might stipulate that, if one agent gets stuck and the other can continue, that first player loses.

This simple modification breaks the game content of the travel game, since players can no longer influence the steps made by other players. Even so, the earlier connection with modal logic remains.

The natural models are now graphs $(\mathbf{G}, s, t)$ with two distinguished points, and the natural modal language is a 'two-dimensional' one, with proposition letters true at single points. Most essential then are two 'travel modalities':

$$(\mathbf{G}, s, t) \models \langle \text{left} \rangle \varphi \text{ iff for some point } s' \text{ with } Rss', (\mathbf{G}, s', t) \models \varphi$$

---

[6]This is not to say that the connection between graph games and 'matching' logics is entirely unproblematic, and we will also raise some conceptual problems in Sect. 4.

[7]Our examples in this paper are mainly cast as sequential games: but it would make sense to also consider versions that allow simultaneous moves for the players.

$$(\mathbf{G}, s, t) \models \langle\text{right}\rangle\varphi \;\text{ iff }\; \text{for some point } t' \text{ with } Rtt', (\mathbf{G}, s, t') \models \varphi$$

These two modalities allow us to express the above observations. For instance, in a given finite graph with starting position (s, t), the first player loses if, for some natural number k, the formula $[\text{left}]^k\bot$ holds at s, while $[\text{right}]^k\bot$ does not hold at t. Moreover, the existence of infinite paths for players can be expressed by μ-calculus formulas of the form $\nu p \cdot \Diamond p$.[8]

The basic modal logic for these models is still decidable, and does not differ all that much from a standard minimal modal logic (Segerberg 1973).[9] One way of seeing this is to note that the language we have introduced can be translated into the two-variable fragment of first-order logic, which is known to be decidable.[10]

A natural variation of this scenario would give both players their own goal regions that they are trying to reach. Similar points can be made about the loss of the earlier game character, but the remaining link with modal logic. In our later versions of goal-reaching scenarios, the game character will be reinstated.

Remark on interaction The two-traveler game raises an interesting general issue. *What makes a game scenario on graphs interactive?* Is it a necessary condition that one player's moves modify or restrict those of the other player? This seems overly strong: even a travel game has interaction in the form of players' strategies influencing optimal play for other players. Another dimension of interaction is the pay-offs. When these are completely independent, there is less interaction than when the goals of the scenario have the players entangled.

Meet/avoid game We investigate the second aspect of interaction, entangled goals, with a slight modification of the travel game with players at two different positions. This time, let A win the game when the two players meet at the same point, while E loses then. So, one player wants to avoid the other, like in children's games of 'Catch' or in 'Cops and Robbers' (Nowakowski and Winkler 1983).

Now there can be more interesting game scenarios, and it is easy to see that E's winning positions are correlated with the existence of suitable cycles in the graph. It is also easy to see that the game is still determined, since it has an open winning condition for A in the sense of the Gale–Stewart Theorem.

As for the logic counterpart of this modified game, we can have the same two-dimensional models as before and the same modalities. However, the new goal now calls for one new propositional constant I such that

$$\mathbf{M}, s, t \models I \;\text{ iff }\; s = t$$

[8]However, there appears to be no single generic formula in the basic modal language, or even in the modal μ-calculus, that expresses the property of reaching the goal first.

[9]There are much more complex modal 'product logics' (Gabbay and Shethman 1998), but these are not needed here.

[10]The crux is that working with just variables x, y for our pairs, bounded existential quantifiers for travel modalities at a point can reuse the variable marking the other point.

In other words, I denotes the identity relation in a graph model $\mathbf{G}$. Then a typical statement about winning positions for E will be of the form

$$[\text{right}]\langle\text{left}\rangle\neg I$$

We believe that this extension of two-dimensional logic is still decidable, even though it gets close to two-dimensional 'compass logics' (Marx and Venema 1998). But there is a catch. It is not clear how to keep the translation of the above language inside the two-variable fragment of first-order logic since we need to compare three points: the originally matched s, t, but also the new point s' reached.[11]

Occupation game Our final travel game lets players occupy parts of graphs, which is reminiscent of many popular board games, such as 'Settlers of Catan'.[12] Each player has his own color, and he automatically colors all points that he visits in his own color if no color was assigned yet. When a player reaches a point with the color of the other player, he loses. Again we can set this up with 'keep going' as a winning condition, or with reaching a specified goal region, one for each player. In what follows, we consider the latter format.

Example Outrun or obstruct?

Here is a graph for a scenario like this.

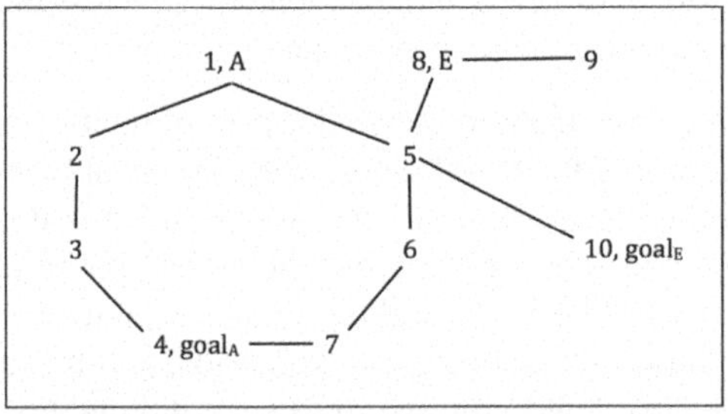

Let the game start with player A moving. Starting in 1, A could reach his goal point 4 in three steps, but that would give the player E enough time to reach her goal point 10. So, what A should do is move to 5, thereby blocking E from ever reaching her goal—even though this is not A's shortest path to his goal.

It would be of interest to analyze winning strategies in occupation games in graph-theoretic terms, as they seem closely related to fundamental results such as Menger's Theorem on the existence of minimal barriers in networks (Diestel 1997).

[11]There is an interesting analogy between the meeting game and modal bisimulation. A bisimulation is a relation Z between points s, t in two models $\mathbf{M}, \mathbf{N}$ such that each step from s to an accessible point s' in $\mathbf{M}$ can be matched by a step from t to an accessible t' in $\mathbf{N}$. Here 'matched' means that $s'Zt'$ holds. The game version of bisimulation is much like our meeting/avoiding game. Now it is known that the notion of bisimulation cannot be expressed with two first-order variables (van Benthem et al. 2009), whence its meta-theory is somewhat complex.

[12]In other widely played parlor games, such as Halma or Chinese Checkers, barriers of occupied points can be jumped over, but we will not consider such variations here.

The occupation game also looks like a modified version of the 'poison game' in the graph-theoretic setting of Duchet and Meyniel (1993) whose corresponding modal logic has been investigated in Mierzewski and Zaffora Blando (2016), Grossi and Rey (2019)—be it that both players can poison positions now.[13]

As for a matching logic, to deal with the dynamics of progressive occupation, we need proposition letters for both players whose extension grows as travel steps are made. This requires introducing, on top of the basic modalities that correspond to steps in a graph, at least a further dynamic modality

$$[+p]$$

which records what holds when we make atom p true at the current point:

$$\mathbf{M}, s \models [+p]\varphi \quad \text{iff} \quad \mathbf{M}\text{-with-}p\text{-set-to-true-at } s, s \models \varphi$$

This local valuation change looks like a simple and harmless addition to basic modal logic. However, using an embedding of a known undecidable hybrid logic, Thompson (2018) shows how basic modal logic extended with $[+p]$ modalities is undecidable.

But the more general, and more explanatory, analogy here is with the 'memory logics' of Areces et al. (2008, 2011), which allows a growing number of points in a graph to be stored in memory. The relevant known theory of these logics includes both general undecidability proofs and decidability results for special fragments.[14]

Having seen how natural variations on the original travel game sustain a logic connection—though one where slight changes may require big logic adjustments—we now turn to some natural modifications of the earlier sabotage game.

2.2 *Modified Sabotage Games*

Sabotage game with localized demon Players in the sabotage game are asymmetric in their powers. Traveler is 'local' in the sense of being restricted to her current position, but Demon is 'global', as his moves can be anywhere in the graph. While this asymmetry of the players is right for modeling some scenarios, such as the general disruption of a communication network, many situations in real life have both players localized in the graph. For concreteness, think of a war game, where Traveler is one party, and Demon the enemy. In that case, we can let the Demon cut at any point where he stands, but he may also make travel steps to get to other

[13]Interestingly, Grossi and Rey (2019) shows how to view graph-changing games as a model for argumentation.

[14]This variety of relevant logics, whose mutual relations are sometimes not so easy to establish, also shows that our matching of games with even just modal logics is not an automatic process, and that there may even be significantly different logics for the same game. For a recent example concerning modal poison logics, cf. Areces, Mierzewski and Zaffora (2018).

positions and cut there. Note that, unlike in the travel scenario, this localization of both players does not necessarily erase the game structure.

Example Localized travelers and demons.

Recall our sabotage game in Sect. 1.2. Assume for concreteness that in each round, Demon first cuts and then makes a step along some edge. For an illustration, if the Demon stands at the top left, he can still win as follows: First cut one link to the right, then go to the exit point E. Now, whatever Traveler does, Demon can cut a connection at E preventing Traveler from reaching the exit point from where he stands, and he has time to cut the third link if the Traveler tries somewhere else. However, if the Demon is localized at the bottom right when the game starts, it is the Traveler who has the winning strategy.

Clearly, this new game restricts the powers of the Demon in the original game. Locality changes things considerably. Zhang (2018, in this volume) gives a polynomial time solution algorithm for local sabotage games, in contrast with the Pspace solution complexity for the global sabotage game. Interestingly, on the logical side, locality induces a significant change in the earlier modal fixed-point definition

$$\nu p \cdot (\gamma \vee \Box\bot \vee \blacksquare \Diamond p)$$

for Traveler's winning positions in the sabotage game. Instead of using the dynamic link deletion modality, we can make do with the following formula

$$\nu p \cdot (\gamma \vee \Box\bot \vee \langle \geqslant 2 \rangle p)$$

Here $\langle \geqslant 2 \rangle \varphi$ is a static graded modality stating the existence of at least two successors of the current point satisfying φ.[15] Proving the equivalence between these two modal fixed-point formulas requires care with comparing winning strategies for Traveler at the same point in different graphs, and the argument typically should fail for the global sabotage game, since the solution complexity for the global version is known to be Pspace-complete, cf. Löding and Rohde (2003). It remains to be seen, of course, whether this particular reduction to static μ-calculi extended with suitable modalities can be generalized to other localized graph games.

Remark: model checking The shifts in complexity considered here need not just be viewed in terms of game solution. In the background lies the fact that the complexity of model checking for sabotage modal logic is Pspace-complete, instead of the Ptime for basic modal logic. And the model checking complexity for modal logics designed to match graph-changing games tends to go up in the same way: van Benthem, Mierzewski and Zaffora (2019) for a recent illustration using the techniques of Löding and Rohde (2003).[16]

---

[15]The number 2 here reflects the out-degree 2 of the 'rich subtrees' of Footnote 3, where local sabotage had the same power for Demon as global sabotage.

[16]There are other natural ways of changing the sabotage game. Say, instead of deleting links, we might also let players delete points and all links leading to them. This makes Demon much more

Fully symmetric sabotage game A still more realistic version makes players symmetric in their powers: they can now both delete links at their current position and decide to move from there, or vice versa, 'burning bridges'. In this symmetric setup, both players can have goal regions, perhaps disjoint ones in special settings such as war games.[17]

Example Local sabotage with two goals.

Consider the four-point graph for the original sabotage game in Sect. 1, with players T and D in various initial positions, and the goal point for T bottom–right, and for D top–left. We first need to stipulate moves. For concreteness, take 'go–cut': first move along an edge, then cut some link there, if you want.

But more interestingly, whatever moves we allow, this new setting obviously invites more refined goals beyond winning and losing. Naturally, both players prefer being in their goal region, while the other is not, slightly worse is both being in their goal region, next down is neither in their goal region, and clearly, worst is having the other in his goal region while you are not.

Refined goal structure yields a significant transition. While going beyond just winning or losing, the above scenario is still a perfect information game, and it can be solved by the standard game-theoretic method of Backward Induction. Now we get non-trivial Nash equilibria, whose definition involves powers of both players, not just one. We give a simple illustration, for convenience, not as a sabotage game, but as an occupation game in our earlier sense.

Example Graph games with preferences.

Consider the following graph, player T is standing at position 1, with 4 as her goal point, player D is standing at 2, with 3 as his goal point. Let us assume this time, for convenience that Traveler starts.

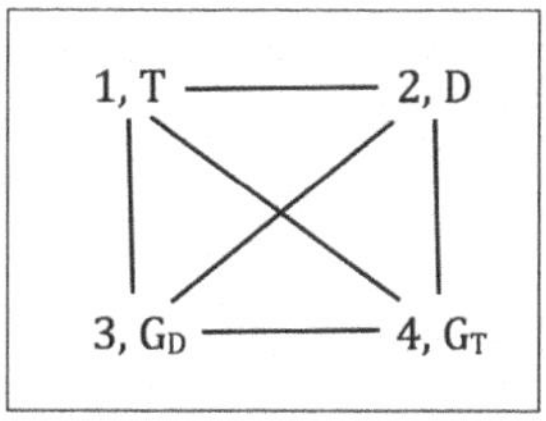

One can easily draw an extensive game tree for the two moves of both players: 'straight down' and 'diagonally down'. It is easy to see that each move by T leaves only one option for D. If T goes straight down, then so does D, and the outcome is

powerful than in the original game. Even so, the logical theory of point deletion shows many similarities with that of link deletion, cf. Chen (2018), and for a broader study connecting up with hybrid logic and first-order logic (van Benthem, Mierzewski and Zaffora 2019). Another natural variation would allow cutting off all links according to a definable rule in a graph. In a uniform global case, this is like link cutting in dynamic-epistemic logic. But the theory of a more realistic local variant shares many features with other complex logics discussed here (Li 2018).

[17]Such variations need not be a priori, they are just as well suggested by actual board games that are reminiscent of the sabotage game, such as 'Quoridor' (Glendenning 2005).

that neither reaches his goal. If T goes diagonally down, then so does D, and both players reach their goal. Following the Backward Induction solution procedure then tells us that the latter outcome will be reached.

This example is extremely simple, but it should be easy to see that much more complex graph games with preferences can be solved in the same way.

We are at a significant point here in the matching of graph games and logics. What resources are needed to analyze the Backward Induction style of reasoning? Our modal logics can still formulate goals, and also preferences, at least, when these are defined by simple enumeration of propositional cases like above, cf. the format of priority graphs in Liu (2011). But defining strategy profiles that form Nash equilibria seems to require further resources, coming from richer game logics than mere graph logics. We will return to this issue in Sect. 4.

3 Parameters for Game Design

Having seen a budget of graph games, let us summarize some main features. There are two parameters for a classification: general game structure (moves, turns, goals), and graph structure: the board on which the games are played.

Modulating moves In choosing moves for graph games, we found options that impact properties of the games, but also of logics describing these games (Aucher et al. 2017; Liu 2017). One pervasive distinction is that between *local versus global moves*: whether players are localized in the graph, like Traveler in the sabotage game, or can range at random, like Demon in that game. A second distinction is whether moves are *arbitrary, or definable*: can Demon delete any links, or should he follow some definition? This is related to another distinction: deletion moves can be *stepwise or uniformly defined*. And in making these comparisons, we even crossed another divide: moves can *stay within a graphs, or jump to a changed graph*. These choice points are not totally independent, but they span a space of design options, whose systematic exploration seems well worthwhile.[18]

Ways of winning and losing In simple graph games, goals for players are *independent unary properties* defining points that they must avoid or regions that they want to be in. However, we can also have *entangled goals*, as in the earlier meeting/avoiding game, itself a variant of so-called 'pursuit-evasion' games on graphs from mathematics and computer science (Parsons 1976). Here Traveler loses if she meets with Demon, Demon then wins: this is not a Boolean combination of separate independent unary goals for Traveler and Demon, but a binary relation between their positions. There may even be still higher-level *procedural goals*, for instance, about how players travel toward the distinguished goal region. For instance, they may not be allowed to visit the same intermediate points twice, or, in occupation games, they may not visit the path visited by the other player. And we also saw how the general

[18]There are many further possible moves in graph games than those discussed here. A quick survey of common parlor games reveals of the host of further types, having to do with how players can move counters, jump across barriers: in single steps or in iterated sequences, and so on.

specification of goals can lead to games that require Backward Induction or other ways of getting at Nash equilibrium outcomes.

Game boards In all this, we may have to loosen our focus on graphs. Our graphs served as game boards, comparable to those of parlor games. But the local states traversed by the above games usually require further structure. At least, we need to indicate the ambient location of players moving through the graph. Moreover, with our more complex occupation scenarios, we need additional propositional information at nodes telling us whether they have been visited already, or even, how often. In general, the game boards are not just graphs, but 'annotated graphs', where these annotations can be game-induced, but might also be game-external properties of vertices used in defining moves.

Several more general issues are suggested by a classification like this.

Changes and thresholds in behavior Localizing players, or changing their available moves has obvious effects on our simple examples of graph games. However, what are their general effects, and how do we measure these? Say, localizing Demon seems a major shift in favor of Traveler in sabotage games, but how can one make such intuitions precise?

One way of doing this might be to determine independent graph-theoretic properties that correspond to winning positions, such as the association of winning positions for Traveler with 'semi-kernels' in poison games (Duchet and Meyniel 1993).

Another approach would involve a general analysis of computational complexity: which design changes cross a threshold, making solution complexity a step harder, or simpler, in the complexity hierarchy?

Finally, there can also be other parameters that determine the behavior of games. For instance, winning conditions are often 'monotonic': some player only gets better off when we increase the goal region or the supply of accessible links. Can we also get general insights by restricting to monotonic goals?[19]

When are two graph games equivalent? However, perhaps the most obvious general issue behind our discussion is this. Are there hidden reductions running underneath the playground of designing new games that we have entered? How much variety is there really? *When are two graph games equivalent?*

Many of our cases illustrate this issue. For instance, travel games with occupation make nodes inaccessible to players, and so they are also sabotage games deleting points. Can this Gestalt switch be turned into a precise equivalence?

And when all is said and done, even our prima facie clear distinction between travel moves that stay within a given graph and sabotage moves that change the graphs can be questioned. Both moves stay inside a richer 'meta-graph' of different graphs and distinguished points, a perspective that is, in fact, quite congenial to logics of sabotage games.[20]

---

[19]There are precedents for this restriction in other logical literature on game design (Ågotnes and van Ditmarsch 2011).

[20]In its turn, this meta-perspective might suggest a generalized 'protocol semantics' where models may constrain available graph changes (van Benthem et al. 2009). This widening of the standard

4 Logics for Graph Games

Having seen the design choices for graph games, we now turn to matching logics.

Modal logics We started with modal languages having enough expressive power for describing simple games, and then moved to more expressive formalisms, sometimes rather mild ones such as two-dimensional modal logic. However, not all these logics were standard, and we found interesting open problems. In particular, logics for graph games always contained the basic standard modalities, the difference usually lies in the dynamic modalities. However, there are unresolved issues here, since the expressive surplus of dynamic modalities can often be calibrated in terms of hybrid logics (Areces et al. 2015). And in the limit, first-order logic itself could serve as a language of graph change (Aucher et al. 2017).

Moreover, these modal languages are not enough to capture game-theoretic notions such as winning positions uniformly. For that, we had to turn to modal fixed-point languages such as the μ-calculus—either in its basic format or in the more subtle formulation needed when adding modalities for graph change. We have seen how these raise interesting new technical issues, such as when winning positions can be defined alternatively in suitably extended static μ-calculi.[21]

However, toward the end of our discussion, we hit a significant borderline. When goals or preferences become more sophisticated than winning or losing, real game-theoretic considerations enter concerning various equilibria, and our emphasis on logics for graphs reaches its limits. Richer formalisms are needed.

From graph logics to logics for explicit game structure The other obvious candidate for describing graph games, now in more procedural detail, are standard 'game logics' for analyzing game structure (van Benthem 2014). These languages are rich, involving moves, preferences, and strategies, and they are usually interpreted on games, either in extensive form or strategic form—or in some cases, on associated models of powers that players have in the game.[22] A typical benchmark for such logics is whether they can describe standard game-theoretic solution procedures such as Backward Induction or Iterated Removal of Strictly Dominated Strategies. And indeed, we encountered the need for the former toward the end of Sect. 2. And contemporary game logics do not even stop here, they can also deal with more complex scenarios such as players of different abilities playing the same graph game, using type spaces or even richer devices.

---

models has been used to lower the complexity of dynamic-epistemic logics, and it might also improve our earlier treatment by lowering the current high complexity of logics of graph change.

[21] Even so, this is just one line, and the emphasis on modal logics in this paper may be a bit biased. If our focus had been on infinite graph games, then an alternative attractive formalism would have been that of *temporal logics*. These, too, can be very useful for defining complex behavior and strategies over time.

[22] For an up to date survey, see the Stanford Encyclopedia entry (van Benthem 2019).

In our view, graph games with complex preferences and procedural conventions eventually need to be studied using game logics.[23] However, this still leaves something to be explained. For a surprisingly long time, our discussion got by with austere logics for graphs. What was going on?

Graph invariants for game logics Graphs may be considered game boards associated with graph games. Now sometimes, the strategic properties of games can be expressed purely in terms of the game board. Game theory has many examples of such 'invariants': think of the simple numerical invariant indicating who has the winning strategy in the game of Nim (Osborne and Rubinstein 1994). Now, our logics for graphs may be viewed as defining invariants that capture the strategic structure of at least simply defined graph games.[24]

It is a very interesting question when such reductions are possible. A deeper study would lead us into results in computational logic on 'memory-free strategies', such as the Positional Determinacy Theorem for parity games (Mostowski 1991).

Here is another well-known example of such a reduction, now inside standard logic. Consider evaluation games for logics, say, Hintikka games for first-order logic (van Benthem 2014). The typical adequacy result one proves about such games is that the 'truth player' Verifier has a winning strategy for the game concerning a formula φ on a model $\mathbf{M}$ if and only if φ is true at $\mathbf{M}$. But reversing directions, the result tells us that the strategic structure of evaluation games is captured by game-free invariants purely at the level of the model.[25]

It is not our aim to solve all technical problems concerning modal logics of graph games and richer game logics. But hopefully, our discussion did show the interest of taking the joint design stance that we have pursued here. We conclude with a few more general perspectives on this match.

Uses of logic Using logic to describe games has obvious benefits. One can use the logic to express general qualitative reasoning patterns about the interactive scenarios that the games are intended to model. One can use some standard techniques such as model checking to verify that given games have specified properties. But these are generic benefits. In the present setting, we would be most interested in new connections with a special thrust toward games.

One interesting example of this kind concerns the general strategic situation in graph games. It is a significant question concerning many games which player has the systematic advantage, say, the opening player or the responding player. In our present setting, one way of making this issue more precise is as follows:

What happens to the winning positions, say, in the sabotage game, as the size of the finite graphs goes to infinity? Now there is a link with the logic here. The *Zero-One Law* for first-order logic says that for each first-order sentence in the language

---

[23] A simple illustration is that a position in a graph does not tell us how the game will go until we know which player has the turn (a game-internal notion) and needs to move there.

[24] Slightly richer invariants would arise on annotated graphs that encode some procedural game-internal information by means of additional unary properties of nodes.

[25] To see how remarkable this is, it suffices to make some obvious variations on well-known evaluation games, and observe how an immediate match with standard logics vanishes.

with a binary relation plus unary predicates, its probability of being true goes to either 0 or 1 as graph size goes to infinity. It can even be decided effectively, for any given formula, which of the two options occurs. And this Zero-One Law has been extended to LFP(FO): first-order logic with fixed-point operators, in Gurevich and Shelah (1986). Now, we were able to express the winning positions in the sabotage game in the sabotage μ-calculus, which translates into LFP(FO). Therefore, anything we can formulate in that language satisfies the Zero-One Law. Now, this does not immediately apply to winning positions, as these refer not just to graphs, but to graphs with distinguished points, for which the Zero-One Law fails. However, here are some assertions that do fall under pure LFP(FO): existence of winning positions for both players, the statement that each winning graph position for Traveler must be connected to at least one other winning position for Traveler, and so on. We believe that this offers a fresh, and to our knowledge, unexplored perspective on the logical analysis of graph games.[26]

The match viewed from both sides The connection between logics and games advocated in this paper can be looked at from two sides. So far, we have gone *from games to logics*, presenting a number of ways of achieving this. Here one obvious question is whether this can be made more systematic, say in the form of automatic methods for turning our earlier definitions for graph games into matching logics. There is a system to how one creates these matches, but we have not been able to see the full generality yet.

But there is also an opposite direction going *from logics to games*. In fact, there has long existed a well-established tradition of designing games for various logical purposes that goes back to the 1950s in mathematical logic, and to many centuries earlier in the dialogical debate tradition in various logical cultures. This historical tradition exemplifies the opposite direction to this paper: using logics to design games.[27] As in going from games to logics, there need not be a unique correlation here. Logical languages are associated with evaluation games, but also with proof search games, or with Ehrenfeucht–Fraïssé games for comparing models. The latter may throw light on graph games, too.

In general, it is a delicate issue how these two perspectives of 'logics for games' and 'logics as games' are related (van Benthem 2014), and many open problems remain. But as noted already in Sect. 1, our graph games have a bit of both since they suggest logics for their description whose evaluation games, at least for certain well-chosen formulas, are those very graph games.

---

[26]After the first version of this paper started circulating, a solution to the question posed here has been found using the countable random graph, where truth coincides with truth almost everywhere in finite models. Mierzewski (2018) shows that, if the goal region has at least two points, all positions in the random graph are winning for Traveler—making the sabotage game massively favorable to Traveler. For special model classes, however, the result remains open.

[27]In a broader perspective, the logic and games connection pursued here would also need a third component, namely, the study of games in computer science. These games go back a long time, cf. Aigner and Fromme (1984), Grädel et al. (2002), and they would form a natural complement to our discussion here.

Thus, the graph games discussed in this paper are one special case for the general question what happens when we pursue cycles of the form

'game, logic of game, game for logic of game', or
'logic, game for logic, logic for game of logic'.

These cycles need not always stabilize, but graph games may be an attractive arena where they do. Getting a bit ahead of the evidence that we presented in this paper, we would like to think of graph games as a natural place where the usual tension between game logics and logic games disappears. One take on this might be that our graph-changing games really are already logical evaluation games, but then games where the model changes in the process of evaluation, as has been suggested for first-order semantics in terms of drawing objects with or without replacement (van Benthem, Mierzewski and Zaffora 2019), and as seems even realistic in logical modeling for quantum-mechanical measurement.[28]

A caveat: logical machinery We conclude by pointing out that our matching also has its problems or limitations. Our logical languages are much richer than the set of assertions one would normally make about graph games or just the particular graph game we started with. This surplus may just be ballast.[29] Have we been pursuing a machinery that is much too heavy for graph games?

For a start, it is easy to overstate this objection. To a student of some particular inference patterns, logicians usually offer a complete formal system. This seems largely harmless, like visiting a store where one buys a lot of medical supplies even when there is just one bandage needed for now. Presumably, the logical system can handle lots of games similar to the original one.

But we can go one step further in the present setting. It may be possible to find a game interpretation for the whole logical language that is congenial to the original graph games. For instance, evaluation games can be defined for arbitrary formulas of sabotage modal logic. These games represent an infinity of *related graph games*, with different schedules and winning conditions. Say, the game for $\Box \Diamond \blacksquare \Diamond \gamma$ defines a simple mixed travel-sabotage game where Traveler must first move to some safe point for her against a sabotaging Demon.

Even so, the question remains whether this infinite class of formula-induced games contains interesting ones beyond the original motivating scenarios.

Finally, related to the preceding point, we are not claiming that every property of our logics has immediate import for graph games. For instance, while logicians tend to emphasize computational complexity of logical systems, this merely tells us that the *theory of* certain graph games may be complex—not that solving these games itself is a complex algorithmic task (van Benthem 2011).

---

[28]There is much more to the match between graph games and evaluation games than what we have probed here, especially, when one tries to correlate equivalent formulations of the same graph game with equivalent formulas in matching fixed-point logics, cf. Grossi (2013).

[29]In some cases, the ballast may even be misleading. For instance, it is emphasized in Aucher et al. (2017) how the valid equivalence $(\Diamond\top \wedge \blacksquare \Diamond p) \leftrightarrow \langle\geqslant 2\rangle\, p$ of the modal logic for local sabotage is a fluke, in being not schematically valid. But as we saw in Sect. 2, for just defining the winning positions of Traveler, this equivalence was the right observation to make.

5 Digression: Imperfect Information

One further test on the perspective offered here would be how well our analysis stands up to one very realistic feature of many games that we have ignored so far. Many games involve *imperfect information*, not necessarily in the sense of bounded rationality, which is, of course, unavoidable in practice, but in the sense that even ideal players may still have limited powers of observation in the graph. In contrast, our whole discussion has been about perfect information games.

In this section, that can be skipped without loss of continuity, we briefly discuss some graph games or game-like scenarios that involve imperfect information, and the resulting partial information flows during play.

Catch game with questions Suppose that a Traveler moves through a graph in which each point has a successor, one step at the time, and an Observer tries to catch him. In each round, the Observer can ask one question of the form 'Are you at point x in the graph?'. If the Traveler is at that point, the game is over, otherwise, Traveler makes a step again, and so on.

Depending on the graph, either the Traveler or Observer will have a winning strategy. For instance, evidently, in a three-point graph with the universal relation, Observer can never be sure to catch Traveler.

But there can be more interesting cases, as shown in the following figure.

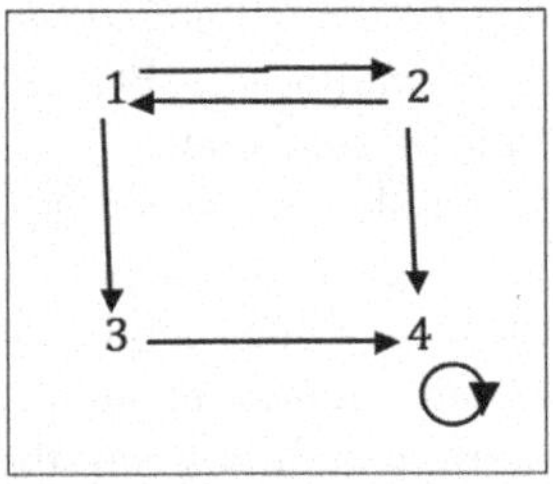

Here Observer has a winning strategy. How many questions does he need? Say, he asks in the following order: 1, 4, 3. The answer to the first question might be negative, so the Traveler is in {2, 3, 4} and the next hidden move then takes the Traveler into {1, 4}. Suppose the answer to the second question is negative again: then the Traveler is at 1, and his next move takes him into {2, 3}. If the answer to the third question is negative once more, Traveler is at 2, so his next move takes him into {1, 4}. Observer still does not know where Traveler is.

However, there is a much better strategy for Observer: just ask *twice* whether Traveler is at 1. As is easy to see, this will pinpoint Traveler's location.

This question scenario looks like a standard information-theoretic one, but the dynamics of moving in graph games add an unexpected flavor. This can be described in a logical framework combining a modal logic for travel steps with an *epistemic logic* for what Observer knows. Worlds in the models for this epistemic logic consist of the above graph with various locations for Traveler. Information flow can then be modeled by dynamic-epistemic techniques (van Benthem 2011).

Games like this resemble 'fugitive problems' from graph theory and computer science: cf. Goranko (2018) for background and structural characterization results.

Sabotage with imperfect observation The sabotage game, too, suggests natural variations with imperfect information. We can make Demon ignorant of where Traveler is, or vice versa, or we can let both players be ignorant of the precise location of the other. For a very simple example, merely to show what additional things can happen in contrast with the perfect information version, suppose that the graph is as follows: where 3 is the goal point of Traveler.

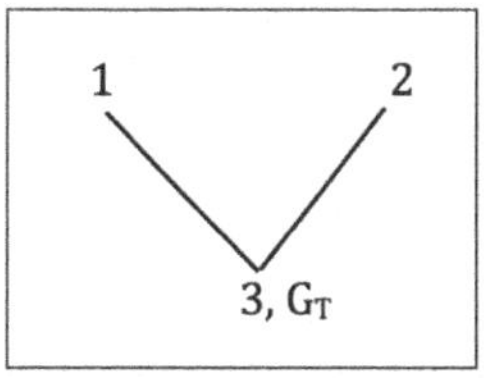

Demon does not know where Traveler is, but does get some information in each round: Traveler must announce whether he is in his goal point. Suppose that no such announcement is made at the start. Then Traveler must be in point 1 or 2. Demon does not know which of the two—so, whichever link he cuts, there is the possibility that Traveler may reach the goal. Clearly, no player has a winning strategy. Demon might happen to cut the right link, so Traveler is stuck. But cutting is not his winnting strategy.

All this is typical for games of imperfect information. To be sure, there is still a tight connection with the logic here. The information flow in sabotage games with imperfect information of various sorts, generated by many different scenarios, can be described completely in dynamic-epistemic logics (van Benthem 2011), with again, graphs with various positions for Traveler and various cuts made by Demon connected by epistemic uncertainty relations.[30]

But what changes is that equilibria can now be probabilistic, involving mixed strategies. For instance, in the above game, one equilibrium consists of Demon choosing a link to be cut with probability 0.5, while Traveler moves to his goal point if he can. Thus, our graph logics must be supplemented with probabilistic considerations when studying definability and computability (van Benthem and Klein 2019).

These are just a few examples of the pervasiveness of imperfect information. Many further natural variations on our graph games deserve investigation.[31] Also, as we have noted, the logic connection does not disappear, but appropriate game

[30]One concrete way in which players might have limited observational powers in graph games is through 'short sight' in the sense of Grossi and Turrini (2012). We might assume that localized players only see the graph up to a certain reachability depth around them. Again, these dynamics can be described in dynamic-epistemic terms (Liu et al. 2015).

[31]Other scenarios with imperfect information that may be connected to graph games are the Boolean Network Games of Seligman and Thompson (2015), with their connections with the epistemic friendship logic of Girard et al. (2014). The social network dynamics in this work is not the same as the graph dynamics in the present paper, since agents do not move, but can only change local properties triggered by those of their neighbors—but a serious comparison seems in order.

logics now require more complex languages and models bringing in techniques from dynamic-epistemic logic. The connection with logic games does not go away either, but one would now have to look at logic games with imperfect information, such as those proposed in frameworks like IF logic (Mann et al. 2011).

6 Conclusion

This paper has presented a way of thinking about graph games as a natural meeting place with logic. In doing so, we are not claiming any exclusive rights: graph games can equally well be studied in terms of game theory, probability theory, computer science, experimental cognitive science, and so on.[32] But a logical perspective adds a special flavor, as our discussion has tried to show, which leads to many new questions. Even so, we acknowledged that the matching of graph games to logics advocated here is not unique, and more importantly, that a complete logical system may be a heavy machinery to associate with a single concrete game. That machinery should be used with care and taste.

To back up our proposed methodology, we have shown by example how one can analyze and design games at the same time in tandem with ideas from logic. This unity of analysis and design is by no means new. It is a classical theme in computer science, in constructive mathematics, and in Euclid's geometry. We are merely taking a grand tradition to a compact, but hopefully rich arena.

Finally, if we had to point out one omission in our program, it is our a priori stance toward design. As we have noted a number of times, our graph games look like actual games that are played in reality, or like simulation games for practical purposes. In fact, society is one big laboratory where new games are designed all the time. A natural complementary challenge to what we have proposed is linking up with the real world of games and gaming. Obviously, this world is not quite that of this paper, since real games are played on fixed boards, and the variety in their play comes from other factors, such as chance moves, or cognitive bounds on players. But this does not mean that no significant links can be made: it is just that we have nothing to offer at this stage.

Acknowledgements This research is supported by Tsinghua University Initiative Scientific Research Program (2017THZWYX08).

References

Ågotnes, T., and H. van Ditmarsch. 2011. What will they say? Public announcement games. *Synthese KRA* 179 (1): 57–85.

[32]These approaches can also be combined in felicitous ways, witness the joint methodology of logical analysis of agent types, simulations, and cognitive experiments pursued in Ghosh (2018).

Aigner, M., and M. Fromme. 1984. A game of cops and robbers. *Discrete Applied Mathematics* 8 (1): 1–12.
Areces, C., F. Carreiro, S. Figueira, and S. Mera. 2011. Expressive power, and decidability for memory logics. In *Proceedings WoLLIC, 2008. Lecture notes in computer science*, vol. 6642, 20–34. Berlin: Springer.
Areces, C., R. Fervari, and G. Hoffmann. 2015. Relation-changing modal operators. *Logic Journal of the IGPL* 23: 601–627.
Areces, C., D. Figueira, S. Figueira, and S. Mera. 2008. Basic model theory, and for memory logics. In *Proceedings WoLLIC, 2011*. Lecture notes in computer science, vol. 5110, 56–68. Berlin: Springer.
Areces, C., K. Mierzewski, and F. Zaffora Blando. 2018 (in this volume). Poison logics and memory logics. Manuscript, Department of Philosophy, Stanford University.
Aucher, G., J. van Benthem, and D. Grossi. 2017. Modal logics of sabotage revisited. *Logic and Computation* 28 (2): 269–303.
van Benthem, J. 2011. *Logical dynamics of information and interaction*. Cambridge: Cambridge University Press.
van Benthem, J. 2014. *Logic in games*. Cambridge: The MIT Press.
van Benthem, J., K. Mierzewski, and F. Zaffora Blando. 2019. The modal logic of stepwise removal. To appear in Reports on Symbolic Logic.
van Benthem, J., B. ten Cate, and J. Väänänen. 2009. Lindström theorems for fragments of first-order logic. *Logical Methods in Computer Science* 5 (3): 1–27.
van Benthem, J., J. Gerbrandy, T. Hoshi, and E. Pacuit. 2009. Merging frameworks for interaction. *Journal of Philosophical Logic* 38 (5): 491–526.
van Benthem, J., and D. Klein. 2019. Logics for analyzing games. *Stanford on-line encyclopedia of philosophy*.
Chen, Y. 2018. Model theory for modal logics of definable point deletion. Department of Philosophy, Tsinghua University.
Diestel, R. 1997. *Graph theory*. Heidelberg: Springer.
Duchet, P., and H. Meyniel. 1993. Kernels in directed graphs: A poison game. *Discrete Mathematics* 115: 273–276.
Gabbay, D., and V. Shethman. 1998. Products of modal logics. *Logic Journal of the IGPL* 6: 73–146.
Ghosh, S. 2018. Strategizing: a meeting of methods. Chennai: Indian Institute of Statistics. Under submission.
Girard, P., F. Liu, and J. Seligman. 2014. Logical dynamics of belief change in the community. *Synthese* 191: 2403–2431.
Glendenning, L. 2005. Mastering quoridor. D.Sc. Thesis, University of New Mexico, Albuquerque.
Goranko, V. 2018. A characterization of observer-fugitive games. Philosophical Institute, University of Stockholm.
Grädel, E., W. Thomas, and T. Wilke (eds.). 2002. Automata, logics, and infinite games: A guide to current research. *Lecture notes in computer science*, vol. 2500. Berlin: Springer.
Grossi, D. 2013. Abstract argument games via modal logic. *Synthese* 190 (5): 5–29.
Grossi, D., and P. Turrini. 2012. Short sight in extensive games. In *Proceedings AAMAS 12*, Valencia, 805–812.
Grossi, D., and S. Rey. 2019. Credulous acceptability, poison games, and modal logic. In *Proceedings AAMAS 2019*, Montreal.
Gurevich, Y., and S. Shelah. 1986. Fixed-point extensions of first-order logic. *Annals of Pure and Applied Logic*, 32.
Li, D. 2018. Losing Connections: Modal logics of definable link deletion. Department of Philosophy, Tsinghua University. Presented at LOFT 13, Milano.
Liu, F. 2011. *Reasoning about preference dynamics*. Dordrecht: Springer Science Publishers.
Liu, F. 2017. Choice points for graph games. Seminar Note. Department of Philosophy, Tsinghua University.

Liu, C., F. Liu, and K. Su. 2015. A dynamic-logical characterization of solutions in sight-limited extensive games. In *Proceedings PRIMA*, 467–480.

Löding, C. and Ph. Rohde. 2003. Model checking and satisfiability for sabotage modal logic. In *Proceedings FSTTCS 2003*, Springer lecture notes in computer science, 302–313.

Mann, A., G. Sandu, and M. Sevenster. 2011. *Independence-friendly logic: A game- theoretic approach*. Cambridge: Cambridge University Press.

Marx, M., and Y. Venema. 1998. *Multi-dimensional modal logic*. Dordrecht: Springer Science Publishers.

Mierzewski, K. 2018. Sabotage in the random graph. Department of Philosophy, Stanford University.

Mierzewski, K., and F. Zaffora Blando. 2016 (in this volume). The modal logic(s) of poison games. Department of Philosophy, Stanford University. New version with co-author Carlos Areces to appear in *Proceedings Fourth Asian Workshop on Philosophical Logic*, Beijing. 2018.

Mostowski, A. 1991. Games with forbidden positions. Technical Report 78, University of Danzig, Institute of Mathematics and Informatics.

Nowakowski, R., and P. Winkler. 1983. Vertex-to-vertex pursuit in a graph. *Discrete Mathematics* 43: 235–239.

Osborne, M., and A. Rubinstein. 1994. *A course in game theory*. Cambridge: The MIT Press.

Parsons, T. 1976. Pursuit-evasion in a graph. *Theory and applications of graphs*. Springer lecture notes in mathematics, vol. 642, 426–441.

Segerberg, K. 1973. Two-dimensional modal logic. *Journal of Philosophical Logic* 2 (1): 77–96.

Seligman, J., and D. Thompson. 2015. Boolean network games and iterated boolean games. In *Proceedings LORI Taiwan*. Springer lecture notes in computer science, vol. 9394, 353–365.

Thompson, D. 2018. A modal logic of local graph change. Department of Philosophy, Stanford University. To appear in *Proceedings fourth Asian workshop on philosophical logic*, Beijing. 2018.

Zhang, T. 2018 (in this volume). *Localized sabotage algorithm*. Manuscript: Tsinghua University.

De Re, De Dicto, and Binding Modalities

Melvin Fitting

Abstract In classical logic, the move from propositional to quantificational is profound but essentially takes one route, following a direction we are all familiar with. In modal logic, such a move shoots off in many directions at once. One can quantify over things or over intensions. Quantifier domains can be the same from possible world to possible world, shrink or grow as one moves from a possible world to an accessible one, or follow no pattern whatsoever. A long time ago, Kripke showed us how shrinking or growing domains are related to validity of the Barcan and the converse Barcan formulas, bringing some semantic order into the situation. But when it comes to proof theory, things get somewhat strange. Nested sequents for shrinking or growing domains, or for constant domains or completely varying domains, are relatively straightforward. But axiomatically some oddities are quickly apparent. A simple combination of propositional modal axioms and rules with standard quantificational axioms and rules proves the converse Barcan formula, making it impossible to investigate its absence. Kripke showed how one could avoid this, at the cost of using a less common axiomatization of the quantifiers. But things can be complicated and even here an error crept into Kripke's work that wasn't pointed out until 20 years later, by Fine. Justification logic was created by Artemov, with a system called LP which is related to propositional S4. This was extended to a quantified version by Artemov and Yavorskaya, for which a possible world semantics was supplied by Fitting. Subsequently Artemov and Yavorskaya introduced what they called *binding modalities*, by transferring ideas back from quantified LP to S4. In this paper, we continue the investigation of binding modalities, but for K, and show that they provide a natural intuition for Kripke's non-standard axiomatization, and relate directly to the distinction between *de re* and *de dicto*. Unlike in Kripke's

This chapter is in its final form and it is not submitted to publication anywhere else. I want to thank Felipe Salvatore, whose joint work with myself, Fitting (2017), and accompanying discussions helped lead to the present paper.

M. Fitting (✉)
Department of Philosophy, Graduate Center, City University of New York,
New York City, NY, USA
e-mail: melvin.fitting@gmail.com
URL: https://www.melvinfitting.org

F. Liu et al. (eds.), *Knowledge, Proof and Dynamics*,
Logic in Asia: Studia Logica Library,
https://doi.org/10.1007/978-981-15-2221-5_8

treatment, the heavy lifting is done through a generalization of the modal operator, instead of a restriction on quantifier axiomatizations.

1 Introduction

In his famous 1963 paper Kripke (1963), Saul Kripke analyzed quantificational modal logic. His paper introduced the now familiar connection between the Barcan formula and monotonicity of domains, and the converse Barcan formula and anti-monotonicity of domains. While the analysis was primarily semantic, it was important to say something about connections with axiomatic approaches to quantified modal logics, which were already well-established in the literature. But here a problem arose. Prior had shown in (1956) that the Barcan formula was provable in first-order S5, and the converse Barcan formula was provable in first-order versions of modal logics more generally, meaning that Kripke's semantics apparently characterized some logical systems not corresponding to axiomatizations. Kripke traced the source of the problem to the innocuous looking universal instantiation axiom, $\forall x\varphi(x) \supset \varphi(y)$, and its interplay with the usual rule of necessitation in modal logics. He proposed that using a less common way of axiomatizing classical quantificational logic would solve the problem—work with a system in which only closed formulas can appear in proofs, and hence only closed formulas are provable. As he noted, such a system had already appeared in Quine's well-known (1940). With a correction added later by Kit Fine (1983), Kripke's proposal did work successfully for modal varying domain logics, as well as for first-order S5.

Remarkably, Kripke's solution was re-discovered independently many years later, albeit in disguise. This time, the source was justification logic, which is a kind of explicit counterpart of modal logic. It has had a long and substantial development—see Artemov and Fitting (2019, 2011) for details but, except for this introduction, they will not actually be needed in this paper. The idea behind justification logics is that the simple assertion that some formula is necessary, $\Box\varphi$, is replaced by an *explicit* version, $t{:}\varphi$, informally asserting that φ is necessary for reason t, or that t is a justification for the necessity of φ. Most modal logics are now known to have justification logic counterparts, and there is a precise characterization of what it means to be a counterpart. The use of a justification logic counterpart for S4, known as LP for *logic of proofs*, allowed Artemov to complete Gödel's program of finding an arithmetical semantics for propositional intuitionistic logic (Artemov 2001).

All justification logics were propositional until 2011, Artemov and Yavorskaya (2011), when Sergei Artemov and Tatiana Yavorskaya created a quantified version of LP, allowing them to extend the work on arithmetic semantics for intuitionistic logic. A possible world semantics for their system was introduced in Fitting (2014). A central problem that Artemov and Yavorskaya solved was how to formulate a quantified system so that what is called *internalization* would be provable, where internalization is a justification logic analog of the necessitation rule for modal logics. The key lays in discovering how to handle internalization for applications of the universal generalization rule. Following a proof theoretical motivation, they separated two different roles that free variables play in first-order proofs. On the one hand, they serve as place holders, so that a proof can be seen as a kind of proof template from which one can manufacture proof after proof by substituting different terms for instances of a free variable throughout the template. The other role played by a free variable is that of a formal symbol subject to binding using the universal generalization rule. The two roles are, in fact, incompatible. Artemov and Yavorskaya introduced syntactic machinery to distinguish the two roles.

It was clear that the justification logic machinery Artemov and Yavorskaya introduced could be modified to apply to modal logic itself, and this was done in Artemov and Yavorskaya (2013), where the machinery acquired the name *binding modalities*. They provided a binding modality version of quantified $\mathsf{S4}$, including monotonicity. They worked with $\mathsf{S4}$ because this logic is the modal counterpart of the first-order LP justification logic they had created earlier. As it happens, $\mathsf{S4}$ has a very well behaved binding modalities version, so the connection with Kripke's ideas was missed. Here we will see that binding modalities for K need ideas that exactly mirror those of Kripke. But for both Kripke and Artemov/Yavorskaya, the approach taken was basically a proof theoretic one. We will see that when approached semantically, binding modalities have a very natural meaning in their own right. As was, in fact, noted in Artemov and Yavorskaya (2013), binding modalities provide an intuitively plausible way of capturing aspects of the familiar *de re*/*de dicto* distinction. In this paper, the collection of ideas and machinery just discussed will be examined, with *de re* and *de dicto* as the pegs on which everything hangs. We work specifically with varying domain K. The choice of K itself plays no special role other than being the weakest normal modal logic and hence at the base of all the others. We discuss briefly how other domain conditions can be brought in. We also provide an axiomatization and prove soundness and completeness for varying domain K.

2 Binding Modalities—*De Re*/*De Dicto*

The familiar *de re*/*de dicto* distinction has been a long-standing bone of contention in intentional logic. It is a problem with a four-valued solution space: accept the distinction, reject it, or reduce one notion to the other, which could be done in either of two directions. We present one way of embracing the distinction, hopefully with minimum disruption but with some nice technical benefits. We begin with a discussion of one of the problems orbiting around *de re*/*de dicto*.

We use Kripke modal models somewhat informally for now, with formal details later on. The models are first-order, and we will think of them as having varying domains. No special assumptions are made about accessibility, so our models are appropriate for K. We write $\mathcal{M}, \Gamma \Vdash \varphi$ to mean formula φ is true at possible world Γ of modal model $\mathcal{M}$. Our problem comes from the presence of free variables in φ; how should we understand $\mathcal{M}, \Gamma \Vdash \varphi(x)$? Of course the usual solution is to assume we have a valuation in mind, so free variables get assigned domain members, and the problem is avoided. But still, the question makes sense as asked. For instance, one commonly understands the assertion that $\varphi(x)$ is true in a *classical* model as meaning it is true with respect to every assignment of values to x from the domain of the model. Very loosely, this corresponds to the rule of universal generalization. Modally we could do a similar thing, provided there are no modal operators present. But how should we understand something like $\mathcal{M}, \Gamma \Vdash \Box\varphi(x)$? (For simplicity in the discussion let us suppose that in $\Box\varphi(x)$ the initial $\Box$ is the only modal operator present.) The formula is ambiguous, with two plausible meanings.

Meaning 1: $\mathcal{M}, \Gamma \Vdash \Box\varphi(x)$ is to be taken as true provided that for every assignment of a value, say a, from the quantificational domain associated with Γ, and for every accessible possible world Δ, $\mathcal{M}, \Delta \Vdash \varphi(x)$ is true when x is assigned the value a. Briefly, $\mathcal{M}, \Delta \Vdash \varphi(a)$ is true for all accessible Δ and all a in the domain of Γ. This amounts to a *de re* usage. We choose an individual, a, in Γ, and ask how that individual behaves in Δ.

Meaning 2: $\mathcal{M}, \Gamma \Vdash \Box\varphi(x)$ is to be taken as true provided $\mathcal{M}, \Delta \Vdash \varphi(x)$ is true for every accessible world Δ (just as we understand things in the propositional modal case). Since no further modal operators are present, this simply means $\mathcal{M}, \Delta \Vdash \varphi(a)$ is true for every assignment of a value a to x, where a is taken from the domain of Δ. This is a *de dicto* reading since we have reduced things to the truth of a formula at Δ, with no individuals identified across worlds.

The two meanings are not identical, since the domains associated with Γ and with the various accessible worlds Δ need not be the same.

Binding Modalities were introduced in Artemov and Yavorskaya (2012, 2013). Here is the idea, very informally. Suppose φ is a formula and X *is a finite set of variables*. Then $\Box_X\varphi$ counts as a formula. To say it is valid is to say it holds universally when free variables in φ that appear in X are understood in a *de re* sense and free variables in φ that are not in X are understood in a *de dicto* sense. Now it is time for the formal details.

3 Binding Modalities—Syntax

We assume a mostly standard first-order modal language. We have relation symbols of all arities, typically written $P, Q, \ldots$. We do not have function symbols or constants. We have a countable set of individual variables $\mathbf{V} = \{x_1, x_2, x_3, \ldots\}$, and we will generally write $x, y, \ldots$ for arbitrary members of $\mathbf{V}$. An atomic formula is a relation symbol followed by the appropriate number of individual variables separated by commas, or is $\bot$, representing falsehood. Formulas are built up from atomic formulas using the logical connectives, $\wedge, \vee, \supset$, and the universal quantifier (we take negation and the existential quantifier to be defined). We write quantified formulas as $\forall x\varphi$, where x is an individual variable and φ is a formula. The central new rule of formation is the following.

Modal Formation Rule: For every finite subset X of $\mathbf{V}$ and for every formula φ, the modality $\Box_X\varphi$ is a formula.

An objection might be raised that, since finite sets are involved, $\Box_X\varphi$ is not really syntax. This is actually a minor point. Every finite set of variables can be coded syntactically, say by a list of its members separated by commas. We would also need axioms to say that order change and repetition of members produces equivalent results. This would not be hard to do, but would seriously obscure the central points. We will use finite sets as honorary syntactic objects, in the same spirit that they are commonly used when working with Gentzen sequents.

Notation Convention: For a finite set of free variables X and a single free variable y, we write $\Box_{Xy}\varphi$ as shorthand for $\Box_{X\cup\{y\}}\varphi$, and *it is assumed that* y *does not occur in* X.

For free and bound variable occurrences, we use the standard definition, augmented with the following.

Free 1: All variable occurrences in the finite set X are free occurrences in $\Box_X\varphi$.
Free 2: All free variable occurrences in the formula φ involving variables that are in the set X are free occurrences in $\Box_X\varphi$.

For example, suppose $\varphi(x, z)$ has free occurrences of x and z. Let ψ be the formula $\Box_{\{x,y\}}\varphi(x, z)$. In ψ, the occurrences of x and y in the subscript of $\Box$ are free. Also in ψ, all free occurrences of x in $\varphi(x, z)$ are free, and no occurrences of z are free. Note that the *set* of variables occurring free in a formula $\Box_X\varphi$ is simply X.

We also need the notion of a variable y being *free for* occurrences of x in a formula φ, which informally means that replacing free x occurrences by occurrences of y won't lead to any accidental variable binding. The definition has its usual clauses, see Mendelson (1997) for instance. In addition, we have the following.

Free For: An individual variable y is free for x in $\Box_X\varphi$ provided y is free for x in φ and, either y does not occur free in φ or $y \in X$.

We will write $\varphi(x/y)$ to denote the result of substituting variable y for free occurrences of x in $\varphi(x)$, where it is understood that y is free for x in φ. We often write $\vec{x}, \vec{y}, \vec{z}, \ldots$ for finite sequences of individual variables, where sequence length can be inferred from context, and we write $\varphi(\vec{x}/\vec{y})$ to denote the result of substituting terms of $\vec{y}$ for free occurrences of the corresponding terms of $\vec{x}$ in φ, all with appropriate free for conditions understood. If no confusion will result, we may write $\varphi(\vec{x}/\vec{y})$ as $\varphi(\vec{y})$ provided the formula $\varphi(\vec{x})$ is understood from context.

4 Binding Modalities—Semantics

First-order modal models are generally familiar, and are what we use even though binding modalities are being considered. We give the basic varying domain semantics definitions, partly to establish notation. In Sect. 5, we connect conditions involving binding modalities to conditions on the domain function, with monotonicity and anti-monotonicity playing their familiar role.

Definition 4.1 (*Skeleton*) A skeleton is a structure $\langle \mathcal{G}, \mathcal{R}, \mathcal{D} \rangle$ where $\mathcal{G}$ is a non-empty set (of possible worlds or states), $\mathcal{R}$ is a binary relation on $\mathcal{G}$ (of accessibility), and $\mathcal{D}$ is a function that maps members of $\mathcal{G}$ to non-empty sets (domains). The *domain of the skeleton* is defined to be $\cup_{w \in \mathcal{G}} \mathcal{D}(w)$.

Definition 4.2 (*Model*) A (first-order modal) model is a structure $\mathcal{M} = \langle \mathcal{G}, \mathcal{R}, \mathcal{D}, \mathcal{I} \rangle$ where $\langle \mathcal{G}, \mathcal{R}, \mathcal{D} \rangle$ is a skeleton (on which the model is *based*) and $\mathcal{I}$ is an *interpretation function*, assigning to each n-ary relation symbol P and to each possible world $\Gamma \in \mathcal{G}$ an n-ary relation $\mathcal{I}(P, \Gamma)$ on the domain of the skeleton. In this context, we also refer to the domain of the skeleton as *the domain of the model* and denote it by $\mathcal{D}_{\mathcal{M}}$.

Note that the interpretation function is not restricted, at a world, only to members of the domain of that world. For instance, for a one-place relation symbol P, we could have $a \in \mathcal{I}(P, \Gamma)$ while $a \notin \mathcal{D}(\Gamma)$, though we must have $a \in \mathcal{D}_{\mathcal{M}}$, and so $a \in \mathcal{D}(\Delta)$ for some $\Delta \in \mathcal{G}$. Conditions can be placed on accessibility relations but we do not do so now, which means that at the propositional level, our modal logic is K. Also conditions can be placed on domain functions but we do not do this either (yet).

It is standard to use *valuations* when working with first-order models. Valuations map free variables to members of a domain, and we commonly talk about the truth value of a formula φ with respect to a valuation v. In this paper, we follow an alternate convention, allowing members of the domain of a model to appear directly in formulas as if they were constants; in fact we call them *model constants*. We call formulas containing these model constants *model formulas*. Very strictly speaking, a model formula is not a formula, but any assertion about it can easily be rephrased using proper formulas and valuations. In a model formula, members of the domain of the model can only appear in what would be free variable positions. A formula, as usually understood, is simply a model formula with no members of the domain of a model appearing as constants. The following says all this more formally.

Definition 4.3 (*Model Formula*) Let $\mathcal{M}$ be a model. A *model formula of* $\mathcal{M}$, or more simply an $\mathcal{M}$ *formula*, is the result of substituting members of $\mathcal{D}_{\mathcal{M}}$ for some or all of the free variable occurrences in a formula. We refer to members of $\mathcal{D}_{\mathcal{M}}$ as *model constants* in their appearance in a modal formula. If the context makes it clear which model is involved, we may refer to a model formula of $\mathcal{M}$ simply as a model formula.

We extend the notation for substitution, from Sect. 3. Let $\mathcal{M} = \langle \mathcal{G}, \mathcal{R}, \mathcal{D}, \mathcal{I} \rangle$ be a model. We write $\vec{a}, \vec{b}, \vec{c}, \ldots$ for finite sequences of members of the domain $\mathcal{D}_{\mathcal{M}}$ of the model. The notation $\varphi(\vec{x}/\vec{a})$ denotes the model formula in which members of $\vec{a}$ replace free occurrences of corresponding members of $\vec{x}$ in φ. Since model constants are not variables, we can think of a model constant as being automatically free for any variable in a formula.

It is convenient to write $\vec{a} \in \mathcal{D}_{\mathcal{M}}$ instead of $\vec{a} \in \mathcal{D}^n_{\mathcal{M}}$. Likewise we write $\{\vec{x}\}$ for the set whose members are the terms of $\vec{x}$, and similarly for $\{\vec{a}\}$. If $\varphi(\vec{a})$ contains no free individual variables, we say it is a *closed model formula*. A closed model formula with no members of the model domain appearing as constants is simply a closed formula in the usual sense.

Definition 4.4 (*Truth in a Model*) Let $\mathcal{M} = \langle \mathcal{G}, \mathcal{R}, \mathcal{D}, \mathcal{I} \rangle$ be a model, and let φ be an $\mathcal{M}$ formula. We write $\mathcal{M}, \Gamma \Vdash \varphi$ to indicate that φ is true at possible world Γ of model $\mathcal{M}$. The conditions defining this are as follows.

(0) Let $\vec{x}$ be all the free variables of $\varphi(\vec{x})$; $\mathcal{M}, \Gamma \Vdash \varphi(\vec{x})$ if and only if $\mathcal{M}, \Gamma \Vdash \varphi(\vec{a})$ for every $\vec{a} \in \mathcal{D}(\Gamma)$. (This reduces truth for formulas with free variables to truth for closed model formulas. These, in turn, are covered by the following.)
(1) Let $P\vec{a}$ be an atomic closed model formula, where $\vec{a} \in \mathcal{D}_{\mathcal{M}}$. $\mathcal{M}, \Gamma \Vdash P\vec{a}$ if and only if $\vec{a} \in \mathcal{I}(P, \Gamma)$
(2) $\mathcal{M}, \Gamma \nVdash \bot$.
(3) Let $\varphi \supset \psi$ be a closed model formula; $\mathcal{M}, \Gamma \Vdash \varphi \supset \psi$ if and only if $\mathcal{M}, \Gamma \nVdash \varphi$ or $\mathcal{M}, \Gamma \Vdash \psi$ (and similarly for the other propositional connectives).
(4) Let $\forall x \varphi(x)$ be a closed model formula; $\mathcal{M}, \Gamma \Vdash \forall x \varphi(x)$ if and only if $\mathcal{M}, \Gamma \Vdash \varphi(a)$ for each $a \in \mathcal{D}(\Gamma)$.
(5) Let $\Box_{\{\vec{a}\}}\varphi(\vec{x}, \vec{a})$ be a closed model formula; $\mathcal{M}, \Gamma \Vdash \Box_{\{\vec{a}\}}\varphi(\vec{x}, \vec{a})$ if and only if $\mathcal{M}, \Delta \Vdash \varphi(\vec{x}, \vec{a})$ for every $\Delta \in \mathcal{G}$ with $\Gamma\mathcal{R}\Delta$. (Note that this reduces things back to case 0.)

Definition 4.5 (*Validity*) For a model $\mathcal{M}$ and an $\mathcal{M}$ formula φ, we say φ is *valid in* $\mathcal{M}$ if it is true at every possible world Γ of $\mathcal{M}$. φ is simply *valid* if it is valid in every model $\mathcal{M}$ such that φ is an $\mathcal{M}$ formula.

Example 4.6 Suppose $\varphi(\vec{a}, \vec{x})$ is an $\mathcal{M}$ formula, where $\mathcal{M} = \langle \mathcal{G}, \mathcal{R}, \mathcal{D}, \mathcal{I} \rangle$. Assume $\vec{a}$ are all the model constants, and $\vec{x}$ are all the free variables in the formula. Invoking case (0) from Definition 4.4, to say $\varphi(\vec{a}, \vec{x})$ is valid in $\mathcal{M}$ is to say that, at each world $\Gamma \in \mathcal{G}$, and for each $\vec{b} \in \mathcal{D}(\Gamma)$, $\mathcal{M}, \Gamma \Vdash \varphi(\vec{a}, \vec{b})$.

Example 4.7 Let $\mathcal{M} = \langle \mathcal{G}, \mathcal{R}, \mathcal{D}, \mathcal{I} \rangle$ be a model where $\mathcal{G} = \{\Gamma, \Delta\}$, under $\mathcal{R}$ each possible world is accessible from the other, $\mathcal{D}(\Gamma) = \{a\}$, $\mathcal{D}(\Delta) = \{b\}$, $\mathcal{I}(P, \Gamma) = \{a\}$, and $\mathcal{I}(P, \Delta) = \{b\}$ (where $a \neq b$). Schematically we have what is shown in Fig. 1.

In this model, $\mathcal{M}, \Gamma \Vdash Px$ because every instance from the domain of Γ is true at Γ, namely Pa. Similarly at Δ, so Px is valid in this model. Further, $\mathcal{M}, \Gamma \Vdash \Box_{\emptyset} Px$ because Px is true at all worlds accessible from Γ, that is, at Δ. On the other hand, if we had $\mathcal{M}, \Gamma \Vdash \Box_{\{x\}} Px$, by Definition 4.4 case (0) we should have every Γ instance of Px true at Δ, and hence $\mathcal{M}, \Delta \Vdash Pa$, and we do not have this. Thus $\mathcal{M}, \Gamma \nVdash \Box_{\{x\}} Px$, and similarly $\mathcal{M}, \Delta \nVdash \Box_{\{x\}} Px$.

Fig. 1 A simple model

$\Gamma\ \boxed{a} \Vdash Pa \longleftrightarrow \Delta\ \boxed{b} \Vdash Pb$

5 *De Re/De Dicto* Conversions

As we have said, the idea is that binding modalities distinguish between individual variables playing a *de dicto* role and those playing a *de re* role. More specifically, in $\Box_{\{\vec{x}\}}\varphi(\vec{x}, \vec{y})$, the variables of $\vec{x}$ are *de re* while those of $\vec{y}$ are *de dicto*. Various people have, from time to time, made a case that *de dicto* usages could be converted into *de re*, or conversely. In this section, we look at what direct conversions amount to, in terms of binding modalities.

We get a trivial case out of the way first. In considering issues of *de re* and *de dicto* conversion, we have vacuous occurrences for free. The proof is quite simple, but will serve to set up notation and machinery for subsequent use.

Theorem 5.1 $\Box_{Xy}\varphi \equiv \Box_X\varphi$ *is valid provided y is a free variable that does not occur free in* φ.

Proof Let $\mathcal{M} = \langle \mathcal{G}, \mathcal{R}, \mathcal{D}, \mathcal{I} \rangle$ and assume $\Box_{Xy}\varphi \equiv \Box_X\varphi$ is an $\mathcal{M}$ formula. To keep notation a bit simpler, we assume X contains no model constants. Let $\Gamma \in \mathcal{G}$; we show $\mathcal{M}, \Gamma \Vdash \Box_{Xy}\varphi \equiv \Box_X\varphi$. Let $X = \vec{x}$, and let $\vec{z}$ be the free variables in φ other than those in $\vec{x}$, so $\varphi = \varphi(\vec{x}, \vec{z})$ (since y is not free in φ).

Suppose $\mathcal{M}, \Gamma \Vdash \Box_{\{\vec{x}\}y}\varphi(\vec{x}, \vec{z})$. Then for each $\vec{a}, b \in \mathcal{D}(\Gamma)$, $\mathcal{M}, \Gamma \Vdash \Box_{\{\vec{a}\}b}\varphi(\vec{a}, \vec{z})$, and so for each $\Delta \in \mathcal{G}$ with $\Gamma\mathcal{R}\Delta$, $\mathcal{M}, \Delta \Vdash \varphi(\vec{a}, \vec{z})$. Since Δ was arbitrary, $\mathcal{M}, \Gamma \Vdash \Box_{\{\vec{a}\}}\varphi(\vec{a}, \vec{z})$, and since $\vec{a}$ was arbitrary, $\mathcal{M}, \Gamma \Vdash \Box_{\{\vec{x}\}}\varphi(\vec{x}, \vec{z})$. The converse implication is similar. ■

Now we move on to the serious cases, where vacuous occurrences are not guaranteed.

Definition 5.2 (*De Re/De Dicto Conversion*) This is first at the level of models and then at the level of skeletons.

Model Level: Let $\mathcal{M} = \langle \mathcal{G}, \mathcal{R}, \mathcal{D}, \mathcal{I} \rangle$ be a model. We say $\mathcal{M}$ admits *de re* to *de dicto* conversion provided, for every $\mathcal{M}$ formula φ, for every finite set X, for every variable y (note, not a model constant), and for every $\Gamma \in \mathcal{G}$, if $\mathcal{M}, \Gamma \Vdash \Box_{Xy}\varphi$ then $\mathcal{M}, \Gamma \Vdash \Box_X\varphi$. Under similar circumstances, $\mathcal{M}$ admits *de dicto* to *de re* conversion if $\mathcal{M}, \Gamma \Vdash \Box_X\varphi$ implies $\mathcal{M}, \Gamma \Vdash \Box_{Xy}\varphi$.

Skeleton Level: Let $\mathcal{S} = \langle \mathcal{G}, \mathcal{R}, \mathcal{D} \rangle$ be a skeleton. We say $\mathcal{S}$ admits *de re* to *de dicto* conversion if every model based on $\mathcal{S}$ admits *de re* to *de dicto* conversion. Likewise, $\mathcal{S}$ admits *de dicto* to *de re* conversion if every model based on $\mathcal{S}$ admits *de dicto* to *de re* conversion.

In propositional modal logic, it is standard to associate frame conditions with the validity of particular modal schemes in all models based on frames meeting those conditions. In the first-order setting, we have the additional machinery of quantifier domains and, as Kripke showed us, these too can be made use of in similar ways, with skeletons from Definition 4.1, replacing frames.

Definition 5.3 (*Monotonicity and Anti-monotonicity*) Let $\mathcal{S} = \langle \mathcal{G}, \mathcal{R}, \mathcal{D} \rangle$ be a skeleton. $\mathcal{S}$ is *monotonic* if, for every $\Gamma, \Delta \in \mathcal{G}$, $\Gamma \mathcal{R} \Delta$ implies $\mathcal{D}(\Gamma) \subseteq \mathcal{D}(\Delta)$. $\mathcal{S}$ is *anti-monotonic* if, for every $\Gamma, \Delta \in \mathcal{G}$, $\Gamma \mathcal{R} \Delta$ implies $\mathcal{D}(\Delta) \subseteq \mathcal{D}(\Gamma)$. A model is monotonic or anti-monotonic if the skeleton on which it is based is monotonic or anti-monotonic.

Theorem 5.4 *Let $\mathcal{S}$ be a skeleton.*

De Dicto *to* De Re*:* $\mathcal{S}$ *admits* de dicto *to* de re *conversion if and only if $\mathcal{S}$ is monotonic,*

De Re *to* De Dicto*:* $\mathcal{S}$ *admits* de re *to* de dicto *conversion if and only if $\mathcal{S}$ is anti-monotonic.*

Proof For the following, assume $\mathcal{S} = \langle \mathcal{G}, \mathcal{R}, \mathcal{D} \rangle$ is a skeleton.

De Dicto to *De Re* $\Leftarrow$ monotonic: Suppose $\mathcal{S}$ is monotonic and $\mathcal{M}$ is built on $\mathcal{S}$. Let $\Gamma \in \mathcal{G}$, and suppose $\mathcal{M}, \Gamma \Vdash \Box_{\{\vec{a}\}}\varphi(\vec{a}, \vec{x}, y)$; we show $\mathcal{M}, \Gamma \Vdash \Box_{\{\vec{a}\}y}\varphi(\vec{a}, \vec{x}, y)$. To show this, let b be an arbitrary member of $\mathcal{D}(\Gamma)$; we show $\mathcal{M}, \Gamma \Vdash \Box_{\{\vec{a}\}b}\varphi(\vec{a}, \vec{x}, b)$. And to show this, let $\Delta \in \mathcal{G}$ be arbitrary with $\Gamma \mathcal{R} \Delta$; we show $\mathcal{M}, \Delta \Vdash \varphi(\vec{a}, \vec{x}, b)$.
Since $b \in \mathcal{D}(\Gamma)$ we have $b \in \mathcal{D}(\Delta)$ by monotonicity. And since $\mathcal{M}, \Gamma \Vdash \Box_{\{\vec{a}\}}\varphi(\vec{a}, \vec{x}, y)$ then $\mathcal{M}, \Delta \Vdash \varphi(\vec{a}, \vec{c}, d)$ for every $\vec{c}, d \in \mathcal{D}(\Delta)$. In particular, $\mathcal{M}, \Delta \Vdash \varphi(\vec{a}, \vec{c}, b)$ for every $\vec{c} \in \mathcal{D}(\Delta)$, and hence $\mathcal{M}, \Delta \Vdash \varphi(\vec{a}, \vec{x}, b)$.

De Dicto to *De Re* $\Rightarrow$ monotonic: Suppose $\mathcal{S}$ is not monotonic; we show it does not admit *de dicto* to *de re* conversion. Since $\mathcal{S}$ is not monotonic, there are $\Gamma, \Delta \in \mathcal{G}$ with $\Gamma \mathcal{R} \Delta$ but $\mathcal{D}(\Gamma) \not\subseteq \mathcal{D}(\Delta)$; say $d \in \mathcal{D}(\Gamma)$ but $d \notin \mathcal{D}(\Delta)$. Let P be a unary predicate symbol, and build a model $\mathcal{M} = \langle \mathcal{G}, \mathcal{R}, \mathcal{D}, \mathcal{I} \rangle$ on $\mathcal{S}$ by setting the interpretation $\mathcal{I}$ so that P is interpreted at every world as being true of exactly the members of the domain of that world, and the interpretation of other predicate symbols doesn't matter. We have $\mathcal{M}, \Gamma \Vdash \Box_{\emptyset}P(y)$ because, for every Ω with $\Gamma \mathcal{R} \Omega$, we have $\mathcal{M}, \Omega \Vdash P(y)$. But we do not have $\mathcal{M}, \Gamma \Vdash \Box_{\{y\}}P(y)$ because $d \in \mathcal{D}(\Gamma)$ but we do not have $\mathcal{M}, \Gamma \Vdash \Box_{\{d\}}P(d)$, and we do not have this because we have $\Gamma \mathcal{R} \Delta$ but we do not have $\mathcal{M}, \Delta \Vdash P(d)$ since $d \notin \mathcal{D}(\Delta)$.

De Re to *De Dicto* $\Leftarrow$ anti-monotonic: This is very similar to the monotonic case. Suppose $\mathcal{S}$ is anti-monotonic and $\mathcal{M}$ is built on $\mathcal{S}$. Let $\Gamma \in \mathcal{G}$, and suppose $\mathcal{M}, \Gamma \Vdash \Box_{\{\vec{a}\}y}\varphi(\vec{a}, \vec{x}, y)$; we show $\mathcal{M}, \Gamma \Vdash \Box_{\{\vec{a}\}}\varphi(\vec{a}, \vec{x}, y)$. To show this, let $\Gamma \mathcal{R} \Delta$, where Δ is arbitrary; we must show that $\mathcal{M}, \Delta \Vdash \varphi(\vec{a}, \vec{x}, y)$. And to show this, let $d \in \mathcal{D}(\Delta)$ be arbitrary; we must show $\mathcal{M}, \Delta \Vdash \varphi(\vec{a}, \vec{x}, d)$. Since we have anti-monotonicity, $d \in \mathcal{D}(\Gamma)$, and since we have $\mathcal{M}, \Gamma \Vdash \Box_{\{\vec{a}\}y}\varphi(\vec{a}, \vec{x}, y)$, we have $\mathcal{M}, \Gamma \Vdash \Box_{\{\vec{a}\}d}\varphi(\vec{a}, \vec{x}, d)$, and $\mathcal{M}, \Delta \Vdash \varphi(\vec{a}, \vec{x}, d)$ follows.

De Re to *De Dicto* $\Rightarrow$ anti-monotonic: We show the contrapositive, with an argument rather similar to the monotonic case. Suppose $\mathcal{S}$ is not anti-monotonic, and $\Gamma, \Delta \in \mathcal{G}$, $\Gamma \mathcal{R} \Delta$, but $\mathcal{D}(\Delta) \not\subseteq \mathcal{D}(\Gamma)$, say $d \in \mathcal{D}(\Delta)$ but $d \notin \mathcal{D}(\Gamma)$. Let P be a unary predicate symbol, and build a model $\mathcal{M} = \langle \mathcal{G}, \mathcal{R}, \mathcal{D}, \mathcal{I} \rangle$ on $\mathcal{S}$ by setting the interpretation $\mathcal{I}$ so that P is interpreted at every world as being true of exactly the members of $\mathcal{D}(\Gamma)$.

We have $\mathcal{M}, \Gamma \Vdash \Box_{\{y\}} P(y)$ because we have $\mathcal{M}, \Gamma \Vdash \Box_{\{e\}} P(e)$ for every $e \in \mathcal{D}(\Gamma)$, and we have this because for every Ω we have $\mathcal{M}, \Omega \Vdash P(e)$ for all $e \in \mathcal{D}(\Gamma)$. But if we had $\mathcal{M}, \Gamma \Vdash \Box_{\emptyset} P(y)$, since $\Gamma \mathcal{R} \Delta$ we must have $\mathcal{M}, \Delta \Vdash P(y)$, and hence we must have $\mathcal{M}, \Delta \Vdash P(d)$ since $d \in \mathcal{D}(\Delta)$, but we do not since $d \notin \mathcal{D}(\Gamma)$ and so d is not in the interpretation of P at Δ. ■

6 Binding Modalities—Varying Domain Axiomatics

A modal logic incorporating binding modalities was first axiomatized in Artemov and Yavorskaya (2012, 2013). It derived from work in Artemov and Yavorskaya (2011), Artemov and Fitting (2019), Chap. 10 on the quantified version of the justification logic LP, which itself is related to propositional S4. Consequently, the binding modalities logic that Artemov and Yavorskaya presented was a quantified version of S4. In their approach, what we called *De Dicto* to *De Re* conversion was built into the axiomatization, and this considerably simplified things. Here we want to be completely general, so we give axioms and rules for a first-order version of modal K with binding modalities, but with no assumptions of *De Re*/*De Dicto* conversion. The intended varying domain semantics is as described in Sect. 4. Since binding modalities are not standard, there is no established naming convention here. We will call this logic FOBMK, for "first-order, binding modalities, K". Because of the use of binding modalities, and the distinctions it allows, the underlying axiomatization of classical first-order logic can be of the more conventional sort, allowing free variables.

The rule of necessitation is a problem, as we might expect. If we were to apply it in its usual form to $\forall x \varphi(x) \supset \varphi(y)$, to get $\Box(\forall x \varphi(x) \supset \varphi(y))$, we have a free variable, y, in the scope of the necessity sign, and should $\varphi(y)$ be understood in the *de re* or *de dicto* sense? Of course this is exactly the issue that binding modalities were designed to address. It is well-known that for propositional modal logics, the necessitation rule can be replaced by the assumption that for each axiom, we also have that axiom with any number of prefixed necessity operators. We will do something similar here. This allows us considerable control because we do not have to treat each axiom the same way, and in fact we do not.

The axioms and rules are given in Fig. 2. The two rules called *axiom necessitation* could really have been included in the axiom list, but they replace a conventional rule of necessitation, so we called them rules to emphasize that.

The usual proof of the distributivity of necessitation over conjunction goes through for FOBMK with no changes. Thus we have $\vdash \Box_X(\varphi \wedge \psi) \equiv (\Box_X \varphi \wedge \Box_X \psi)$ for every finite set X of individual variables.

Since we have two versions of axiom necessitation, we do not get a derived necessitation rule in full generality. Still, the use of binding modalities allows us to get a partial version. This can be strengthened when we take monotonicity and anti-monotonicity into account, as we will see in Sect. 7.

FOBMK–A-1 All tautologies

FOBMK–A-2 $\forall x\varphi(x) \supset \varphi(y)$ provided y is free for x in φ

FOBMK–A-3 $\forall x(\varphi \supset \psi) \supset (\varphi \supset \forall x\psi)$ provided x does not occur free in φ

FOBMK–A-4 $\Box_{Xy}\varphi \equiv \Box_X\varphi$ provided y does not occur free in φ

FOBMK–A-5 $\Box_X\varphi \supset \Box_X\forall y\varphi$ provided $y \notin X$

FOBMK–A-6 $\Box_X(\varphi \supset \psi) \supset (\Box_X\varphi \supset \Box_X\psi)$

FOBMK–R-1 (*modus ponens*) $\vdash \varphi,\ \vdash \varphi \supset \psi \ \Rightarrow \vdash \psi$

FOBMK–R-2 (*universal generalization*) $\vdash \varphi \ \Rightarrow \vdash \forall x\varphi$

FOBMK–R-3 (*axiom necessitation 1*) For every axiom φ except for instances of FOBMK–A-2, $\vdash \Box_{X_1}\Box_{X_2}\ldots\Box_{X_n}\varphi$ for $n > 0$ and arbitrary finite sets $X_1, X_2, \ldots, X_n$

FOBMK–R-4 (*axiom necessitation 2*) For an instance of FOBMK–A-2, $\forall x\varphi(x) \supset \varphi(y)$ where y is free for x in φ, $\vdash \Box_{X_1}\Box_{X_2}\ldots\Box_{X_n}(\forall x\varphi(x) \supset \varphi(y))$ where $n > 0$, $X_1, X_2, \ldots, X_n$ are arbitrary finite sets, and $y \notin X_n$

Fig. 2 FOBMK axioms and rules

Theorem 6.1 (FOBMK Necessitation) *Let X be a finite set of individual variables. Call an* FOBMK *axiomatic proof X* acceptable *provided that whenever $\forall x\varphi(x) \supset \varphi(y)$ occurs in the proof, where y is free for x in φ, then $y \notin X$. In* FOBMK, *if $\vdash \varphi$ and there is an X acceptable proof, then $\vdash \Box_X\varphi$.*

Proof The argument is by induction on proof length. Assume the result is known for proofs whose length is less than i and we now have a proof of length i. Let $\varphi_1, \varphi_2, \ldots, \varphi_i$ be an X acceptable FOBMK proof. We show $\Box_X\varphi_i$ is provable. There are several cases.

If φ_i is an axiom or was added to the proof using FOBMK–R-3 or FOBMK–R-4, we have $\Box_X\varphi_i$ using either FOBMK–R-3 or FOBMK–R-4.

For FOBMK–R-1, suppose φ_i follows from earlier lines of the proof φ_j and $\varphi_j \supset \varphi_i$ using *modus ponens*. By the induction hypothesis, we have $\Box_X\varphi_j$ and $\Box_X(\varphi_j \supset \varphi_i)$. $\Box_X\varphi_i$ follows using axiom FOBMK–A-6.

For FOBMK–R-2, suppose φ_i is $\forall y\varphi_j$ where $j < i$, and we have a proof of $\Box_X\varphi_j$. If $y \notin X$, we construct a proof of $\Box_X\forall y\varphi_j$ using FOBMK–A-5. Now suppose $y \in X$. Let $\widehat{X}$ be X with y removed. Since the proof is also $\widehat{X}$ acceptable, by the induction hypothesis we have a proof of $\Box_{\widehat{X}}\varphi_j$, so by FOBMK–A-5 we have a proof of $\Box_{\widehat{X}}\forall y\varphi_j$. Since y is not free in $\forall y\varphi_j$, we get $\Box_X\forall y\varphi_j$ using FOBMK–A-4. ■

Corollary 6.2 *If $\vdash \varphi$ then $\vdash \Box_\emptyset\varphi$.*

Corollary 6.3 (Quantifier Absorbtion) $\vdash \Box_X\psi \equiv \Box_X\forall y_1\forall y_2\cdots\forall y_n\psi$ *provided $y_1, y_2, \ldots, y_n \notin X$.*

Proof In the following, assume $y_1, y_2, \ldots, y_n \notin X$.

$\Box_X \psi \supset \Box_X \forall y_n \psi$ is axiom FOBMK–A-5, and hence provable. Another instance of FOBMK–A-5 is $\Box_X \forall y_n \psi \supset \Box_X \forall y_{n-1} \forall y_n \psi$, hence $\Box_X \psi \supset \Box_X \forall y_{n-1} \forall y_n \psi$ is provable. And so on.

For the implication in the other direction:

(1) $\forall y_n \psi \supset \psi$ FOBMK–A-2
(2) $\Box_X (\forall y_n \psi \supset \psi)$ FOBMK–R-4 since $y_n \notin X$
(3) $\Box_X (\forall y_n \psi \supset \psi) \supset (\Box_X \forall y_n \psi \supset \Box_X \psi)$ FOBMK–A-6
(4) $\Box_X \forall y_n \psi \supset \Box_X \psi$ *modus ponens*
(5) $\forall y_{n-1} \forall y_n \psi \supset \forall y_n \psi$ FOBMK–A-2
(6) $\Box_X (\forall y_{n-1} \forall y_n \psi \supset \forall y_n \psi)$ FOBMK–R-4 since $y_{n-1} \notin X$
(7) $\Box_X \forall y_{n-1} \forall y_n \psi \supset \Box_X \forall y_n \psi$, using FOBMK–A-6, 6, and *modus ponens*
(8) $\Box_X \forall y_{n-1} \forall y_n \psi \supset \Box_X \psi$ from 4 and 7 by classical logic

And so on. ■

7 Monotonicity and Anti-monotonicity

We briefly discuss how one can add axioms to FOBMK to capture the *De Re*/*De Dicto* conversions discussed in Sect. 5, and thus bring in monotonicity and anti-monotonicity conditions. This section is a small interlude, since we are primarily interested in axiomatizing varying domain semantics as such in this paper.

The monotonicity condition on models is that a move to an accessible world may increase the quantification domain; if $\Gamma \mathcal{R} \Delta$ in a model, then $\mathcal{D}(\Gamma) \subseteq \mathcal{D}(\Delta)$. Kripke showed in (1963) that this semantic condition corresponds to the axiomatic converse Barcan scheme, $\Box \forall y \varphi \supset \forall y \Box \varphi$. The form this takes with binding modalities can be seen in Theorem 5.4. In the Artemov–Yavorskaya paper that introduced binding modalities, Artemov and Yavorskaya (2013), motivation came from thinking of $\Box$ as representing provability, following Gödel in (1933). With provability in mind, the converse Barcan scheme is quite natural: if we have proved that φ holds of everything, then for each particular item, we can prove that φ holds of it. Artemov and Yavorskaya assumed a stronger version of our FOBMK–A-4, breaking it into two implications. First they had $\Box_{Xy} \varphi \supset \Box_X \varphi$ provided y does not occur free in φ, which is half of our axiom. And second they had $\Box_X \varphi \supset \Box_{Xy} \varphi$, with no conditions. In their understanding, based on the familiar machinery of axiomatic provability, one could freely announce that a variable was being "pulled out" to serve as a parameter that could be substituted for in a proof, and thus was no longer subject to quantification. While reasonable in their context, it turns out that this is exactly what corresponds to the converse Barcan formula.

Before starting formalities, we make some comments on the form that the converse Barcan scheme takes when binding modalities are present. Similar comments apply to the Barcan scheme too. In its usual form, the converse Barcan scheme is $\Box \forall y \varphi \supset \forall y \Box \varphi$. In this paper, the $\Box$ operator is replaced with a binding version, and we would like the binding to be as exact as possible. In the antecedent of the converse Barcan scheme, $\Box\forall y\varphi$, the variable y is quantified, and hence bound. Then in a binding modalities version, $\Box_X\forall y\varphi$, it does not matter whether y occurs in X since we have axiom FOBMK–A-4. The narrowest version would assume $y \notin X$. The consequent, $\forall y\Box\varphi$, becomes $\forall y\Box_X\varphi$ when binding modalities are brought in, and if $y \notin X$ then y would not be free in $\Box_X\varphi$, so the quantifier in $\forall y\Box_X\varphi$ would be vacuous. Then here we want to have $y \in X$. So, finally, our binding modalities version of converse Barcan is $\Box_X\forall y\varphi \supset \forall y\Box_{Xy}\varphi$ with the usual condition that $y \notin X$.

Theorem 7.1 (Converse Barcan Equivalents) *In axiomatic* FOBMK, *the following are equivalent, following the convention that* $y \notin X$:

(1) $\Box_X\varphi \supset \Box_{Xy}\varphi$
(2) $\forall y(\Box_X\varphi \supset \Box_{Xy}\varphi)$
(3) $\Box_X\varphi \supset \forall y\Box_{Xy}\varphi$
(4) $\Box_X\forall y\varphi \supset \forall y\Box_{Xy}\varphi$

Proof Items (1) and (2) are equivalent using universal generalization FOBMK–R-2 and universal instantiation FOBMK–A-2. Items (2) and (3) are equivalent because y is not free in $\Box_X\varphi$. Items (3) and (4) are equivalent using Quantifier Absorbtion, Theorem 6.3. ■

We could choose any of the four versions from Theorem 7.1 to add as new axiom schema. We follow Artemov and Yavorskaya, using $\Box_X\varphi \supset \Box_{Xy}\varphi$, which seems simplest. We also need a version of axiom necessitation to go with our new axiom scheme; we add it to the FOBMK–R-3 version. That is, we allow $\vdash \Box_{X_1}\Box_{X_2}\ldots\Box_{X_n}(\Box_X\varphi \supset \Box_{Xy}\varphi)$ for all finite sets X_i. One nice consequence is that the restriction imposed in FOBMK–R-4 vanishes. The result is a formulation of what we call FOBMKmon similar to the one for binding modalities S4 given in Artemov and Yavorskaya (2011, 2013). The following makes all this official.

Definition 7.2 Axiom system FOBMKmon results from FOBMK by adding the axiom scheme $\Box_X\varphi \supset \Box_{Xy}\varphi$ for $y \notin X$, and extending rule FOBMK–R-3 to include this new axiom scheme thus:

$$\vdash \Box_{X_1}\Box_{X_2}\ldots\Box_{X_n}(\Box_X\varphi \supset \Box_{Xy}\varphi) \text{ for any finite sets } X_1, X_2, \ldots, X_n, \text{ where } y \notin X.$$

Theorem 7.3 *In* FOBMKmon *we have full necessitation: if* $\vdash \varphi$ *then* $\vdash \Box_X\varphi$ *for every finite set* X.

Proof Adding the new case to rule FOBMK–R-3 without modifying rule FOBMK–R-4 does not affect the proof of Theorem 6.1 or its Corollary 6.2. So we have that $\vdash \varphi$ implies $\vdash \Box_\emptyset\varphi$. Now $\vdash \Box_X\varphi$ follows by repeated use of instances of the new axiom scheme. ■

Corollary 7.4 *The restriction that* $y \notin X_n$ *in FOBMK–R-4 can be dropped in* FOBMKmon. *That is, FOBMK–R-4 can be folded into FOBMK–R-3.*

Proof $\forall x\varphi \supset \varphi(x/y)$ is provable since it is an axiom. Then by repeated use of the theorem above, $\Box_{X_1}\Box_{X_2}\dots\Box_{X_n}\varphi$ is provable for any finite X_1, ..., X_n without restriction. ■

As Kripke also showed in Kripke (1963), the Barcan formula corresponds to the semantic condition of anti-monotonicity. Binding modalities provide us with fewer options here than in the anti-monotonic case.

Theorem 7.5 (Versions of the Barcan formula) *In axiomatic* FOBMK, *the following are equivalent, where* $y \notin X$:

(1) $\forall y\Box_{Xy}\varphi \supset \Box_X\varphi$
(2) $\forall y\Box_{Xy}\varphi \supset \Box_X\forall y\varphi$

Proof Use Quantifier Absorbtion, Theorem 6.3. ■

Definition 7.6 Axiom system FOBMKantimon results from FOBMK by adding the axiom scheme $\forall y\Box_{Xy}\varphi \supset \Box_X\varphi$ for $y \notin X$, and extending rule FOBMK–R-3 to include this new axiom scheme thus:

$$\vdash \Box_{X_1}\Box_{X_2}\dots\Box_{X_n}(\forall y\Box_{Xy}\varphi \supset \Box_X\varphi) \text{ for any finite sets } X_1, X_2, \dots, X_n, \text{ where } y \notin X.$$

Finally, for constant domains, we simply combine our earlier systems.

Definition 7.7 Axiom system FOBMKcons is the combination of FOBMKmon and FOBMKantimon. Equivalently, add to FOBMK the axiom scheme $\forall y\Box_{Xy}\varphi \equiv \Box_X\varphi$.

8 Soundness

Now we return to a consideration of FOBMK alone, and show our axiomatization matches our semantics. In this section, we show soundness in the usual way: each axiom is valid, and validity is preserved by the rules. But truth in models, Definition 4.4, is directly about model formulas, while axioms are in a language that does not allow members of a model domain to appear. Further, axioms can contain free variable occurrences, while we have given a central role to *closed* model formulas—in Definition 4.5, we characterized validity of a model formula with free variables so that it is equivalent to the validity of its universal closure. (Recall that at a particular possible world, the universal quantifier ranges over only the members of the domain *of that world*.) It is useful now to introduce something stronger, essentially allowing universal closure with respect to the domain of a model, and not just with respect to the domain of a world.

Definition 8.1 *(Strong Validity)* Let $\varphi(\vec{x})$ be a formula not containing members of a domain, but having all its free variables in $\vec{x}$, and let $\mathcal{M} = \langle \mathcal{G}, \mathcal{R}, \mathcal{D}, \mathcal{I} \rangle$ be a model. $\varphi(\vec{x})$ is *strongly valid* in $\mathcal{M}$ provided, for each $\Gamma \in \mathcal{G}$ and each $\vec{a} \in \mathcal{D}_S$, $\mathcal{M}, \Gamma \Vdash \varphi(\vec{a})$.

Note that since $\mathcal{D}(\Gamma) \subseteq \mathcal{D}_S$, strong validity in a model implies validity in that model.

Theorem 8.2 *Each of the axioms* FOBMK–A-1, FOBMK–A-3, FOBMK–A-4, FOBMK–A-5, *and* FOBMK–A-6 *are strongly valid. Also* $\forall y(\forall x\varphi \supset \varphi(x/y))$ *is strongly valid where* y *is free for* x *in* φ, *and hence* FOBMK–A-2 *is valid.*

Proof We check FOBMK–A-4, FOBMK–A-5, and FOBMK–A-2, leaving FOBMK–A-1, FOBMK–A-3, and FOBMK–A-6 to the reader. For the rest of this proof, $\mathcal{M} = \langle \mathcal{G}, \mathcal{R}, \mathcal{D}, \mathcal{I} \rangle$ is an arbitrary model and Γ is an arbitrary world in $\mathcal{G}$.

For FOBMK–A-4, to keep notation simple, we work with a representative special case. Let us assume φ is an $\mathcal{M}$ formula that has only x and w as free variables, so we write it as $\varphi(x, w)$, and assume $X = \{x, z\}$. We show strong validity of $\Box_{\{x,y,z\}}\varphi(x, w) \equiv \Box_{\{x,z\}}\varphi(x, w)$ in $\mathcal{M}$.

Let a, b, c be arbitrary members of $\mathcal{D}_S$ and apply the substitution $(x/a, y/b, z/c)$ getting the closed model formula $\Box_{\{a,b,c\}}\varphi(a, w) \equiv \Box_{\{a,c\}}\varphi(a, w)$. We show this is true at Γ. To do this, we must show $\mathcal{M}, \Gamma \Vdash \Box_{\{a,b,c\}}\varphi(a, w)$ if and only if $\mathcal{M}, \Gamma \Vdash \Box_{\{a,c\}}\varphi(a, w)$. For both, this reduces to the same condition, that for any $\Delta \in \mathcal{G}$ with $\Gamma\mathcal{R}\Delta$, $\mathcal{M}, \Delta \Vdash \varphi(a, w)$, so we are done.

For FOBMK–A-5, again to keep notation simple, assume we have an $\mathcal{M}$ formula $\varphi = \varphi(x, y, w)$ where all free variables are displayed, and $X = \{x, z\}$. We show strong validity of $\Box_{\{x,z\}}\varphi(x, y, w) \supset \Box_{\{x,z\}}\forall y\varphi(x, y, w)$.

Let a, c be arbitrary members of $\mathcal{D}_S$ and apply the substitution $(x/a, z/c)$, getting closed model formula $\Box_{\{a,c\}}\varphi(a, y, w) \supset \Box_{\{a,c\}}\forall y\varphi(a, y, w)$. We show this is true at Γ. Assume $\mathcal{M}, \Gamma \Vdash \Box_{\{a,c\}}\varphi(a, y, w)$. Let Δ be an arbitrary member of $\mathcal{G}$ such that $\Gamma\mathcal{R}\Delta$. Using Definition 4.4 part 5, $\mathcal{M}, \Delta \Vdash \varphi(a, y, w)$. Then by Definition 4.4 part 0, for every $b, d \in \mathcal{D}(\Delta)$, $\mathcal{M}, \Delta \Vdash \varphi(a, b, d)$. Rephrasing this, for each $d \in \mathcal{D}(\Delta)$, $\mathcal{M}, \Delta \Vdash \varphi(a, b, d)$ for every $b \in \mathcal{D}(\Delta)$, so by Definition 4.4 part 4, $\mathcal{M}, \Delta \Vdash \forall y\varphi(a, y, d)$ for each $d \in \mathcal{D}(\Delta)$. Since Δ was arbitrary, $\mathcal{M}, \Gamma \Vdash \Box_{\{a,c\}}\forall y\varphi(a, y, w)$ by Definition 4.4 part 5 again.

Now assume y is free for x in $\mathcal{M}$ formula $\varphi(x)$. We show strong validity of $\forall y(\forall x\varphi(x) \supset \varphi(y))$ in $\mathcal{M}$. Let φ' be the result of substituting members of $\mathcal{D}_S$ for individual variables of φ other than x and y. We want to show that $\forall y(\forall x\varphi'(x) \supset \varphi'(y))$ is true at Γ. But this is immediate using Definition 4.4 part 4.

It now follows that, for each $a \in \mathcal{D}(\Gamma)$, $\mathcal{M}, \Gamma \Vdash \forall x\varphi'(x) \supset \varphi'(a)$, and so we have validity of FOBMK–A-2. ■

Note that, while we proved validity of $\forall x\varphi(x) \supset \varphi(y)$, it is easy to construct examples to show it is not *strongly* valid.

Corollary 8.3 *Let* $\mathcal{M} = \langle \mathcal{G}, \mathcal{R}, \mathcal{D}, \mathcal{I} \rangle$ *be a model,* X *be a finite set of individual variables, and* ψ *be a model formula. If* ψ *is strongly valid in* $\mathcal{M}$ *so is* $\Box_X\psi$.

Proof Assume $\psi(\vec{x}, \vec{y})$ is a strongly valid model formula in $\mathcal{M}$, where all the free individual variables are shown. To show $\Box_{\{\vec{x}\}}\psi(\vec{x}, \vec{y})$ is strongly valid in $\mathcal{M}$ it must be shown that $\mathcal{M}, \Gamma \Vdash \Box_{\{\vec{a}\}}\psi(\vec{a}, \vec{y})$ for every $\Gamma \in \mathcal{D}$ and every $\vec{a} \in \mathcal{D}_S$. Let Δ be an arbitrary member of $\mathcal{G}$ with $\Gamma\mathcal{R}\Delta$; we must show $\mathcal{M}, \Delta \Vdash \psi(\vec{a}, \vec{b})$ for every $\vec{b} \in \mathcal{D}_S$. But this is the case because $\psi(\vec{x}, \vec{y})$ is strongly valid in $\mathcal{M}$. ■

Theorem 8.4 *The rules* FOBMK–R-1 *and* FOBMK–R-2 *preserve weak (and strong) validity, and rules* FOBMK–R-3 *and* FOBMK–R-4 *add strong validity formulas.*

Proof We leave FOBMK–R-1 and FOBMK–R-2 to the reader. For FOBMK–R-3, all axioms except for FOBMK–A-2, are strongly valid by Theorem 8.2, and we can repeatedly apply Corollary 8.3.

For FOBMK–R-4 suppose φ is $\varphi(x, \vec{z}, \vec{w})$, where the $\vec{z}$ and $\vec{w}$ are disjoint and do not contain x. Also suppose y is free for x in this formula, and $X = \{\vec{z}\}$. We first show $\Box_X(\forall x\varphi \supset \varphi(x/y))$ is strongly valid, or using current notation, $\Box_{\{\vec{z}\}}(\forall x\varphi(x, \vec{z}, \vec{w}) \supset \varphi(y, \vec{z}, \vec{w}))$ is strongly valid. Let $\mathcal{M} = \langle\mathcal{G}, \mathcal{R}, \mathcal{D}, \mathcal{I}\rangle$ be a model, $\Gamma \in \mathcal{G}$, and $\vec{a} \in \mathcal{D}_S$. We must show $\mathcal{M}, \Gamma \Vdash \Box_{\{\vec{a}\}}(\forall x\varphi(x, \vec{a}, \vec{w}) \supset \varphi(y, \vec{a}, \vec{w}))$. This means that for each $\Delta \in \mathcal{G}$ with $\Gamma\mathcal{R}\Delta$, we must show $\mathcal{M}, \Delta \Vdash \forall x\varphi(x, \vec{a}, \vec{w}) \supset \varphi(y, \vec{a}, \vec{w})$. To do this, we must show $\mathcal{M}, \Delta \Vdash \forall x\varphi(x, \vec{a}, \vec{b}) \supset \varphi(c, \vec{a}, \vec{b})$ for every $\vec{b} \in \mathcal{D}(\Delta)$ and for every $c \in \mathcal{D}(\Delta)$. This is equivalent to showing $\mathcal{M}, \Delta \Vdash \forall y(\forall x\varphi(x, \vec{a}, \vec{b}) \supset \varphi(y, \vec{a}, \vec{b}))$ for every $\vec{b} \in \mathcal{D}(\Delta)$, and we have this because $\forall y(\forall x\varphi(x, \vec{z}, \vec{w}) \supset \varphi(y, \vec{z}, \vec{w}))$ is strongly valid by Theorem 8.2. Now, as in the previous case, we can repeatedly apply Corollary 8.3. ■

9 Completeness

Completeness proofs for axiomatizations of varying domain modal logics are rather rare, compared to the monotonic version. The proof we give is similar to the very general construction in Corsi (2002). Of course we have binding modalities and that paper does not, but that does allow a kind of free logic, with inner and outer domains, both of which vary, and we do not have this here. A trace of the dual domains can be seen, however, in our use of what we call *specification pairs*, in which one member plays the role of an inner domain and the other plays that of the outer version. But here it is simply a technical device, forced on us by circumstances. The proof in Corsi (2002) can be used as a guide to extend the present argument to the monotonic and anti-monotonic cases, but this would be too much detail for here.

Completeness is proved Henkin style so the language is enlarged with witnesses for existential statements. These are sometimes taken to be formal constants, but we prefer to use variables that are restricted so they never occur bound (obviously they behave like constants in this respect). This keeps the variety of machinery down and simplifies things in some respects.

Definition 9.1 *(Variable Extension)* The language, as defined in Sect. 3, has a set $\mathbf{V} = \{x_1, x_2, x_3, \ldots\}$ of individual variables. We now add a countable set of new variables $\mathbf{W} = \{w_1, w_2, w_3, \ldots\}$ *that are not allowed to occur bound in formulas.* For this section, we call members of $\mathbf{V} \cup \mathbf{W}$ *individual variables*, members of $\mathbf{V}$ (our original variables) *basic variables*, and the newly added members of $\mathbf{W}$ *witness variables.* For a formula φ in our extended language, by $\mathsf{v}(\varphi)$ we mean the set of *witness* variables occurring in φ. The axiom system of Sect. 6 is enlarged to allow formulas containing witness variables, but in free positions only. By *basic language and basic system*, we mean the language and the axiomatic system as originally given, without the extension by witness variables. We call a formula with witness variables *closed* if no *basic* variable occurrences are free. (Thus in a closed formula, the free variables are all witness variables present and the bound variables are all basic variables present.)

Allowing witness variables that can only occur in free positions brings with it some peculiarities that should be noted. Axiom FOBMK–A-2 allows for $\forall x\varphi(x) \supset \varphi(w)$ where w is a witness variable (and x is basic). But Rule FOBMK–R-4 is inapplicable. In particular we cannot even have the $n = 1$ case which would allow us to conclude $\Box_X(\forall x\varphi(x) \supset \varphi(w))$, because w cannot occur in X by the conditions on FOBMK–R-4, but if it does not, its occurrence as part of $\varphi(w)$ in $\Box_X(\forall x\varphi(x) \supset \varphi(w))$ would be a bound occurrence, precisely because it does not occur in X. Also if w is a witness variable, a new instance of axiom FOBMK–A-4 is $\Box_{Xw}\varphi \equiv \Box_X\varphi$ provided w does not occur in φ. The proviso, as actually stated with FOBMK–A-4, is that w does not occur *free* in φ, but since it cannot occur bound, this is equivalent to requiring that it does not occur at all.

We will use witness variables to supply domains of quantification at possible worlds of a canonical model. Different worlds can have different domains, so we will work with many subsets of $\mathbf{W}$ as quantificational domains. We will require that such subsets omit countably many witness variables, so "new" witness variables are always available. Since the countable set $\mathbf{W}$ can always be split up into countably many disjoint countable subsets, this is not a problem and we omit the details.

Our models, as defined in Sect. 4, allow us to evaluate possible world formulas that contain members of the model domain that are not in the domain of that possible world itself. In constructing our Henkin-style model, we need to be careful with this. The overall domain of the model we construct will be the entire set of witness variables, but possible worlds will be sets of formulas *that are required to omit countably many witness variables*, so we should not expect a truth lemma to match up semantic behavior at a possible world with membership in that world for every formula. We need some restrictions on the set of formulas for which we can expect such a match-up to work out. For these purposes, for each possible world in the model that we will construct we specify what are, informally speaking, inner and outer sets of witness variables, where the inner set is the one we quantify over, and the outer set is the one for which truth lemma behavior is well behaved. Here are the formalities.

Definition 9.2 *(Specification Pair)* Let $\mathbf{W}_1 \subseteq \mathbf{W}_2$ be sets of witness variables. We call $\langle \mathbf{W}_1, \mathbf{W}_2 \rangle$ a *specification pair*, and use it to restrict formulas and proofs as follows.

A $\langle \mathbf{W}_1, \mathbf{W}_2 \rangle$ *formula* is any formula allowing basic variables free and bound, and witness variables in free positions but with all witness variables coming from $\mathbf{W}_2$.

A $\langle \mathbf{W}_1, \mathbf{W}_2 \rangle$ *axiomatic proof* is any proof using the system of Sect. 6, allowing witness variables but in which only $\langle \mathbf{W}_1, \mathbf{W}_2 \rangle$ formulas are allowed, and in addition if an instance of axiom FOBMK–A-2 occurs, say $\forall x\varphi \supset \varphi(x/w)$ where w is a witness variable, then w must be in $\mathbf{W}_1$. A proof proves its last line. We write $\vdash_{\langle \mathbf{W}_1, \mathbf{W}_2 \rangle} \varphi$ if φ is a $\langle \mathbf{W}_1, \mathbf{W}_2 \rangle$ formula having a $\langle \mathbf{W}_1, \mathbf{W}_2 \rangle$ proof.

Since witness variables cannot occur bound, a substitution (x/y) in which y is a witness variable is always a free substitution, which is pertinent to axiom FOBMK–A-2 and rule FOBMK–R-4. This simplifies some things. But on the other hand, we cannot apply the generalization rule FOBMK–R-2 if we have a proof of $\varphi(w)$, where w is a witness variable. Fortunately we will have something almost as good in Corollary 9.4.

Theorem 9.3 *Let $\varphi_1, \varphi_2, \ldots, \varphi_n$ be a $\langle \mathbf{W}_1, \mathbf{W}_2 \rangle$ proof, let $w \in \mathbf{W}_2$, and let x be a basic variable that does not appear in the proof. Then $\varphi_1(w/x), \varphi_2(w/x), \ldots, \varphi_n(w/x)$ is also a $\langle \mathbf{W}_1, \mathbf{W}_2 \rangle$ proof.*

Proof This is an easy argument on proof length, which we leave to the reader. ■

Corollary 9.4 (Generalization) *Suppose $\vdash_{\langle \mathbf{W}_1, \mathbf{W}_2 \rangle} \varphi(w)$ where w is a witness variable in $\mathbf{W}_2$. Then $\vdash_{\langle \mathbf{W}_1, \mathbf{W}_2 \rangle} \forall x\varphi(x)$ for some basic variable x not appearing in $\varphi(w)$.*

Proof Assume we have a $\langle \mathbf{W}_1, \mathbf{W}_2 \rangle$ proof of $\varphi(w)$. Let x be any basic variable that does not occur in the proof (and hence not in $\varphi(w)$), and replace occurrences of w throughout the proof with occurrences of x. By Theorem 9.3, this will still be a proof, but of $\varphi(x)$. Now rule FOBMK–R-2 can be applied and $\forall x\varphi(x)$ is thus provable. ■

The next few items are rather technical, but play an important role in our completeness proof.

Corollary 9.5 *Let φ be a $\langle \mathbf{W}_1, \mathbf{W}_2 \rangle$ formula. Recall from Definition 9.1 that $\mathsf{V}(\varphi)$ is the set of witness variables of φ. If $\mathsf{V}(\varphi) \cap \mathbf{W}_1 = \emptyset$ and $\vdash_{\langle \mathbf{W}_1, \mathbf{W}_2 \rangle} \varphi$ then $\vdash_{\langle \mathbf{W}_1, \mathbf{W}_2 \rangle} \Box_{\mathsf{V}(\varphi)}\varphi$.*

Proof Assume $\mathsf{V}(\varphi) \cap \mathbf{W}_1 = \emptyset$, and consider a specific $\langle \mathbf{W}_1, \mathbf{W}_2 \rangle$ proof of φ. We begin by eliminating some possible trouble spots from the proof. Suppose an instance of Axiom FOBMK–A-2, $\forall x\psi \supset \psi(x/w)$, appears in the proof, where w is a witness variable. Using Theorem 9.3, this witness variable can be replaced by a new basic variable throughout the proof and the result will still be a $\langle \mathbf{W}_1, \mathbf{W}_2 \rangle$ proof. Further, since the original proof was a $\langle \mathbf{W}_1, \mathbf{W}_2 \rangle$ proof, w must come from $\mathbf{W}_1$ and since $\mathsf{V}(\varphi) \cap \mathbf{W}_1 = \emptyset$, w cannot occur in φ, and so φ is unchanged by the replacement. In this way we can replace, one by one, all witness variables that occur in the original proof in instances of Axiom FOBMK–A-2 by basic variables without changing what is being proved, while keeping it a $\langle \mathbf{W}_1, \mathbf{W}_2\rangle$ proof. Let us assume this has been done.

We now have a $\langle \mathbf{W}_1, \mathbf{W}_2\rangle$ proof $\varphi_1, \varphi_2, \ldots, \varphi_n = \varphi$ in which no instantiation introduced by FOBMK–A-2 involves a witness variable, only a basic variable. We show that, for each i, $\vdash_{\langle \mathbf{W}_1, \mathbf{W}_2\rangle} \Box_{\mathsf{V}(\varphi_i)}\varphi_i$, and hence $\vdash_{\langle \mathbf{W}_1, \mathbf{W}_2\rangle} \Box_{\mathsf{V}(\varphi)}\varphi$.

The proof is inductive. Suppose it is known that $\vdash_{\langle \mathbf{W}_1, \mathbf{W}_2\rangle} \Box_{\mathsf{V}(\varphi_j)}\varphi_j$ for all $j < i$; we show $\vdash_{\langle \mathbf{W}_1, \mathbf{W}_2\rangle} \Box_{\mathsf{V}(\varphi_i)}\varphi_i$.

Case FOBMK–A-2: φ_i is $\forall x\psi \supset \psi(x/y)$. The variable y cannot be a witness variable since these have been replaced, so it must be a basic variable. As a basic variable, y cannot be in $\mathsf{V}(\varphi_i)$, so we have $\vdash_{\langle \mathbf{W}_1, \mathbf{W}_2\rangle} \Box_{\mathsf{V}(\varphi_i)}(\forall x\varphi \supset \varphi(x/y))$ using rule FOBMK–R-4.

Case all other axioms: φ_i is a an axiom other than FOBMK–A-2. Use Rule FOBMK–R-3.

Case FOBMK–R-1: Suppose $j, k < i$, $\varphi_k = \varphi_j \supset \varphi_i$, and we have $\vdash_{\langle \mathbf{W}_1, \mathbf{W}_2\rangle} \Box_{\mathsf{V}(\varphi_k)}\varphi_k$ and $\vdash_{\langle \mathbf{W}_1, \mathbf{W}_2\rangle} \Box_{\mathsf{V}(\varphi_k \supset \varphi_i)}(\varphi_k \supset \varphi_i)$. Using FOBMK–A-4, $\vdash_{\langle \mathbf{W}_1, \mathbf{W}_2\rangle} \Box_{\mathsf{V}(\varphi_k \supset \varphi_i)}\varphi_k$. Using FOBMK–A-6 we can get $\vdash_{\langle \mathbf{W}_1, \mathbf{W}_2\rangle} \Box_{\mathsf{V}(\varphi_k \supset \varphi_i)}\varphi_k \supset \Box_{\mathsf{V}(\varphi_k \supset \varphi_i)}\varphi_i$. By *modus ponens*, $\vdash_{\langle \mathbf{W}_1, \mathbf{W}_2\rangle} \Box_{\mathsf{V}(\varphi_k \supset \varphi_i)}\varphi_i$. Then by FOBMK–A-4 again, $\vdash_{\langle \mathbf{W}_1, \mathbf{W}_2\rangle} \Box_{\mathsf{V}(\varphi_i)}\varphi_i$.

Case FOBMK–R-2 Using the induction hypothesis, FOBMK–A-5, and the fact that only basic variables can be quantified while members of $\mathsf{V}(\varphi_i)$ are witness variables.

Case FOBMK–R-3 By FOBMK–R-3 itself.

Case FOBMK–R-4 Using FOBMK–R-4 itself.

■

Definition 9.6 Let S be a set of $\langle \mathbf{W}_1, \mathbf{W}_2\rangle$ formulas and φ be a single such formula. We write $S \vdash_{\langle \mathbf{W}_1, \mathbf{W}_2\rangle} \varphi$ if $\vdash_{\langle \mathbf{W}_1, \mathbf{W}_2\rangle} (\psi_1 \wedge \ldots \wedge \psi_n) \supset \varphi$ for some $\psi_1, \ldots, \psi_n \in S$. We say S is $\langle \mathbf{W}_1, \mathbf{W}_2\rangle$ *inconsistent* if $S \vdash_{\langle \mathbf{W}_1, \mathbf{W}_2\rangle} \bot$, and S is $\langle \mathbf{W}_1, \mathbf{W}_2\rangle$ *consistent* if it is not $\langle \mathbf{W}_1, \mathbf{W}_2\rangle$ inconsistent.

Theorem 9.7 *Let* $\langle \mathbf{W}_1, \mathbf{W}_2\rangle$ *and* $\langle \mathbf{W}_3, \mathbf{W}_4\rangle$ *be specification pairs where* $\mathbf{W}_2 \cap \mathbf{W}_3 = \emptyset$ *and* $\mathbf{W}_2 \subseteq \mathbf{W}_4$*, Also let S be a set of* $\langle \mathbf{W}_1, \mathbf{W}_2\rangle$ *formulas and* φ *be a single such formula. If* $S \vdash_{\langle \mathbf{W}_3, \mathbf{W}_4\rangle} \varphi$ *then* $S \vdash_{\langle \mathbf{W}_1, \mathbf{W}_2\rangle} \varphi$.

Proof The members of S and φ are all $\langle \mathbf{W}_1, \mathbf{W}_2\rangle$ formulas. Since $\mathbf{W}_2 \subseteq \mathbf{W}_4$ these are also $\langle \mathbf{W}_3, \mathbf{W}_4\rangle$ formulas. Assume $S \vdash_{\langle \mathbf{W}_3, \mathbf{W}_4\rangle} \varphi$. Then there is a $\langle \mathbf{W}_3, \mathbf{W}_4\rangle$ proof $\varphi_1, \varphi_2, \ldots, \varphi_n$, where $\varphi_n = (\psi_1 \wedge \ldots \wedge \psi_k) \supset \varphi$ for some $\psi_1, \ldots, \psi_k \in S$. Let $w_1, w_2, \ldots, w_m$ be all the witness variables in $\mathbf{W}_4 - \mathbf{W}_2$ that occur in this proof, let $x_1, \ldots, x_m$ be distinct basic variables that do not occur in the proof, and let σ be the substitution $(w_1/x_1, \ldots, w_m/x_m)$. By repeated application of Theorem 9.3, $\varphi_1\sigma$, …, $\varphi_n\sigma$ is a $\langle \mathbf{W}_3, \mathbf{W}_4\rangle$ proof. We show it is also a $\langle \mathbf{W}_1, \mathbf{W}_2\rangle$ proof.

Throughout $\varphi_1, \ldots, \varphi_n$, σ replaces all witness variables from $\mathbf{W}_4$ except those that are also in $\mathbf{W}_2$, so the only witness variables in $\varphi_1\sigma$, …, $\varphi_n\sigma$ are in $\mathbf{W}_2$. Since

$\varphi_1, \ldots, \varphi_n$ was a $\langle \mathbf{W}_3, \mathbf{W}_4 \rangle$ proof, if some φ_i is an instance of FOBMK–A-2 and introduced a witness variable, that witness variable must have come from $\mathbf{W}_3$. Since $\mathbf{W}_3 \subseteq \mathbf{W}_4$, that witness variable must be in $\mathbf{W}_4$. Since $\mathbf{W}_2 \cap \mathbf{W}_3 = \emptyset$, that witness variable cannot be in $\mathbf{W}_2$. Then σ replaced it with a basic variable. So, all witness variables in $\varphi_1, \ldots, \varphi_n$ introduced by FOBMK–A-2 have been replaced by basic variables in $\varphi_1\sigma, \ldots, \varphi_n\sigma$, and thus vacuously all are in $\mathbf{W}_1$. It follows that $\varphi_1\sigma, \ldots, \varphi_n\sigma$ is a $\langle \mathbf{W}_1, \mathbf{W}_2 \rangle$ proof.

Finally φ_n is $(\psi_1 \wedge \ldots \wedge \psi_k) \supset \varphi$ which is a $\langle \mathbf{W}_1, \mathbf{W}_2 \rangle$ formula, thus containing no members of $\mathbf{W}_4$ that are not in $\mathbf{W}_2$. Then $\varphi_n\sigma = (\psi_1 \wedge \ldots \wedge \psi_k) \supset \varphi$, and $\varphi_1\sigma, \ldots, \varphi_n\sigma$ is a $\langle \mathbf{W}_1, \mathbf{W}_2 \rangle$ proof for it. ∎

Corollary 9.8 *Under the conditions of Theorem 9.7, if S is $\langle \mathbf{W}_1, \mathbf{W}_2 \rangle$ consistent then S is $\langle \mathbf{W}_3, \mathbf{W}_4 \rangle$ consistent.*

Proof Take φ to be $\bot$ in Theorem 9.7. ∎

Now we are ready to introduce the things that will function as possible worlds in our varying domain canonical models.

Definition 9.9 We say $\langle S, \mathbf{W}_1, \mathbf{W}_2 \rangle$ is *world-like* if:

1. $\langle \mathbf{W}_1, \mathbf{W}_2 \rangle$ is a specification pair (Definition 9.2) where $\mathbf{W}_2$ omits countably many witness variables,
2. S is a set of closed $\langle \mathbf{W}_1, \mathbf{W}_2 \rangle$ formulas (Definition 9.1),
3. S is $\langle \mathbf{W}_1, \mathbf{W}_2 \rangle$ consistent (Definition 9.6),
4. if S' is a set of closed $\langle \mathbf{W}_1, \mathbf{W}_2 \rangle$ formulas that is $\langle \mathbf{W}_1, \mathbf{W}_2 \rangle$ consistent, and $S \subseteq S'$ then $S = S'$ (that is, S is maximal),
5. S contains witnesses from $\mathbf{W}_1$, meaning that if $\neg\forall x\varphi(x) \in S$ then $\neg\varphi(w) \in S$ for some $w \in \mathbf{W}_1$.

Here is the key item, an elaboration of the usual Henkin construction.

Theorem 9.10 (Extension Theorem) *Let $\mathbf{W}_1$ and $\mathbf{W}_2$ be disjoint sets of witness variables and let S be a set of closed formulas whose witness variables are all in $\mathbf{W}_2$. Assume $\mathbf{W}_1$ is countable, $\mathbf{W}_1 \cup \mathbf{W}_2$ omits countably many witness variables, and S is $\langle \mathbf{W}_1, \mathbf{W}_1 \cup \mathbf{W}_2 \rangle$ consistent. Then S can be extended to a set S' so that $\langle S', \mathbf{W}_1, \mathbf{W}_1 \cup \mathbf{W}_2 \rangle$ is world-like.*

Proof Assume the hypotheses. Enumerate the countable set of closed $\langle \mathbf{W}_1, \mathbf{W}_1 \cup \mathbf{W}_2 \rangle$ formulas, as $\varphi_1, \varphi_2, \ldots$, and enumerate the members of $\mathbf{W}_1$, as $w_1, w_2, \ldots$. Construct a sequence of $\langle \mathbf{W}_1, \mathbf{W}_1 \cup \mathbf{W}_2 \rangle$ consistent sets as follows.

$$S_1 = S$$

$$S_{n+1} = \begin{cases} S_n & \text{if } S_n \cup \{\varphi_n\} \text{ is not } \langle \mathbf{W}_1, \mathbf{W}_1 \cup \mathbf{W}_2 \rangle \text{ consistent} \\ S_n \cup \{\varphi_n\} & \text{if } S_n \cup \{\varphi_n\} \text{ is } \langle \mathbf{W}_1, \mathbf{W}_1 \cup \mathbf{W}_2 \rangle \text{ consistent} \\ & \text{and } \varphi_n \text{ is not a negated universal} \\ S_n \cup \{\varphi_n\} \cup \{\neg\psi(w)\} & \text{if } S_n \cup \{\varphi_n\} \text{ is } \langle \mathbf{W}_1, \mathbf{W}_1 \cup \mathbf{W}_2 \rangle \text{ consistent} \\ & \text{and } \varphi_n = \neg\forall x \psi(x) \\ & \text{and } w \text{ is the first term in the sequence } w_1, w_2, \ldots \\ & \text{that does not occur in } S_n \cup \{\varphi_n\} \end{cases}$$

$S_1 = S$ involves no members of $\mathbf{W}_1$. It follows that every S_n involves only a finite number of members of $\mathbf{W}_1$, so in the third clause of the definition of S_{n+1}, some w_i must be available.

Each member of S_1, S_2, S_3, ... will be consistent. Two of the cases are simple; the only one that needs checking is the negative universal case. To keep notation less cluttered, we write $\vdash$ for $\vdash_{\langle \mathbf{W}_1, \mathbf{W}_1 \cup \mathbf{W}_2 \rangle}$ throughout the following paragraph.

Suppose $S_n \cup \{\varphi_n\}$ is $\langle \mathbf{W}_1, \mathbf{W}_1 \cup \mathbf{W}_2 \rangle$ consistent, $\varphi_n = \neg\forall x \psi(x)$, w does not occur in $S_n \cup \{\varphi_n\}$, and $S_{n+1} = S_n \cup \{\varphi_n\} \cup \{\neg\psi(w)\}$ is not $\langle \mathbf{W}_1, \mathbf{W}_1 \cup \mathbf{W}_2 \rangle$ consistent. We derive a contradiction. If S_{n+1} is not $\langle \mathbf{W}_1, \mathbf{W}_1 \cup \mathbf{W}_2 \rangle$ consistent, for some finite subset F of $S_n \cup \{\varphi_n\}$, $\vdash \bigwedge F \supset \psi(w)$. Now w is a witness variable and cannot be quantified, but we can use Corollary 9.4. For some basic variable y that does not occur in $\bigwedge F \supset \psi(w)$, $\vdash \forall y(\bigwedge F \supset \psi(w))(w/y)$. Since w was a new witness variable when introduced, it does not occur in F, so we have $\vdash \forall y(\bigwedge F \supset \psi(y))$ and hence also $\vdash \bigwedge F \supset \forall y \psi(y)$. But $F \subseteq S_n$ and $S_n \cup \{\neg\forall x \psi(x)\}$ is consistent. This is a contradiction in classical logic.

Of course $S_1 \subseteq S_2 \subseteq S_3 \subseteq \ldots$. Define $S = S_1 \cup S_2 \cup S_3 \cup \ldots$. Familiar arguments, which we omit, show that S is $\langle \mathbf{W}_1, \mathbf{W}_1 \cup \mathbf{W}_2 \rangle$ consistent, appropriately maximal, and we have just shown that there are witnesses in $\mathbf{W}_1$ for negated universals. ■

Definition 9.11 Let S be a set of closed formulas allowing witness variables. By $S^\sharp$ we mean $\{\vec{\forall}\varphi \mid \Box_{\mathsf{v}(\varphi)}\varphi \in S\}$, where $\vec{\forall}$ indicates universal closure and v is from Definition 9.1.

Now we construct our quantified varying domain canonical model. It is here that we use what we called world-like sets in Definition 9.9.

Definition 9.12 *(Varying Domain Canonical Model)* The *binding modality varying domain canonical model* $\mathcal{M} = \langle \mathcal{G}, \mathcal{R}, \mathcal{D}, \mathcal{I} \rangle$ is defined as follows.

(1) $\mathcal{G}$ is the collection of all sets $\langle S, \mathbf{W}_1, \mathbf{W}_2 \rangle$ that are world-like.
(2) Let $\Gamma, \Delta \in \mathcal{G}$, where $\Gamma = \langle S, \mathbf{W}_1, \mathbf{W}_2 \rangle$ and $\Delta = \langle T, \mathbf{W}_3, \mathbf{W}_4 \rangle$. $\Gamma \mathcal{R} \Delta$ iff $\mathbf{W}_2 \subseteq \mathbf{W}_4$ and $S^\sharp \subseteq T$.
(3) Let $\Gamma = \langle S, \mathbf{W}_1, \mathbf{W}_2 \rangle \in \mathcal{G}$. $\mathcal{D}(\Gamma) = \mathbf{W}_1$.

(4) Let $\Gamma = \langle S, \mathbf{W}_1, \mathbf{W}_2 \rangle \in \mathcal{G}$. For an n-place relation symbol P, $\mathcal{I}(P, \Gamma)$ is the set of all $\vec{a}$ where $\vec{a} \in \mathbf{W}_2$ and $P(\vec{a}) \in S$.

We have a somewhat more complicated version of the familiar Truth Lemma than usual.

Theorem 9.13 (Truth Lemma) *Let $\mathcal{M} = \langle \mathcal{G}, \mathcal{R}, \mathcal{D}, \mathcal{I} \rangle$ be the binding modality varying domain canonical model, $\Gamma = \langle S, \mathbf{W}_1, \mathbf{W}_2 \rangle \in \mathcal{G}$, and φ be a closed formula whose witness variables are all in $\mathbf{W}_2$. Then we have the following.*

$$\mathcal{M}, \Gamma \Vdash \varphi \textit{ if and only if } \varphi \in \qquad (*)$$

Proof The proof is by induction on formula complexity. All the cases but one are straightforward and are left to the reader. We consider only the modal case. Assume $\varphi = \Box_X \psi(\vec{x})$ where φ is a closed model formula and $\vec{x}$ are all the free basic variables of ψ. As our induction hypothesis we assume (*) is known for arbitrary possible worlds and for formulas simpler than φ, in particular for closed instances of $\psi(\vec{x})$ involving witness variables. Let $\Gamma = \langle S, \mathbf{W}_1, \mathbf{W}_2 \rangle \in \mathcal{G}$ and assume all witness variables of φ are in $\mathbf{W}_2$. We show (*) for φ itself.

Suppose $\varphi \in S$, that is, $\Box_X \psi(\vec{x}) \in S$. Since all witness variables in φ are in $\mathbf{W}_2$, the same is true of ψ. Now let $\Delta \in \mathcal{G}$ be arbitrary, with $\Gamma \mathcal{R} \Delta$. Δ must be $\langle T, \mathbf{W}_3, \mathbf{W}_4 \rangle$ for some T, $\mathbf{W}_3$, and $\mathbf{W}_4$, and by definition of $\mathcal{R}$, $S^\sharp \subseteq T$ and $\mathbf{W}_2 \subseteq \mathbf{W}_4$. Then $\forall \vec{x} \psi(\vec{x}) \in T$ and all witness variables in ψ are in $\mathbf{W}_4$. Since T meets maximally condition 4 from Definition 9.9, using FOBMK–A-2, $\psi(\vec{a}) \in T$ for all $\vec{a} \in \mathbf{W}_3$. Then by the induction hypothesis (*) we have $\mathcal{M}, \Delta \Vdash \psi(\vec{a})$ for all $\vec{a} \in \mathcal{D}(\Delta)$, and so $\mathcal{M}, \Delta \Vdash \psi(\vec{x})$. Since Δ was arbitrary we have $\mathcal{M}, \Gamma \Vdash \Box_X \psi(\vec{x})$, that is, $\mathcal{M}, \Gamma \Vdash \varphi$.

Now suppose $\varphi \notin S$ so by maximal consistency $\neg\varphi \in S$, that is, $\neg\Box_X \psi(\vec{x}) \in S$. Let $\mathbf{W}_0$ be a countable set of witness variables disjoint from $\mathbf{W}_2$ chosen so that $\mathbf{W}_2 \cup \mathbf{W}_0$ still omits countably many witness variables.

We first show the set $S^\sharp \cup \{\neg\forall\vec{x}\psi(\vec{x})\}$ is $\langle \mathbf{W}_0, \mathbf{W}_0 \cup \mathbf{W}_2 \rangle$ consistent. Suppose not. Then using Definition 9.11 there are $\forall\vec{y}_1 G_1(\vec{y}_1), \ldots, \forall\vec{y}_n G_n(\vec{y}_n) \in S^\sharp$ such that $\vdash_{\langle \mathbf{W}_0, \mathbf{W}_0 \cup \mathbf{W}_2 \rangle} (\forall\vec{y}_1 G_1(\vec{y}_1) \wedge \ldots \wedge \forall\vec{y}_n G_n(\vec{y}_n) \wedge \neg\forall\vec{x}\psi(\vec{x})) \supset \bot$, where for each $i = 1, \ldots, n$, $\Box_{\mathsf{v}(G_i)} G_i(\vec{y}_i) \in S$. Equivalently, $\vdash_{\langle \mathbf{W}, \mathbf{W}_2 \cup \mathbf{W} \rangle} (\forall\vec{y}_1 G_1(\vec{y}_1) \wedge \ldots \wedge \forall\vec{y}_n G_n(\vec{y}_n)) \supset \forall\vec{x}\psi(\vec{x})$. Let $Z = \mathsf{v}(\psi) \cup \mathsf{v}(G_1) \cup \ldots \cup \mathsf{v}(G_n)$. Note that Z is the set of witness variables appearing in $(\forall\vec{y}_1 G_1(\vec{y}_1) \wedge \ldots \wedge \forall\vec{y}_n G_n(\vec{y}_n)) \supset \forall\vec{x}\psi(\vec{x})$. Of course the witness variables in Z must be in $\mathbf{W}_2$. But in a $\langle \mathbf{W}_0, \mathbf{W}_0 \cup \mathbf{W}_2 \rangle$ proof of $(\forall\vec{y}_1 G_1(\vec{y}_1) \wedge \ldots \wedge \forall\vec{y}_n G_n(\vec{y}_n)) \supset \forall\vec{x}\psi(\vec{x})$, witness variables introduced by instances of axiom FOBMK–A-2 must come from $\mathbf{W}_0$. Since $\mathbf{W}_0$ and $\mathbf{W}_2$ are disjoint, in any such proof no witness variable introduced by FOBMK–A-2 can be in Z. So by Corollary 9.5, $\vdash_{\langle \mathbf{W}_0, \mathbf{W}_0 \cup \mathbf{W}_2 \rangle} \Box_Z[((\forall\vec{y}_1 G_1(\vec{y}_1) \wedge \ldots \wedge \forall\vec{y}_n G_n(\vec{y}_n)) \supset \forall\vec{x}\psi(\vec{x})]$. But then, by Theorem 9.7, $\vdash_{\langle \mathbf{W}_1, \mathbf{W}_2 \rangle} \Box_Z[((\forall\vec{y}_1 G_1(\vec{y}_1) \wedge \ldots \wedge \forall\vec{y}_n G_n(\vec{y}_n)) \supset \forall\vec{x} \psi(\vec{x})]$. It follows that $\vdash_{\langle \mathbf{W}_1, \mathbf{W}_2 \rangle} (\Box_Z \forall\vec{y}_1 G_1(\vec{y}_1) \wedge \ldots \wedge \Box_Z \forall\vec{y}_n G_n(\vec{y}_n)) \supset \Box_Z \forall\vec{x}\psi (\vec{x})$. Using FOBMK–A-4, $\vdash_{\langle \mathbf{W}_1, \mathbf{W}_2 \rangle} (\Box_{\mathsf{v}(G_1)} \forall\vec{y}_1 G_1(\vec{y}_1) \wedge \ldots \wedge \Box_{\mathsf{v}(G_n)} \forall\vec{y}_n G_n(\vec{y}_n)) \supset \Box_X \forall\vec{x}\psi(\vec{x})$, so by Theorem 6.3, $\vdash_{\langle \mathbf{W}_1, \mathbf{W}_2 \rangle} (\Box_{\mathsf{v}(G_1)} G_1(\vec{y}_1) \wedge \ldots \wedge \Box_{\mathsf{v}(G_n)} G_n(\vec{y}_n)) \supset$

$\Box_X\psi(\vec{x})$. But for each i, $\Box_{\mathsf{v}(G_i)}G_i(\vec{y}_i) \in S$, and so $\Box_X\psi(\vec{x}) \in S$, contradicting the consistency of S. This contradiction establishes that $S^\sharp \cup \{\neg\forall\vec{x}\psi(\vec{x})\}$ is $\langle \mathbf{W}_0, \mathbf{W}_0 \cup \mathbf{W}_2\rangle$ consistent.

Now we apply Theorem 9.13; $S^\sharp \cup \{\neg\forall\vec{x}\psi(\vec{x})\}$ extends to S' so that $\langle S', \mathbf{W}_0, \mathbf{W}_0 \cup \mathbf{W}_2\rangle$ is world-like, and hence is in $\mathcal{G}$. Call this Δ. We have $\Gamma\mathcal{R}\Delta$ because $S^\sharp \subseteq S'$ and $\mathbf{W}_2 \subseteq \mathbf{W}_0 \cup \mathbf{W}_2$. By item 5 of Definition 9.9, for some $\vec{w} \in \mathbf{W}_0 = \mathcal{D}(\Delta)$, $\neg\psi(\vec{w}) \in S'$, so by consistency, $\psi(\vec{w}) \notin S'$. Then by the induction hypothesis, $\mathcal{M}, \Delta \not\Vdash \psi(\vec{w})$. Hence $\mathcal{M}, \Gamma \not\Vdash \Box_{\mathsf{v}(\psi)}\psi(\vec{x})$ and, using **FOBMK–A-4**, $\mathcal{M}, \Gamma \not\Vdash \Box_X\psi(\vec{x})$, that is, $\mathcal{M}, \Gamma \not\Vdash \varphi$. ■

Now strong completeness follows by the usual argument, somewhat fancied up.

Theorem 9.14 (Completeness) *Let S be a set of closed formulas and φ be a single closed formula, all in the basic language. If $S \not\vdash \varphi$ in the basic language then, in the binding modality varying domain canonical modality $\mathcal{M} = \langle \mathcal{G}, \mathcal{R}, \mathcal{D}, \mathcal{I}\rangle$, there is some $\Gamma \in \mathcal{G}$ such that $\mathcal{M}, \Gamma \Vdash \psi$ for every $\psi \in S$ but $\mathcal{M}, \Gamma \not\Vdash \varphi$.*

Proof Suppose $S \not\vdash \varphi$ in the basic language. Then using the terminology of Definitions 9.2 and 9.6, $S \cup \{\neg\psi\}$ is $\langle\emptyset, \emptyset\rangle$ consistent. Let $\mathbf{W}_0$ be a countable set of witness variables that also omits countably many witness variables. By Corollary 9.8, $S \cup \{\neg\psi\}$ is $\langle \mathbf{W}_0, \mathbf{W}_0 \cup \emptyset\rangle$ consistent. Then by Theorem 9.13, S extends to a set S' so that $\Gamma = \langle S', \mathbf{W}_0, \mathbf{W}_0 \cup \emptyset\rangle$ is world-like. $\Gamma \in \mathcal{G}$ and, by Theorem 9.13, $\mathcal{M}, \Gamma \Vdash \psi$ for every $\psi \in S$ but $\mathcal{M}, \Gamma \not\Vdash \varphi$. ■

10 Conclusion

Binding modalities machinery has a plausible interpretation, is straightforward to work with, and helps elucidate the *de re/de dicto* dichotomy. What remains is to say something about the connection between binding modalities and the familiar necessity operator and, in particular, with the axiomatic approach that Kripke took in Kripke (1963).

As noted earlier, Kripke used an axiom system that did not allow free variable occurrences in formulas appearing in a proof. To do this, he worked with his version of the "closure" of a formula, defined as follows.

> "If *A* is a formula containing free variables, we define a *closure* of *A* to be any formula without free variables obtained by prefixing universal quantifiers and necessity signs, in any order, to *A*."

He then presented a standard classical first-order axiomatization using schemas, allowing free variable occurrences, except that universal instantiation had the form $\forall y(\forall x\varphi(x) \supset \varphi(y))$. Of course he also adopted appropriate modal axiom schemes, again allowing free variables. But all this was actually preliminary to his official axiom system. For this he took *all closures of instances of the schemes just mentioned*, together with modus ponens.

It is not hard to see that Kripke's use of his notion of closures corresponds to our two versions of axiom necessitation. As an illustrative example, consider the following, where $\varphi(x_1, x_2, x_3)$ is an instance of one of our axiom schemes other than FOBMK–A-2, and all free variables are shown. Using FOBMK–R-3, Axiom Necessitation 1, the following might occur in a proof in FOBMK.

$$\Box_{\{x_1\}}\Box_{\{x_1,x_2\}}\varphi(x_1, x_2, x_3)$$

Using Quantifier Absorption, Corollary 6.3, this is equivalent to the following.

$$\Box_{\{x_1\}}\forall x_2\Box_{\{x_1,x_2\}}\forall x_3\varphi(x_1, x_2, x_3)$$

And by Universal Generalization and Universal Instantiation, it is further equivalent to the following, which has no free variables.

$$\forall x_1\Box_{\{x_1\}}\forall x_2\Box_{\{x_1,x_2\}}\forall x_3\varphi(x_1, x_2, x_3)$$

Suppose we write $\Box\psi$ for $\Box_X\psi$ where X consists of all the variables occurring free in ψ. Then the formula above rewrites to the following.

$$\forall x_1\Box\forall x_2\Box\forall x_3\varphi(x_1, x_2, x_3)$$

In this, the prefix has exactly the form of Kripke's closure.

We call attention to our rule FOBMK–R-4, which we called Axiom Necessitation 2. It covers formulas of the form $\forall x\varphi(x) \supset \varphi(y)$, which are exactly the ones that Kripke singled out as the source of provability problems. In our necessitation rule formulation, we can conclude $\Box_{X_1}\Box_{X_2}\ldots\Box_{X_n}(\forall x\varphi(x) \supset \varphi(y))$, *provided* $y \notin X_n$. Kripke didn't allow the free variable version of universal instantiation we use, but replaced it with $\forall y(\forall x\varphi(x) \supset \varphi(y))$. The prefixed universal quantifier corresponds directly to our condition that $y \notin X_n$.

As was noted in Artemov and Yavorskaya (2013), the binding modalities formulation "is as expressive as traditional FOS4 since the binding modality $\Box_X F$ can be semantically encoded by $\Box\forall\vec{y}F$ where $\Box$ is the FOS4 modality and $\vec{y}$ is the list of all free variables of F that are not in X." Exact connections with Kripke's axiomatization require some delicacy however. In Kripke (1963), Kripke observed that "all the laws of quantification theory—modified to admit the empty domain—hold." The empty domain must be brought in since $\forall x\varphi(x) \supset \varphi(y)$ is disallowed in favor of its closure. Such details need care, but are not fundamentally significant. Our point is, simply, that relationships between the binding modalities system and Kripke's system are easily seen. The important difference is that in Kripke's approach the complexities he adds are essentially proof theoretic devices to avoid the possibility of proving something undesirable, while with binding modalities, the extra machinery has a direct semantic interpretation, enhances the natural expressiveness of the system, and relates to issues of long-standing philosophical interest.

References

Artemov, Sergei, and Melvin Fitting. 2019. *Justification logic: reasoning with reasons*. Cambridge tracts in mathematics book 216. Cambridge: Cambridge University Press.

Artemov, Sergei N. 2001. Explicit Provability and Constructive Semantics. *Bulletin of Symbolic Logic* 7 (1): 1–36.

Artemov, Sergei N., and Melvin C. Fitting. 2011. Justification logic. In ed. Edward N. Zalta. Revised 2015. http://plato.stanford.edu/entries/logic-justification/.

Artemov, Sergei N., and Tatiana Yavorskaya (Sidon). 2011. First-order logic of proofs. Technical report, TR-2011005. City University of New York.

Artemov, Sergei N., and Tatiana Yavorskaya (Sidon). 2012. Binding modalities. Technical report, TR-2012011. City University of New York.

Artemov, Sergei N., and Tatiana Yavorskaya (Sidon). 2013. Binding modalities. *Journal of Logic and Computation* 26 (1): 451–461. Published online 07 Oct 2013.

Corsi, Giovanna. 2002. A unified completeness theorem for quantified modal logics. *Journal of Symbolic Logic* 67 (4): 1483–1510.

Feferman, Solomon, et al. (eds.). 1986–2003. *Kurt Gödel collected works*, vol. 5. Oxford: Oxford University Press.

Fine, Kit. 1983. The permutation principle in quantificational logic. *Journal of Philosophical Logic* 12: 33–37.

Fitting, Melvin. 2014. Possible world semantics for first-order logic of proofs. *Annals of Pure and Applied Logic* 165 (1): 225–240. Published online in Aug 2013.

Fitting, Melvin C., and Felipe Salvatore. 2017. First-order justification logic with constant domain semantics. *To appear in Annals of Pure and Applied Logic.*

Göodel, Kurt. 1933. Eine Interpretation des intuistionistischen Aussagenkalküls. In *Ergebnisse eines mathematischen Kolloquiums*, vol. 4. Translated as An interpretation of the intuitionistic propositional calculus in [8] I, 296–301, 39–40.

Kripke, Saul. 1963. Semantical considerations on modal logics. In *Acta philosophica fennica, modal and many-valued logics*, 83–94.

Mendelson, Elliott. 1997. *Introduction to mathematical logic*, 4th ed. London: Chapman and Hall.

Prior, Arthur. 1956. Modality and quantification in S5. *Journal of Symbolic Logic* 21: 60–62.

Quine, Willard Van Orman. 1940. *Mathematical logic*, 2nd ed. Cambridge: Harvard University Press. Revised 1951.

Refutation Systems: An Overview and Some Applications to Philosophical Logics

Valentin Goranko, Gabriele Pulcini and Tomasz Skura

Abstract Refutation systems are systems of formal, syntactic derivations, designed to derive the non-valid formulas or logical consequences of a given logic. Here we provide an overview with comprehensive references on the historical development of the theory of refutation systems and discuss some of their applications to philosophical logics.

1 Introduction

The idea of refutation systems goes back to Aristotle's approach to systematic rejection of all invalid syllogisms. Aristotle showed that some syllogisms are invalid by providing suitable examples (in modern terms, counter-models) demonstrating that they can derive false conclusions from true premises. Other syllogisms he then rejected by applying some deductive reasoning, showing that assuming their validity would imply validity of already rejected syllogisms. Essentially, he applied the inference rule of Modus Tollens: '*If* B *is rejected and* $A \to B$ *is accepted (valid), then* A *must be rejected, too.*' This approach worked quite well, enabling Aristotle to classify most of the invalid syllogisms, though his classification was shown to be incomplete by Jan Łukasiewicz, but only much later, in the first half of the twentieth

This chapter is in its final form and it is not submitted to publication anywhere else.

V. Goranko (✉)
Stockholm University, Stockholm, Sweden
e-mail: valentin.goranko@philosophy.su.se

University of Johannesburg (Visiting Professorship), Johannesburg, South Africa

G. Pulcini
University of Amsterdam, Amsterdam, The Netherlands
e-mail: g.pulcini@uva.nl

T. Skura
University of Zielona Góra, Zielona Góra, Poland
e-mail: T.Skura@ifil.uz.zgora.pl

F. Liu et al. (eds.), *Knowledge, Proof and Dynamics*,
Logic in Asia: Studia Logica Library,
https://doi.org/10.1007/978-981-15-2221-5_9

century. Łukasiewicz embraced Aristotle's idea and formalized it in the concept of refutation system in Łukasiewicz (1939) and applied it to Aristotle's syllogistic in his seminal book (Łukasiewicz 1951), though a much earlier important pre-cursor of the development of his ideas on refutation systems was Łukasiewicz (1921) (see English translation in Borkowski 1970), where he introduced the notion of formal rejection of a false proposition.[1] In Łukasiewicz (1951), Łukasiewicz wrote '...*Out of the two intellectual acts, acceptance and rejection of a statement, only the first has been taken into account in modern formal logic. Gottlob Frege introduced into logic the idea of assertion, and the sign of assertion* ($\vdash$), *accepted afterwards by the authors of Principia Mathematica. The idea of rejection, however, so far as I know, has been neglected up to the present day.*'

Since Łukasiewicz's foundational work in Łukasiewicz (1951), the concept of refutation systems has been developing at a slow but steady pace, with various types of refutation systems being designed both for classical logic and for a number of non-classical logics, and several technical and conceptual applications of these have been proposed in the literature.

In this paper, we provide a concise, but comprehensively referenced overview of the literature on refutation systems and discuss some of their applications to philosophical logics.

2 Refutation Systems: Basic Concepts

2.1 Semantic Versus Deductive Refutability

Consider a logical system L with a given semantics, defining L-validity, and, more generally, L-logical consequence, both denoted by $\models_{\mathsf{L}}$. Following a common tradition in non-classical logics, we will often identify the logic L with the set of its valid formulas.

An L-formula A is **(semantically) refutable** in L iff it is not valid, i.e. $\not\models A$. That is, there is an L-model falsifying A. Likewise, for logical consequence: $\Phi \not\models A$ means that there is an L-model satisfying all formulas in Φ, but falsifying A.

Now, consider also a deductive system **D** for L, with a derivation relation $\vdash_{\mathbf{D}}$. Then, for any L-formula A, provability of A in **D** is denoted $\vdash_{\mathbf{D}} A$. If **D** is sound and complete for L, then $\vdash_{\mathbf{D}} A$ corresponds to validity $\models_{\mathsf{L}} A$.

However, one can argue that, in general, the matching syntactic notion to semantical refutability is *not* non-provability of A, i.e. $\nvdash_{\mathbf{D}} A$. It is rather the notion of '*deductive refutability*', i.e. existence of a formal derivation of A in a suitable *derivation system for* L*-refutable formulas*.

[1] As noted in Wybraniec-Skardowska (2018), for introducing the concept of formal rejection, Łukasiewicz was apparently influenced by Brentano, who was probably the first to consider on a par the two kinds of judgment, viz., *acceptance* and *rejection*, and to use a pair of symbols, + and −, to syntactically distinguish between them.

A new symbol is needed for that notion, for which we will use $\dashv_{\mathsf{L}}$, following Łukasiewicz, where

$$\dashv_{\mathsf{L}} A$$

means '*A* **is deductively refutable in** L', i.e. '*the non-validity of A in* L *can be formally derived*', or '*the refutation of A in* L *is derivable/derived*'. When the logic L is fixed by the context, we will write simply $\dashv$.

The above naturally extends to deductive refutations of logical consequences.

Thus, the notion of '**(deductive) refutation systems**' arises.

2.2 *Refutation Rules and Systems*

Pure rule of refutation inference:

$$\frac{\dashv B_1, \ldots, \dashv B_n}{\dashv C}$$

A typical example is Łukasiewicz's rule **Reverse substitution**:

$$\frac{\dashv \sigma(A)}{\dashv A}$$

where σ is a uniform substitution.

The intuitive meaning: if $\sigma(B)$ is derived as non-valid, then B is derived as non-valid, too.

Usually, pure refutation rules do not suffice to capture adequately semantic refutability, so we also consider the following, more general refutation rules.

(Mixed) rule of refutation inference (based on a deductive system **D**):

$$\frac{\vdash_{\mathbf{D}} A_1, \ldots, \vdash_{\mathbf{D}} A_m, \ \dashv B_1, \ldots, \dashv B_n}{\dashv C}$$

The intuitive interpretation: if each A_i is derived by **D** as valid and each B_j is derived as non-valid, then C is derived as non-valid.

A typical example is Łukasiewicz's rule **Reverse modus ponens**:

$$\frac{\vdash A \to B, \dashv B}{\dashv B}$$

A **refutation system** is a set R of refutation rules. Refutation rules with no premises are called **refutation axioms** and we write them simply as $\dashv A$.

A **refutation derivation in** R, or just an **R-derivation** for a formula A, is a sequence S_1, ..., S_t, where S_t is $\dashv A$ and every S_i is either a refutation axiom or has the form $\vdash B$ or is obtained from some preceding formulas by applying a refutation rule from R. Note that when R contains mixed refutation rules, then the notion of R-refutability is relativised to the underlying deductive system **D**. We now say that a formula A is **refutable in** R (or, just R-**refutable**) iff there is a refutation derivation for A in R.

Given a logical system L, we say that a refutation system R is:

- **refutation-sound**, or **Ł-sound, for** L, if only non-valid in L formulae (more generally, logical consequences) are R-refutable.
- **refutation-complete**, or **Ł-complete, for** L, if all non-valid in L formulae (more generally, L logical consequences) are R-refutable.

These are readily relativized to any fixed set Φ of formulas in L.

3 Refutation Systems: An Overview

3.1 Beginning and Early Work on Refutation Systems

The notion of formal rejection was introduced by Łukasiewicz in his early paper (Łukasiewicz 1921), but his first publication on formal refutation systems were Łukasiewicz (1939), where he introduced the concept and proposed a complete refutation system for the classical propositional calculus PL, based on the refutation rules of reverse substitution and Modus Tollens (reverse Modus Ponens).

Later, in his seminal book (Łukasiewicz 1951), Łukasiewicz showed that Aristotle's refutation system for syllogisms is 'refutationally incomplete', i.e. not deriving all non-valid syllogisms. He then added two more non-valid syllogisms to Aristotle's system, to make it (refutationally) complete for all non-valid syllogisms, but not for all non-valid expressions of the language.

Later, Łukasiewicz's student J. Słupecki showed in Słupecki (1955) that no finite set of refutation axioms would suffice for that. In that work, he proposed an additional refutation rule to make Łukasiewicz's system complete. In that work, Słupecki also developed further the concept of refutation system, introduced the notions of rejection consequence relation, Ł-decidability, and Ł-consistency, and solved the problem, posed by Łukasiewicz, of proving decidability of Aristotle's Syllogistics. During the 1970s, Słupecki together with his students Bryll and Wybraniec-Skardowska developed a general systematic theory of rejected propositions, in the style of Tarski's theories of consequence relations and deductive systems, see in Słupecki et al. (1971) and Słupecki et al. (1972). Follow up and related works along that line include (Staszek 1971, 1972), and the more recent (Wybraniec-Skardowska 2016), where two notions of axiomatic refutation systems, dual to Tarski's concept of deductive system for PL were constructed and proved to be equivalent.

A recent comprehensive historical overview and discussion of the early work on refutation systems by Łukasiewicz, Słupecki and his school can be found in Wybraniec-Skardowska (2018). For a nice discussion on Łukasiewicz's development of the concept of refutation system see also Tamminga (1994).

3.2 Overview of Refutation Systems in Classical Logic

In parallel with the fundamental work of Polish school, several other, essentially independent works developing variants of refutation systems for classical logic were published since the 1960s:

- In (1960) Härtig studied Łukasiewicz's refutation system for CPC and established some necessary and sufficient conditions for Ł-completeness of axiomatic refutation systems for the not-provable formulae in the classical propositional calculus CPC.
- In (1961) Hailperin, apparently unaware of Łukasiewicz's notion of refutation system and related work, developed probably the first explicit axiomatic refutation system for first-order logic FOL, for languages without equality and function symbols. That system was proved there to be sound and complete for falsifiability in finite FOL models.
- Later, in (1971) Staszek considered Łukasiewicz's refutation system for Aristotle's syllogistic and extends it with rules for the quantifiers.
- In (1958) Stahl introduced the notion of 'opposite system' for axiomatic derivation of all logically false (unsatisfiable) sentences of a given logic and proposed two axiomatizations for such system for CPC. Later, Morgan in (1973) independently presented a finitely axiomatized refutation system for all logically false (unsatisfiable) sentences of CPC.
- Independently from all previous work on refutation systems mentioned so far, Caicedo proposed in Caicedo (1978) a finitary formal system for deriving all non-theorems of CPC, consisting only of *pure* refutation rules.
- Later, in (1990) Skura proposed a general method for constructing refutation systems with pure refutation rules for sentential logics.
- Another line of relatively independent work started in Japan with Ishimoto's (1981), which studied the method of axiomatic rejection in PL. Inoue discussed and followed on Stahl's work in Inoué (1989) and Ishimoto's work in Inoué (1989). In (1990), Inoue studied unprovability-preserving translations between formal systems for propositional logic and in Inoué (1994) he discussed the atomic formula property of Härtig's refutation calculus in Härtig (1960).
- In (1990), Varzi introduced a simplified Ł-complete refutation system for PL, based on the single axiom: $\bot$, and the two rules of inference: reverse substitution and replacement of equivalent subformulas. That system was further refined in Varzi (1992).
- In (2000, 2002) Kulicki discusses applications of axiomatic rejection, for refuting the non-valid syllogisms of Aristotle's Syllogistic and, more generally, as syntactic decision procedures.
- Morgan et al. (2007) presents a sound and complete axiomatic proof system for deriving for the logical contingencies in PL, that is the PL-formulas that are both satisfiable and falsifiable.

Lastly, we note that refutation systems are close in design and purpose to tableaux-based methods, as both can be used to derive non-validity or satisfiability of the input formula. A number of more syntactically presented tableau-based methods for FOL have been developed, of which we only mention here some, which relate more explicitly with the concept of deductive refutation.

- In (1989) Dutkiewicz developed a tableau-based Ł-complete refutation system for IPL (also mentioned further).
- In (1998) Bry and Torge present a tableaux-based algorithmic deduction method for FOL, for which they prove that it is refutation-complete for finite satisfiability in FOL. They also discuss several areas of applications of their methods to computer science, incl. databases, planning, natural language understanding, program verification, and theorem proving.
- In (2005), Skura introduced 'invertible reduction rules', reversing refutation rules and preserving validity, for IPL.

3.3 Gentzen-Style Refutation Systems

The first constructed refutation systems were of axiomatic type, but gradually the other classical types of deductive systems, such as tableaux, sequent calculi, and systems of natural deduction, were adapted as refutation systems. We mention here those that were constructed for classical logic, while others, specifically designed for non-classical logics, are discussed further in this section.

- In (1988), Tiomkin constructed a sequent-style refutation calculus for FOL without function symbols and sketched a proof of its Ł-completeness for the formulas refutable in finite models.
  Independently, a few years later both Bonatti in Bonatti (1993)[2] and Goranko in Goranko (1994) developed an Ł-complete sequent refutation calculi for PL, completing Tiomkin's calculus by including (the same) left- and right-introduction rules for the full propositional language. (See also Bonatti and Varzi 1995 containing the rules for implication and negation.) The sequent refutation calculus constructed in Goranko (1994) was then extended in the same paper to some important normal modal logics.
- In (1994) Tamminga developed a system of natural deduction for deriving the non-theorems of PL, which he proved to be (Ł-sound and) Ł-complete. He also presented a subsystem of that system, Ł-complete for the set of all contradictions of PL and illustrated the application of both systems with some examples.

[2]This reference is imported from other sources, albeit none of the authors have seen this technical report and it appears to be currently inaccessible in the internet.

- In (2013) Tiomkin constructed an Ł-complete sequent calculus for the contingencies of PL, i.e. the propositional formulas that are neither valid nor contradictory, and also proved cut-elimination and the subformula property for that calculus.
- In Carnielli and Pulcini (2017), the sequent refutation systems of Tiomkin and Goranko were enriched with two admissible unary cut-rules which come shaped as reverse Weakenings. Then the authors designed a normalization procedure and studied some related computational properties.
- In Pulcini and Varzi (2019), the authors propounded a proof-nets system sound and complete with respect to the set of classically invalid (right-sided) sequents. They also provided a normalization procedure that allows for the improvement of the basic computational properties already studied in Carnielli and Pulcini (2017).

3.4 Refutation Systems for Non-validity in the Finite

Ł-complete refutation systems can be naturally associated with logics determined by effectively definable classes of finite models, as the non-validities in such logics are effectively enumerable, and therefore possible to capture by deductive refutation systems. Cases of particular importance are first-order logic on the class of all finite structures, hereafter denoted $\mathsf{FOL}^{\mathsf{fin}}$, as well as modal and other non-classical logics with finite model property (FMP). Several such refutation systems have been proposed, including the following.

- We have already mentioned the axiomatic refutation systems in Hailperin (1961) and in Bry and Torge (1998), as well as the sequent refutation calculus in Tiomkin (1988) for FOL without function symbols, each proved to be Ł-sound and Ł-complete for $\mathsf{FOL}^{\mathsf{fin}}$.
- In (1991) Skura discussed refutation systems as decision procedures and in Skura (1992) he developed a general method for constructing Ł-complete refutation systems for intermediate logics with the FMP, as well as for some logics without the FMP. Further, in Skura (1994), he extended that method to modal logics with the FMP.
- In (2004b) Skura proposed Scott-style refutation rules for the intermediate logics of finite n-ary trees.
- In (1994) Goranko constructed Ł-complete refutation systems for several modal logics that are complete for classes of finite (intransitive or transitive) trees.
- In (2018) Goranko and Skura develop generic refutation systems for modal logics with finite model property and for FOL theories determined by classes of finite models. In particular, they construct an Ł-complete refutation system for $\mathsf{FOL}^{\mathsf{fin}}$ in arbitrary FOL languages (incl. function symbols).

3.5 Refutation Systems for Intuitionistic and Modal Logics

Research on extending refutation systems to non-classical logics began already with Łukasiewicz, who conjectured in Łukasiewicz (1952) that, to derive all intuitionistically non-valid formulae it suffices to add to CPC ($\dashv$) the Disjunction Rule (DR):

$$\frac{\dashv A, \ \dashv B}{\dashv A \vee B}$$

This conjecture was refuted later (in 1957) by Kreisel and Putnam, who showed in Kreisel and Putnam (1957) that DR is admissible in a proper extension of the intuitionistic propositional calculus IPC, now known as Kreisel–Putnam's logic, KP = IPC + $(\neg A \to (B \vee C)) \to (\neg A \to B) \vee (\neg A \to C)$.

Several complete refutation systems for IPL and its extensions were developed since then, including

- In (1957) D. Scott constructed an Ł-complete refutation system for IPL by introducing an infinite family of non-structural refutation rules (now called Scott-style refutation rules), involving Wajsberg/Mints-style normal forms. These go back to Wajsberg (1938) and Mints (1990); see also Mints (1992) for further details.
- In (1989), Skura improved Scott's system by constructing an Ł-complete refutation system for IPL involving an infinite family of structural rules.
- In (1989) Dutkiewicz developed another, tableau-based infinitary Ł-complete refutation system for IPL.
- In (1992) Skura constructed refutation calculi for some intermediate propositional logics.
- Essential connections between refutations and admissibility of rules have been analysed by Goudsmit in the context of IPL and intermediate logics in Goudsmit (2014, 2015), where using some techniques for admissible rules he constructs refutation systems (involving structural refutation rules) for the Gabbay–de Jongh logics.

Refutation systems are very close in purpose to counter-model producing procedures. Of the many and various such methods developed in the literature, we mention just a few that make that relation explicit and make an essential use of it.

- In (1995) Pinto and Dyckhoff use the contraction-free sequent calculi LJT for IPL to build (as they say, 'for expository reasons') a sequent 'Calculus for Refutation of Intuitionistic Propositions' CRIP, that is Ł-complete for IPL.
- Related in spirit and purpose is Goré and Postniece (2008) where Goré and Postniece present a cut-free sequent calculus for the *bi-intuitionistic logic*, being the union of intuitionistic and dual intuitionistic logic, which also involves refutation calculi for both. See also the related in topic (Pinto and Uustalu 2009).
- In (2014) Negri explores further the duality between proof-search and counter-model search or, more generally, refutation search. She presents a method for unifying proof-search and counter-model construction that puts together several

ideas, incl. sequent calculi with internalized semantics and Tait-Schütte-Takeuti style completeness proofs, combined with procedures for effective finite counter-model construction. The method is applied to a variety of intermediate, modal, and other non-classical logics.
- In (2017) Fiorentini and Ferrari employ Maslov's inverse method (a saturation based theorem proving technique using forward proof-search strategy) to develop a 'forward unprovability calculus' FRJ for IPL. They show how to extract a Kripke counter-model for a formula from a derivation of the formula in FRJ.

Other works on refutation systems, specifically for modal logics include

- The first Ł-complete refutation system for modal logic was the one for S5 constructed by Słupecki and Bryll in Słupecki and Bryll (1973).
- In Goranko (1994) (see also Goranko 1991) Goranko constructed Ł-complete axiomatic and sequent-style refutation systems for several important modal logics, incl. K, T, K4, GL, S4Grz and others.
- In (1995) Skura constructed a refutation system for S4 and proved its Ł-completeness by using algebraic methods. Further, in Skura (1996), he presented Scott-style refutation rules for S4. In Skura (2002), he also constructed Scott-style refutation rules for K4. Recently, in Skura (2019b) Skura explored the relationship between tableaux and refutations in S4. See also his related works (Skura 1994, 1999), and the monograph on refutation systems for modal logics Skura (2013), where, refutation systems and constructions of counter-models from syntactic refutations for several specific modal logics are described.
- In (2015) Fiorentini introduces a refutation calculus RS4 for the modal logic S4 with the subformula property, by using a contraction-free sequent calculus GS4 for S4 with the same property, and shows how to generate an S4-model of any formula refuted by that calculus.
- As already mentioned earlier, refutation systems are very close in purpose to counter-model producing procedures. In addition to the earlier mentioned works exploring that relation, we now also mention (Garg et al. 2012), where Garg et al. present a novel counter-model-producing syntactic decision procedure, based on backwards search in labelled sequent calculi, that applies to several intuitionistic and classical multi-modal logics, thus also obtaining new decidability results.
- One topic not explicitly mentioned yet is the application of algebraic semantics and algebraic methods for the construction and analysis of refutation systems for non-classical logics. Such methods have been applied in Skura (1989, 1992, 1994, 1995) and, more recently, in Citkin (2013), where Citkin uses Jankov-style formulas to construct refutation rules and systems. These ideas were developed further in Citkin (2015a, b).

3.6 Refutation Systems for Other Non-classical Logics

Refutation systems have been developed for several other non-classical logics, too, including the following.

- In (1996) Bonatti developed sequent refutation calculi for default and autoepistemic logics. Following his approach, Egly and Tompits constructed in Egly and Tompits (1997) a similar sequent refutation calculus for the intuitionistic default logic. Later, in Bonatti and Olivetti (2002), the authors extended Bonatti's work to sequent refutation calculi for propositional nonmonotonic logics.
- In (2007) Sochacki studied axiomatic rejection in the implication-negation fragment of Łukasiewicz's many-valued logic. Later, in Sochacki (2011), he constructed refutation systems for Finn's 3-valued 'nonsense-logic'. See also Sochacki (2010) for a general reference on his related work.
- Kulicki (2012), presented a complete axiomatisation of the quantifier-free fragment of Leśniewski's Ontology, called there a 'pure calculus of names', by employing the method of axiomatic rejection.
- In (2011) Oetsch and Tompits developed Gentzen-type refutation systems for three-valued logics and proposed applications of them for disproving strong equivalence.
- In (2013) Berger and Tompits constructed an axiomatic rejection system for the description logic *ALC*.
- Several refutation calculi and related systems for paraconsistent logics have also been developed: In Skura (2017a) Skura used ideas from refutation systems, discussed further here, in Sect. 5.1.1 to prove that Segerberg's logic $JE'E_1^Q$ is the greatest paraconsistent extension of Johansson's logic.
  In (2018), Pulcini and Varzi observed that any decidable logic which semantically circumscribes a set of truth-functional contingent formulas, classical propositional logic *in primis*, allows for a refutational counterpart which is paraconsistent. The authors stress this observation to draw some philosophical remarks about the notion of paraconsistency.
  In (2018), Trybus developed a method for constructing maximal paraconsistent logics, motivated by ideas from refutation systems.
  Other applications to paraconsistent logics can be found in Skura (2004a), also mentioned in Sect. 3.7 and used further in this paper, in Sect. 5.1.1.
- In (2017b), Skura constructs Scott-style refutation rules for Wansing's nonmonotonic logic and uses them to provide an efficient decision procedure for that logic.

3.7 Systems Combining Proofs and Refutations

Another important line of research on refutation systems is the idea of building systems of deduction combining proofs and refutations on a par. Here are several references exploring that idea.

- The idea of developing combined derivation systems for proofs and refutations was discussed at the end of Goranko (1994), where some *hybrid refutation rules* for such systems were suggested. That idea has recently been further developed by designing and exploring combined ('hybrid') proof/refutation systems of natural deduction for classical and intuitionistic logic.
- The idea is also close to the 'complementary systems' for sentential logics proposed in Bonatti and Varzi (1995).
- In (2004a), Skura studies symmetric inference systems (that is, pairs of inference systems), that can be used for characterizing maximal non-classical logics with certain properties. The method is also applied there to paraconsistent logics.
- In (2011) Wybraniec-Skardowska and Waldmajer explore the general theory of deductive systems employing the two dual consequence operators, the standard logical consequence, inferring validities, and the refutation consequence, inferring non-validities.
- In (2008) Caferra and Peltier, motivated by potential applications to automated reasoning, take a unifying perspective on accepting or rejecting propositions by deriving them from other, already accepted or rejected propositions.
- In (2008) Goré and Postniece combine derivations and refutations to obtain cut-free complete systems for bi-intuitionistic logic.
- In (2013), Negri explores the duality of proofs and counter-models in labelled sequent calculi, further developed in the already discussed (Negri 2014).
- Likewise, in (2018), Fiorentini and Ferrari explore the duality between unprovability and provability in forward proof-search for intuitionistic propositional logic, following on their previous, already mentioned work (Fiorentini and Ferrari 2017).

3.8 Further General References on Refutation Systems

More recent further general references on the theory of refutation systems include the following.

- A general theory of refutation systems, involving the concept of multiple-conclusion (or Scott-style) consequence relation is presented in Skura (2009),
- An introduction to refutation systems, describing and discussing a variety of concepts, methods, and tools connected with them, is provided in Skura (2011b).
- A general method for proving completeness with respect to refutation rules involving certain normal forms is presented in Skura (2011a).
- The relationship between refutations, tableaux and counter-models constructions in classical logic is discussed in Skura (2016).

4 Refutation Systems Versus Deductive Systems

Here we discuss and compare briefly refutation systems with some of the main types of traditional deductive systems, viz., tableaux and sequent calculi. We also outline the closely related idea of 'inverse refutation systems'. For axiomatic refutation systems we refer the reader, e.g. to Skura (2011b) and for natural deduction style refutation systems—to Tamminga (1994).

4.1 Tableaux Versus Refutation Systems

Refutation systems and tableau systems[3] are naturally related, as both are used to establish falsifiability of the input formula. If tableaux are regarded as satisfiability testing procedures then, as noted in Skura (2016), tableaux and refutation systems are complementary. Still, there are some clear distinctions, both conceptual and technical, which we discuss here.

Tableau procedures are typically used as *proof-search procedures*: if there is a closed tableau for $\neg A$, then A is proved. When terminating, they can also be used as *validity/satisfiability testing decision procedures*: if there is no closed tableau for $\neg A$, then a (usually finite) counter-model for A can be constructed from open and saturated tableaux for $\neg A$. Note however, that, unlike tableaux (which are derivations in an object language), such model constructions are done on a meta-logical level.

A typical (pure) refutation rule is of the form

$$\frac{\dashv A_1, \ldots, \dashv A_m}{\dashv A}$$

meaning that if every A_i is (derived as) non-valid, then so is A.

If we turn such a rule around (bottom-up), by contraposition we obtain an **inverse refutation rule**, preserving validity:

$$\frac{\vdash A}{\vdash A_1 \mid \ldots \mid \vdash A_m}$$

meaning that if A is (derived as) valid, then so is some A_i. (Rules of this kind were introduced in Skura (2005, 2011b).) Using them will facilitate our further discussion.

In a tableau procedure, in order to prove the validity of A, one assumes that A is non-valid, hence $\neg A$ is satisfiable. Starting with $\neg A$ as input, one then applies (satisfiability preserving) tableau rules, the purpose of which is to search systematically for a satisfying truth-assignment (more generally, satisfying model) for the input formula $\neg A$. The aim, however, is to show that this is not possible, by eventually producing a *closed tableau*: a finite tree in which every branch ends with (the

[3]For background on tableau methods for non-classical logics we refer the reader to Goré (1999).

unsatisfiable) $\bot$. If such a closed tableau is constructed, it means that $\neg A$ cannot be satisfied. This contradiction with the initial assumption implies that A is valid.

Dually, in a refutation procedure, in order to refute A, we can assume that A is valid. Then, by applying (validity preserving) branching inverse refutation rules as above we aim to obtain a finite tree in which every branch ends with a refutation axiom (which is non-valid). If such a tree is constructed, this is a contradiction with the initial assumption, implying that A is non-valid.

Thus, tableau systems (as proof procedures) and refutation systems (as refutation procedures) can be regarded as mutually dual calculi, and they can be studied by using similar formal tools. We illustrate that in the next subsection.

4.2 Inverse Refutation Systems

Given finite sets X, Y, Z of (modal) formulae, we define

$\Box X := \{\Box A : A \in X\}$, $X \longrightarrow Y := \bigwedge X \to \bigvee Y$.

When $Y = \{C\} \cup Z$, we may write simply $X \longrightarrow C, Z$.

Note that an application of an inverse refutation rule to a formula produces $m \geq 1$ immediate successors. These successors can be written just as a set Ψ, representing a multiple-conclusion list $B_1 \mid ... \mid B_k$, where $\Psi = \{B_1, ..., B_k\}$.

An **inverse refutation tree** for a formula A is a finite tree with root labelled by A, whose nodes have labels containing formulas. That tree is expanded by applying inverse refutation rules to the node labels, and the labels of the leaves (end nodes) are *refutation axioms*, specifically defined for the system. An **inverse refutation system DR** consists of a set of refutation axioms and a set of inverse refutation rules. A formula A is **refutable** in an inverse refutation system **DR** iff there is an inverse refutation tree for A in **DR**. A system **DR** is **Ł-complete** for a logic L iff we have $A \notin \mathsf{L}$ iff A is refutable in **DR**.

For example, in modal logic, complete inverse refutation systems can be obtained by reformulating the refutation systems involving normal forms in Skura (2002, 2013). Roughly speaking, such an inverse refutation system for a normal modal logic L contains two rules: MP and a characteristic inverse refutation rule R_{L}. (In fact, usually the whole MP is not necessary, but a few specific cases of MP are sufficient (see Skura 2013)). For a concrete example, the characteristic rule for the basic modal logic **K** looks as follows (cf. Lemmon and Scott 1977, p. 46; Goranko 1994; Skura 1999).

(R_K):

$$\frac{\vdash \Box\Phi \longrightarrow C, \Box\Psi}{\{\vdash \Phi \longrightarrow B : B \in \Psi\}}$$

where C is a $\Box$-free formula that is not (valid) in CL.

It seems that the meta-logical constructions of counter-models for **K** from open tableaux (see Goré 1999) and our syntactic refutations (which are formal derivations) are different descriptions of essentially the same procedure. Both terminate because the number of modal connectives is reduced in each step.

However, in other logics, the procedures may generally be different, for the following reasons. The rule R_K (mirroring the modal model construction) is not a sound refutation rule for any proper extension of **K**, as it suffices to refute every formula that is not in **K**. So, new refutation rules must be provided for other modal logics. It seems that, in transitive logics like **S4**, refutation rules must be restricted to Mints-style normal forms (cf. Skura 2013). But it is in these logics that refutation systems really make a difference because they provide refutations that, at least in some cases, are simpler than the tableau counter-model constructions (cf. Skura 2013, p. 125). Also, unlike tableau procedures for transitive logics, our refutation procedures are reduction procedures; in each step the rank of the Mints-style normal form (a specified natural number) is reduced, so our procedures are cycle-free.

Finally, we remark that counter-models can be obtained directly from inverse refutation trees, by removing the nodes to which MP was applied and extracting a suitable valuation from the normal forms involved. For details, see Skura (2002, 2013, 2017a). A more detailed analysis of the relationship between refutation systems and tableau systems (illustrated by **S4**) is given in Skura (2019b).

4.3 Gentzen-Style Sequent Refutation Systems

4.3.1 The Antisequent System $\overline{\mathsf{LK}}$ for Classical Propositional Logic

We use capital Greek letters $\Gamma, \Delta, \ldots$ to stand for finite multisets of formulas $[A_1, A_2, \ldots, A_n]$. To simplify our notation, we write Γ, A and Γ, Δ to mean the multisets $\Gamma \uplus [A]$ and $\Gamma \uplus \Delta$, respectively. If $\Gamma = A_1, A_2 \ldots, A_n$, then $\bigwedge \Gamma \equiv A_1 \wedge A_2 \wedge \cdots \wedge A_n$ and $\bigvee \Gamma \equiv A_1 \vee A_2 \vee \cdots \vee A_n$. With LK we refer to Gentzen's sequent calculus for classical propositional logic, as introduced in Gentzen (1935), Szabo (1969).

Refutational counterparts of ordinary sequent systems are often called in the literature *antisequent* systems. Antisequents are usually written as $\Gamma \dashv \Delta$ and they differ from standard sequents *à la* Gentzen insofar as they are intended to affirm their own invalidity. In semantical terms, affirming $\Gamma \dashv \Delta$ corresponds to say that there is a valuation v such that $v(\bigwedge \Gamma \rightarrow \bigvee \Delta) = 0$. Notice that, in spite of the use of the inverted turnstile symbol $\dashv$, the validity of $\Gamma \dashv \Delta$—i.e. the *in*validity of $\Gamma \vdash \Delta$ and so its unprovability in LK—does not necessarily imply the validity of its mirror image $\Delta \vdash \Gamma$. For a very simple example, the sequent $p \vdash q$ is invalid, therefore the *anti*sequent $p \dashv q$ is valid, but the sequent $q \vdash p$ is classically invalid as well.

The first antisequent system for classical logic was constructed by Tiomkin in Tiomkin (1988). The propositional part of Tiomkin's calculus consists in one refutation axiom scheme together with left- and right-introduction rules for negation and disjunction. The refutation axiom scheme allows the user to introduce any antisequent $\Gamma \dashv \Delta$ provided that Γ and Δ are *two disjoint multisets of atoms* (example: $p, q \dashv t$). Thus, the empty antisequent '$\dashv$' is readily provable in the system as a limit

Axiom:

$$\frac{}{\Gamma \dashv \Delta}\ \overline{ax}. \quad \text{where } \Gamma \text{ and } \Delta \text{ are two finite multisets of atoms s.t. } \Gamma \cap \Delta = \varnothing$$

Logical rules:

$$\frac{\Gamma, A, B \dashv \Delta}{\Gamma, A \wedge B \dashv \Delta}\ \wedge\dashv \qquad \frac{\Gamma \dashv \Delta, A}{\Gamma \dashv \Delta, A \wedge B}\ \dashv\wedge_{\mathcal{R}} \qquad \frac{\Gamma \dashv \Delta, B}{\Gamma \dashv \Delta, A \wedge B}\ \dashv\wedge_{\mathcal{L}}$$

$$\frac{\Gamma, A \dashv \Delta}{\Gamma, A \vee B \dashv \Delta}\ \vee_{\mathcal{R}}\dashv \qquad \frac{\Gamma, B \dashv \Delta}{\Gamma, A \vee B \dashv \Delta}\ \vee_{\mathcal{L}}\dashv \qquad \frac{\Gamma \dashv A, B, \Delta}{\Gamma \dashv A \vee B, \Delta}\ \dashv\vee$$

$$\frac{\Gamma \dashv A, \Delta}{\Gamma, A \to B \dashv \Delta}\ \to\dashv_{\mathcal{R}} \qquad \frac{\Gamma, B \dashv \Delta}{\Gamma, A \to B \dashv \Delta}\ \to\dashv_{\mathcal{L}} \qquad \frac{\Gamma, A \dashv B, \Delta}{\Gamma \dashv A \to B, \Delta}\ \dashv\to$$

$$\frac{\Gamma \dashv A, \Delta}{\Gamma, \neg A \dashv \Delta}\ \neg\dashv \qquad \frac{\Gamma, A \dashv \Delta}{\Gamma \dashv \neg A, \Delta}\ \dashv\neg$$

Fig. 1 The $\overline{\mathsf{LK}}$ sequent calculus

$$\cfrac{\cfrac{\cfrac{\cfrac{\cfrac{}{\dashv q, p, p}\ \overline{ax}.}{\neg p \dashv q, p}\ \neg\dashv}{p \to q, \neg p \dashv q}\ \to\dashv_{\mathcal{R}}}{p \to q \dashv \neg p \to q}\ \dashv\to}{\dashv (p \to q) \to (\neg p \to q)}\ \dashv\to$$

Fig. 2 An example of a $\overline{\mathsf{LK}}$-proof

case of the refutation axiom wherein $\Gamma = \Delta = \varnothing$. As noted in Sect. 3.3, antisequent calculi for the full language of PL were independently constructed a few years later in Goranko (1994) and (reportedly) in Bonatti (1993).

From now on we will write $\overline{\mathsf{LK}}$ to refer to the full antisequent system, as formulated in Goranko (1994) (cf. Fig. 1). Notice that $\overline{\mathsf{LK}}$'s logical rules are all unary so that $\overline{\mathsf{LK}}$-proofs turn out to be simple chains of antisequents. From semantic viewpoint, logical rules are designed in a way that the very valuation that falsifies the premise, also falsifies the conclusion. In Fig. 2 we give an example of a $\overline{\mathsf{LK}}$-derivation ending in the truth-functional contingency $(p \to q) \to (\neg p \to q)$.

In general we have that

- $\overline{\mathsf{LK}}$ proves $\dashv A$ if and only if A is *not* a tautology.
- $\overline{\mathsf{LK}}$ proves $A \dashv$ if and only if A is *not* a contradiction.

Therefore, by combining the two previous points, we can conclude that $\overline{\mathsf{LK}}$ proves both $\dashv A$ and $A \dashv$ if and only if A is a truth-functional contingency (Tiomkin 2013).

In Carnielli and Pulcini (2017), the authors observed how $\overline{\mathsf{LK}}$ logical rules can be algorithmically generated by a bottom-up reading of the tableau rules for classical logic once: (i) formulas labelled as true are placed on the left-hand side of the sequent symbol, (ii) formulas labelled as false are placed on the right-hand side, and (iii) context variables are added on both sides of the sequents. The guiding idea is that branching tableau rules generate two distinct $\overline{\mathsf{LK}}$ logical rules, whereas non-branching rules generate a single $\overline{\mathsf{LK}}$ logical rule. This procedure is similar to the one employed in D'Agostino (1999) to derive the rules of sequent systems for classical logic dealing with implicit structural rules (cfr. Kleene's system G4 or its variant G3 Kleene 1967; Negri et al. 2008) from tableau rules.

As an illustrative example, consider the tableau rules for classical conjunction. First, consider the case in which the formula $A \wedge B$ is true, namely,

$$\begin{array}{c} \mathrm{T} : A \wedge B \\ | \\ \mathrm{T} : A \\ | \\ \mathrm{T} : B \end{array}$$

In this case we proceed by putting all the true formulas on the left-hand side of the antisequent symbol, distinguishing the conclusions from the premises as follows: $\dfrac{A, B \dashv}{A \wedge B \dashv}$. Then we add context variables on both sides so as to finally achieve the following $\overline{\mathsf{LK}}$ rule:

$$\frac{\Gamma, A, B \dashv \Delta}{\Gamma, A \wedge B \dashv \Delta} \wedge \dashv.$$

When we focus, instead, on the case in which $A \wedge B$ is taken to be false, we have to consider the following branching rule

$$\begin{array}{c} \mathrm{F} : A \wedge B \\ \diagup \quad \diagdown \\ \mathrm{F} : A \quad \mathrm{F} : B \end{array}$$

which is expected to produce two different logical rules, one for each branch. Consider, for instance, the left-most branch, the one claiming the falsity of the first conjunct. In this case we rewrite the rule as follows: $\dfrac{\dashv A}{\dashv A \wedge B}$. Then we add contexts variables on both sides so as to get the following logical rule:

$$\frac{\Gamma \dashv \Delta, A}{\Gamma \dashv \Delta, A \wedge B} \dashv \wedge_{\mathcal{R}}.$$

Each of the other logical rules can be obtained by applying the procedure in the same way.

Two interesting facts are worth observing at this point of our discussion. The first is that $\overline{\mathsf{LK}}$ is paraconsistent, in the sense that, for any truth-functional contingency A,

$\overline{\mathsf{LK}}$ proves both the antisequents $\dashv A$ and $\dashv \neg A$. This peculiarity of the calculus has been emphasized in Pulcini and Varzi (2018) to draw some philosophical remarks and reflect light on the logical nature of the notion of *paraconsistency*, cf. Priest (2002), Carnielli and Coniglio (2016). The other fact is that $\overline{\mathsf{LK}}$ can be classified as a substructural logic insofar as it can be proved sound and complete with respect to the set of invalid sequents without resorting to the structural rules Weakening and Contraction, neither explicitly nor implicitly.

4.3.2 Cut-Elimination and Related Computational Properties

As is well-known, there are two ways to formulate the cut-rule. The context-mixing version is termed *multiplicative* cut:

$$\frac{\Gamma \vdash \Delta, A \qquad \Gamma', A \vdash \Delta'}{\Gamma, \Gamma' \vdash \Delta, \Delta'}\;; \tag{MC}$$

whereas the context-sharing rule is called *additive* cut:

$$\frac{\Gamma \vdash \Delta, A \qquad \Gamma, A \vdash \Delta}{\Gamma \vdash \Delta}\;. \tag{AC}$$

In presence of the structural rules Weakening and Contraction, the two versions MC and AC can be easily proved to be equivalent. However, as already observed, the $\overline{\mathsf{LK}}$ calculus is a completely substructural system and so this equivalence cannot be imported from the affirmative to the refutational side. As a matter of fact, whereas the additive version proves admissible in $\overline{\mathsf{LK}}$ (namely, adding AC to the $\overline{\mathsf{LK}}$ sequent calculus does not entail new provable antisequents), its multiplicative counterpart is not. The following is an easy counterexample to the admissibility of MC in $\overline{\mathsf{LK}}$.

$$\frac{p \dashv q \qquad q \dashv p}{p \vdash p}\;\mathrm{MC}$$

Tiomkin proposes the two following 'hybrid' structural rules that he calls 'cuts for the unprovability' (Tiomkin 1988).

$$\frac{\Gamma \dashv \Delta \qquad \Gamma \vdash \Delta, A}{\Gamma, A \dashv \Delta} \qquad\qquad \frac{\Gamma \dashv \Delta \qquad \Gamma, A \vdash \Delta}{\Gamma \dashv \Delta, A}$$

The problem with these rules is that they do not violate the subformula property so, from a strict proof-theoretic point of view, they can be hardly defined as 'cuts' for all intent and purposes: where is the cut-formula? what is to be cut?

As shown in Carnielli and Pulcini (2017), the additive cut, though admissible, is unnecessarily general for $\overline{\mathsf{LK}}$, in the sense that it can be replaced by the two following simpler unary cuts.

$$\frac{\Gamma, A \dashv \Delta}{\Gamma \dashv \Delta}\;\mathit{cut}\dashv \qquad\qquad \frac{\Gamma \dashv \Delta, A}{\Gamma \dashv \Delta}\;\dashv\mathit{cut}$$

Once these rules are added to the system, AC turns out not only admissible, but also trivially derivable.

In Carnielli and Pulcini (2017), the authors provide a fully syntactical proof of cut-elimination, i.e. any antisequent provable by means of a proof that resorts to cut applications is also provable by means of a cut-free proof. Such a result is, of course, implicit in the completeness proof provided, for instance, in Tiomkin (1988). However, the fact of focussing on the dynamics of a normalization procedure instead of dealing with the static concept of cut-redundancy provides the full setting for the study of the computational properties of $\overline{\mathsf{LK}}$. In particular, the normalization procedure outlined in Carnielli and Pulcini (2017) has been shown to enjoy the weak normalization property as well as the uniqueness of the normal form (provided that contexts in antisequents are taken as sequences of formulas). Computational properties related to normalization in $\overline{\mathsf{LK}}$ have been further optimized in Pulcini and Varzi (2019) by resorting to the proof-nets technique.

In the traditional terminology of proof theory, a proof is said to be *analytic* in case any formula displayed in the proof is a subformula of some formula among those displayed in the end-sequent. To take an example, the proof presented in Fig. 2 is analytic. In both LK and $\overline{\mathsf{LK}}$, cut-elimination implies the fact that any provable (anti)sequent is provable analytically, i.e. by means of an analytic proof. However, in LK, analyticity does not necessarily imply simplicity since cut-elimination may increase the size of LK-proofs. On the contrary, the implementation of cut-elimination in the complementary system $\overline{\mathsf{LK}}$ always returns one of the simplest proofs (intended as one having minimal length) for any provable antisequent (Carnielli and Pulcini 2017).

5 Some Applications to Non-classical Logics

5.1 *Defining Non-classical Logics via Refutability*

5.1.1 Positive and Non-negative Ways of Defining a Logic

Propositional logics are usually defined in a positive way, as the least sets of formulas containing certain given formulas (axioms, deemed a priori acceptable/valid) and closed under given inference rules (usually, *uniform substitution* SUB, MP, and possibly others). However, non-classical logics are often motivated by negative requirements saying that certain principles are not acceptable, hence any formulas that formalize or imply them should be rejected. For example, Intuitionistic Logic can be obtained from some systems of Classical Logic CL by rejecting the law of excluded middle (LEM) $p \vee \neg p$. Or, paraconsistent logics (see Skura 2004a) can be obtained from CL by rejecting the law of explosion:

$$\text{(XP)} \qquad \neg p \to (p \to q)$$

A positive way to define such a logic would be to

1. Specify a set POS of formulas (axioms) that we regard as acceptable.
2. Ensure that the rejected law is not derivable from POS by the inference rules.
3. Define the new logic as the least set of formulas containing POS and closed under the inference rules.

In Skura (2004a), a new, 'non-negative' method of defining a logic was introduced. It can be described as follows.

- In addition to POS, specify a set NEG of rejected formulas.
- Make sure that no $A \in$ NEG is derivable from POS.
- Define the set Ref(POS, NEG) (of *refutable formulas*) as follows: $A \in \mathsf{Ref}$(POS, NEG) iff some $B \in$ NEG is derivable from A by using POS and the inference rules.
- Define the new logic to be the set of non-refutable formulas (that is, the complement $-\mathsf{Ref}$(POS, NEG) of Ref(POS, NEG)).

Metaphorically speaking, in the positive approach we want to keep in what is good, and in the 'non-negative' approach we want to keep out what is bad.

We now briefly discuss two examples of non-classical logics obtained from CL in the non-negative way.

5.1.2 Paraconsistent Logics

Let FOR($\rightarrow, \wedge, \vee, \bot$) be the set of all formulas generated from the set VAR of propositional variables by the binary connectives $\rightarrow, \wedge, \vee$ and the constant $\bot$ (falsum). We define $\neg A := A \rightarrow \bot$.

We take as POS the axioms of Positive Logic, i.e. the fragment of the intuitionistic logic IPL with the connectives $\rightarrow, \wedge, \vee$ (cf. e.g. Odintsov 2008). As NEG we take $\{\mathsf{XP}\}$.

Johansson's logic J (cf. Odintsov 2008) is the least set of formulas of FOR($\rightarrow, \wedge, \vee, \bot$) containing POS and closed under SUB and MP. We note that adding XP to **J** results in IPL.

For any $A \in$ FOR, the symbol $\mathbf{J}A$ denotes the extension of **J** with the additional axiom A, i.e. the least set of formulas containing $\mathbf{J} \cup \{A\}$ and closed under uniform substitution and MP. For example, the logic $\mathbf{J}E$, where

$$(\mathsf{E}) \qquad p \vee (p \rightarrow q) \qquad \text{(the extended law of excluded middle)}$$

is called '*the logic of classical refutability*' (see Odintsov 2008).

Then, by using some results in Skura (2004a, 2017a) and in Odintsov (2008), it can be shown that

$$-\mathsf{Ref}(\text{POS}, \text{NEG}) = \mathbf{J}E.$$

As an interesting corollary, we obtain the following. Let L be an (axiomatic) extension of **J**. We say that L is a **paraconsistent analogue of** CL iff $\mathsf{L} \subseteq \mathsf{CL}$ and $\mathsf{XP} \notin \mathsf{L}$. (Hence, $\mathsf{L} \subset \mathsf{CL}$.) It turns out that $-\mathsf{Ref}$(POS, NEG) is the greatest paraconsistent analogue of CL (see Skura 2004a, Corollary 3.2).

5.1.3 Implicational Relevance Logics

A logic L is said to have the **variable-sharing property** (VSP) if

$A \to B$ is non-valid in L*, whenever A and B share no variables.*

For instance, *relevance logics* are logics with the VSP (see Anderson and Belnap 1975).

The VSP is readily presented as a negative property. It turns out (see Skura 2019a) that, in a large class of *implicational logics* (i.e. logics with $\to$ as the only logical connective), the VSP is equivalent to the simpler property that the following formula PP, regarded as unacceptable in relevance logics, is non-valid:

$$(\mathsf{PP}) \qquad p \to (q \to p) \qquad \text{(Positive Paradox)}$$

We now take the axioms of the implicational fragment $\mathbf{R}_{\to}$ of the relevance logic **R** as POS and $\{\mathsf{PP}\}$ (or all formulas $A \to B$, where A and B share no variables) as NEG.

Characterizing the set $-\mathsf{Ref}$(POS, NEG) turns out to be a hard technical problem (see Skura 2019a for some open problems). However, extending the basic logic to $\mathbf{RMO}^{\top}_{\to}$, defined below, makes the task feasible.

Let FOR$(\to, \top)$ be the set of all formulas generated from VAR by the connective $\to$ and the constant $\top$. The logic $\mathbf{RMO}^{\top}_{\to}$ is defined as the least set of formulas in FOR$(\to, \top)$ closed under the rules SUB and MP, and containing the following axioms (cf. Avron 1984, 1990; Dunn and Restall 2002).

- $p \to \top$
- $p \to (p \to p)$
- $(p \to q) \to ((q \to r) \to (p \to r))$
- $(p \to (q \to r)) \to (q \to (p \to r))$
- $(p \to (p \to q)) \to (p \to q)$

Consider the logical matrix (see Skura 2019a for details)

$$\mathbf{3} := (\{-1, 0, 1\}, \{0, 1\}, \to),$$

where the set of designated values is $\{0, 1\}$ and $x \to y = max(-x, y)$ if $x \leq y$; else $x \to y = min(-x, y)$. A **valuation** in the matrix **3** is a function v from VAR to $\{-1, 0, 1\}$ extended with $v(\top) = 1$ and $v(A \to B) = v(A) \to v(B)$.

We say that a formula A is **valid in 3**, denoted $A \in Val(\mathbf{3})$, iff $v(A) \in \{0, 1\}$ for every valuation v. We remark that $\mathbf{RMO}^{\top}_{\to} \subseteq Val(\mathbf{3})$, and $\mathsf{PP} \notin Val(\mathbf{3})$.

Let us now take as POS the set of axioms of $\mathbf{RMO}_{\rightarrow}^{\top}$ listed above, and as NEG we take {PP}. Then, as shown in Skura (2019a), we have

$$-\mathsf{Ref}(\text{POS}, \text{NEG}) = Val(\mathbf{3}).$$

Also, let L be an axiomatic extension of $\mathbf{RMO}_{\rightarrow}^{\top}$. We say that L is a **relevant analogue** of CL iff $\mathsf{L} \subseteq \mathsf{CL}$ and $\mathsf{PP} \notin \mathsf{L}$. (Thus, $\mathsf{L} \subset \mathsf{CL}$.) It follows that $-\mathsf{Ref}(\text{POS}, \text{NEG})$ is the greatest relevant analogue of CL (see Skura 2004a, Corollary 3.2).

6 Concluding Remarks

Refutation systems are not yet explored nearly as deep as traditional logical deductive systems, and there is an abundance of related open questions and potential applications, in addition to those discussed in the present paper. We only briefly mention here one such application, viz. to provide a recursive axiomatization of the non-validities of a logic that may not be (known to be) recursive axiomatizable, but the non-validities of which are known to be effectively enumerable. A prime example is $\mathsf{FOL}^{\mathsf{fin}}$, for which several Ł-complete refutation systems proposed in the literature (referenced in Sect. 3.4) illustrate the idea. More generally, recursive axiomatizatizable Ł-complete refutation systems exist for any logic or theory that is defined in terms of an effectively enumerable class of finite models. For some generic results of that type see Goranko and Skura (2018).

A well-known example of a logic with no known (and, possibly not existing) recursive axiomatization is *Medvedev's logic of finite problems*, for which a surprisingly simple Ł-complete refutation system (though, employing deduction in Kreisel–Putnam's logic KP) was designed in Skura (1992). There are, however, many less-known such logical systems waiting for Ł-complete refutation systems to be designed for them; just one example is the interval temporal logic of reflexive subinterval relation over the integers, proved to be (surprisingly) undecidable in Marcinkowski and Michaliszyn (2011), while it has the finite model property by definition, and therefore an effectively enumerable set of non-validities.

In conclusion, the inherent importance and potential of the idea of refutation systems is yet to be unravelled. That idea is still waiting for new, sufficiently convincing applications that would attract the attention of the wider logic community and would eventually properly address Łukasiewicz's concern about the disparity between deductive acceptance and rejection of a logical statement, raised in Łukasiewicz (1951). Discovering such applications is the so far main challenge in that field.

Acknowledgements The work of Valentin Goranko was partly supported by a research grant 2015-04388 of the Swedish Research Council. Gabriele Pulcini thankfully acknowledges the support from the Dutch Research Council (NWO) through the Open Competition-SSH project 406.18.TW.009

"A Sentence Uttered Makes a World Appear—Natural Language Interpretation as Abductive Model Generation".

References

Anderson, A.R., and N.D. Belnap. 1975. *Entailment*, vol. 1. Princeton: Princeton University Press.

Avron, A. 1984. Relevance entailment: semantics and formal systems. *Journal of Symbolic Logic* 49: 334–342.

Avron, A. 1990. Relevance and paraconsistency - a new approach Part II: The formal systems. *Notre Dame Journal of Formal Logic* 31: 169–202.

Berger, G., and H. Tompits. 2013. On axiomatic rejection for the description logic *ALC*. In *Proceedings of KDPD 2013*, vol. 8439. Lecture notes in computer science, 65–82. Berlin: Springer.

Bonatti, P. 1993. A Gentzen system for non-theorems. Technical report, Technische Universität Wien, Institut für Informationssysteme.

Bonatti, P., and A.C. Varzi. 1995. On the meaning of complementary systems. In *10th international congress of logic, methodology and philosophy of science. Volume of abstracts*, ed. E. Castagli, and M. Konig, 122.

Bonatti, P.A. 1996. Sequent calculi for default and autoepistemic logics. In *Proceedings of TABLEAUX '96*, 127–142.

Bonatti, P.A., and N. Olivetti. 2002. Sequent calculi for propositional nonmonotonic logics. *ACM Transactions on Computational Logic* 3 (2): 226–278.

Borkowski, L. (ed.). 1970. *J. Łukasiewicz: Selected works*. Amsterdam: North-Holland Publishing Co.

Bry, F., and S. Torge. 1998. A deduction method complete for refutation and finite satisfiability. In *Proceedings of JELIA '98*, 122–138.

Caferra, R., and N. Peltier. 2008. Accepting/rejecting propositions from accepted/rejected propositions: A unifying overview. *International Journal of Intelligent Systems* 23 (10): 999–1020.

Caicedo, X. 1978. A formal system for the non-theorems of the propositional calculus. *Notre Dame Journal of Formal Logic* 19 (1): 147–151.

Carnielli, W.A., and M.E. Coniglio. 2016. *Paraconsistent logic: Consistency, contradiction and negation*. Berlin: Springer.

Carnielli, W.A., and G. Pulcini. 2017. Cut-elimination and deductive polarization in complementary classical logic. *Logic Journal of the IGPL* 25 (3): 273–282.

Citkin, A. 2013. Jankov-style formulas and refutation systems. *Reports on Mathematical Logic Logic* 48: 67–80.

Citkin, A. 2015a. Characteristic inference rules. *Logica Universalis* 9 (1): 27–46.

Citkin, A. 2015b. A meta-logic of inference rules: Syntax. *Logic and Logical Philosophy* 24: 313–337.

D'Agostino, M. 1999. Tableau methods for classical propositional logic. In *Handbook of tableau methods*, 45–123. Berlin: Springer.

Dunn, J.M., and G. Restall. 2002. Relevance logic. In *Handbook of philosophical Logic*, vol. 6, ed. D.M. Gabbay, and F. Guenthner, 1–128. Kluwer.

Dutkiewicz, R. 1989. The method of axiomatic rejection for the intuitionistic propositional logic. *Studia Logica* 48 (4): 449–459.

Egly, U., and H. Tompits. 1997. A sequent calculus for intuitionistic default logic. In *Proceedings of the 12th workshop on logic programming WLP'97*, Tech. Rep. PMS-FB-1997-10, 69–79. LMU, München.

Fiorentini, C. 2015. Terminating sequent calculi for proving and refuting formulas in S4. *Journal of Logic and Computation* 25 (1): 179–205.

Fiorentini, C., and M. Ferrari. 2017. A forward unprovability calculus for intuitionistic propositional logic. In *Proceedings of TABLEAUX 2017*, vol. 10501, LNCS, 114–130. Berlin: Springer.

Fiorentini, C., and M. Ferrari. 2018. Duality between unprovability and provability in forward proof-search for intuitionistic propositional logic. CoRR arXiv:abs/1804.06689.

Garg, D., V. Genovese, and S. Negri. 2012. Countermodels from sequent calculi in multi-modal logics. In *Proceedings of the 27th annual IEEE symposium on logic in computer science, LICS 2012, Dubrovnik, Croatia, June 25–28, 2012*, 315–324. IEEE Computer Society.

Gentzen, G. 1935. Unstersuchungen über das logische Schliessen. *Mathematische Zeitschrift* 39: 176–210. English translation in (Szabo, 1969).

Goranko, V. 1991. Proving unprovability in some normal modal logics. *Bulletin of the Section of Logic, PAS* 20 (1): 23–29.

Goranko, V. 1994. Refutation systems in modal logic. *Studia Logica* 53 (2): 299–324.

Goranko, V., and T. Skura. 2018. Refutation systems in the finite. In *Games, Cognition, Logic, College Publications*, ed. M. Urbański, T. Skura, and P. Łupkowski, 2020, 173–184.

Goré, R. 1999. Tableau methods for modal and temporal logics. In *Handbook of Tableau methods*, ed. M. D'Agostino, D. Gabbay, R. Hähnle and J. Posega, 297–396. Kluwer.

Goré, R., and L. Postniece. 2008. Combining derivations and refutations for cut-free completeness in bi-intuitionistic logic. *Journal of Logic and Computation* 20: 233–260.

Goudsmit, J.P. 2014. Admissibility and refutation: some characterisations of intermediate logics. *Archive for Mathematical Logic* 53 (7): 779–808.

Goudsmit, J.P. 2015. Intuitionistic rules : Admissible rules of intermediate logics. PhD thesis, University of Utrecht, Utrecht.

Hailperin, T. 1961. A complete set of axioms for logical formulas invalid in some finite domains. *ZML* 7: 84–96.

Härtig, K. 1960. Zur axiomatisierung der nicht-identitäten des aussagenkalkül. *ZML* 6: 240–247.

Inoué, T. 1989. A note on Stahl's opposite system. *Mathematical Logic Quarterly* 35 (5): 387–390.

Inoué, T. 1989. On Ishimoto's theorem in axiomatic rejection - the philosophy of unprovability, (in Japanese). *Philosophy of Science* 22 (5): 77–93.

Inoué, T. 1990. A note on unprovability-preserving sound translations. *Logique Et Analyse* 33 (31): 243–257.

Inoué, T. 1994. On the atomic formula property of Härtig's refutation calculus. *Bulletin of the Section of Logic, PAS* 23 (4): 146–150.

Ishimoto, A. 1981. On the method of axiomatic rejection in classical propositional logic, (in Japanese). *Philosophy of Science* 14 (5): 45–60.

Kleene, S.C. 1967. *Mathematical logic*. New Jersey: Wiley.

Kreisel, G., and H. Putnam. 1957. Eine Unableitbarkeitsbeweismethode für den intuitionistischen Aussagenkalkül. *Archiv für mathematische Logik und Grundlagenforschung* 3: 74–78.

Kulicki, P. 2000. The use of axiomatic rejection. In *Logica yearbook 1999*. Filosophia.

Kulicki, P. 2002. Remarks on axiomatic rejection in Aristotle's syllogistic. *Studies in Logic and Theory of Knowledge* 5: 231–236.

Kulicki, P. 2012. An axiomatisation of a pure calculus of names. *Studia Logica* 100 (5): 921–946.

Lemmon, E., and D. Scott. 1977. *An introduction to modal logic, in collaboration with D. Scott*. Oxford: Blackwell.

Łukasiewicz, J. 1921. Logika dwuwartościowa (Two-valued logic). *Przeglad Filozoficzny* 23 (5): 189–205.

Łukasiewicz, J. 1939. *O sylogistyce Arystotelesa. Sprawozdania PAU* 44: 220–227.

Łukasiewicz, J. 1951. *Aristotle's syllogistic from the standpoint of modern formal logic*. Oxford: Clarendon.

Łukasiewicz, J. 1952. On the intuitionistic theory of deduction. *Indagationes Mathematicae* 14: 202–212.

Marcinkowski, J., and J. Michaliszyn. 2011. The ultimate undecidability result for the Halpern-Shoham logic. *Proceeding of LICS 2011*, 377–386.

Mints, G. 1990. Gentzen-type systems and resolution rules. *Lecture Notes in Computer Science* 417: 198–231.
Mints, G. 1992. *A short introduction to modal logic*. CSLI.
Morgan, C., A. Hertel, and P. Hertel. 2007. A sound and complete proof theory for propositional logical contingencies. *Notre Dame Journal of Formal Logic* 48 (4): 521–530.
Morgan, C.G. 1973. Sentential calculus for logical falsehoods. *Notre Dame Journal of Formal Logic* 14: 347–353.
Negri, S. 2013. On the duality of proofs and countermodels in labelled sequent calculi. In *Proceedings of TABLEAUX 2013*, 5–9.
Negri, S. 2014. Proofs and countermodels in non-classical logics. *Logica Universalis* 8 (1): 25–60.
Negri, S., J. Von Plato, and A. Ranta. 2008. *Structural proof theory*. Cambridge: Cambridge University Press.
Odintsov, S. 2008. *Constructive negations and paraconsistency*. Berlin: Springer.
Oetsch, J., and H. Tompits. 2011. Gentzen-type refutation systems for three-valued logics with an application to disproving strong equivalence. In *Proceedings of LPNMR 2011*, 254–259.
Pinto, L., and R. Dyckhoff. 1995. Loop-free construction of counter-models for intuitionistic propositional logic. In *Symposia Gaussiana*, 225–232.
Pinto, L., and T. Uustalu. 2009. Proof search and counter-model construction for bi-intuitionistic propositional logic with labelled sequents. In *Proceedings of TABLEAUX 2009*, 295–309.
Priest, G. 2002. Paraconsistent logic. In *Handbook of philosophical logic*, 287–393. Berlin: Springer.
Pulcini, G., and A.C. Varzi. 2018. Paraconsistency in classical logic. *Synthese* 195 (12): 5485–5496.
Pulcini, G., and A.C. Varzi. 2019. Proof-nets for non-theorems. Forthcoming in Studia Logica.
Scott, D. 1957. Completeness proofs for the intuitionistic sentential calculus. In *Summaries of talks presented at the Summer Institute of Symbolic Logic*, 2nd ed, 231–241. Princeton: Cornell University.
Skura, T. 1989. A complete syntactical characterization of the intuitionistic logic. *Reports on Mathematical Logic* 23: 75–80.
Skura, T. 1990. On pure refutation formulations of sentential logics. *Bulletin of the Section of Logic, PAS* 19 (3): 102–107.
Skura, T. 1991. On decision procedures for sentential logics. *Studia Logica* 50 (2): 173–179.
Skura, T. 1992. Refutation calculi for certain intermediate propositional logics. *Notre Dame Journal of Formal Logic* 33 (4): 552–560.
Skura, T. 1994. Syntactic refutations against finite models in modal logic. *Notre Dame Journal of Formal Logic* 35 (4): 595–605.
Skura, T. 1995. A Łukasiewicz-style refutation system for the modal logic S4. *Journal of Philosophical Logic* 24 (6): 573–582.
Skura, T. 1996. Refutations and proofs in S4. In *Proof theory of modal logic*, ed. H. Wansing, 45–51. Dordrecht: Springer.
Skura, T. 1999. *Aspects of refutation procedures in the intuitionistic logic and related modal systems*. Wrocław: Wydawnictwo Uniwersytetu Wrocławskiego.
Skura, T. 2002. Refutations, proofs, and models in the modal logic K4. *Studia Logica* 70 (2): 193–204.
Skura, T. 2004a. Maximality and refutability. *Notre Dame Journal of Formal Logic* 45: 65–72.
Skura, T. 2004b. Rules and refutation rules for the logic of finite n-ary trees. *Journal of Logic and Computation* 14 (3): 429–435.
Skura, T. 2005. Intuitionistic Socratic procedures. *Journal of Applied Non-Classical Logics* 15 (4): 453–464.
Skura, T. 2009. A refutation theory. *Logica Universalis* 3 (2): 293–302.
Skura, T. 2011a. On refutation rules. *Logica Universalis* 5 (2): 249–254.
Skura, T. 2011b. Refutation systems in propositional logic. In *Handbook of philosophical logic*, vol. 16, ed. D.M. Gabbay, and F. Guenthner, 115–157. Berlin: Springer.
Skura, T. 2013. *Refutation methods in modal propositional logic*. Warszawa: Semper.

Skura, T. 2016. What is a refutation system? In *Let's be logical*, vol. 22, ed. A. Moktefi, A. Moretti, and F. Schang, 189–198. College Publications.
Skura, T. 2017a. The greatest paraconsistent analogue of Intuitionistic Logic. In *Proceedings of the 11th Panhellenic logic symposium*, 71–76.
Skura, T. 2017b. Refutations in Wansing's logic. *Reports on Mathematical Logic* 52: 83–99.
Skura, T. 2019a. Implicational logic, relevance, and refutability. *Logic and Logical Philosophy*. forthcoming.
Skura, T. 2019b. Refutation rules and refined model constructions in the modal logic S4. *In Preparation*.
Słupecki, J. 1955. On Aristotelian syllogistic. *Studia Philosophica* 4: 275–300.
Słupecki, J., and G. Bryll. 1973. Proof of Ł-decidability of Lewis system S5. *Studia Logica* 32 (1): 99–105.
Słupecki, J., G. Bryll, and U. Wybraniec-Skardowska. 1971. Theory of rejected propositions. *I. Studia Logica* 29 (1): 75–123.
Słupecki, J., G. Bryll, and U. Wybraniec-Skardowska. 1972. The theory of rejected propositions. *II. Studia Logica* 30 (1): 97–145.
Sochacki, R. 2007. Axiomatic rejection in the implicational-negational invariant sentential calculi of Łukasiewicz. *Bulletin of the Section of Logic, PAS* 36 (1/2): 1–6.
Sochacki, R. 2010. *Refutation methods in the study of logical systems*. Opole: Wydawnictwo Uniwersytetu Opolskiego.
Sochacki, R. 2011. Refutation systems for a system of nonsense-logic. *Logic and Logical Philosophy* 20 (3): 233–239.
Stahl, G. 1958. An opposite and an expanded system. *ZML* 4: 244–247.
Staszek, W. 1971. On proofs of rejection. *Studia Logica* 29 (1): 17–25.
Staszek, W. 1972. A certain interpretation of the theory of rejected propositions. *Studia Logica* 30 (1): 147–152.
Szabo, M. (ed.). 1969. *The collected papers of Gerhard Gentzen*. Amsterdam: North-Holland.
Tamminga, A. 1994. Logics of rejection: Two systems of natural deduction. *Logique Et Analyse* 146: 169–208.
Tiomkin, M.L. 1988. Proving unprovability. In *Proceeding of LICS'88)*, 22–26.
Tiomkin, M.L. 2013. A sequent calculus for a logic of contingencies. *Journal of Applied Logic* 11 (4): 530–535.
Trybus, A. 2018. A generalisation of a refutation-related method in paraconsistent logics. *Logic and Logical Philosophy* 27: 235–261.
Varzi, A.C. 1990. Complementary sentential logics. *Bulletin of the Section of Logic, PAS* 19: 112–116.
Varzi, A.C. 1992. Complementary logics for classical propositional languages. *Kriterion. Zeitschrift für Philosophie* 4: 20–24.
Wajsberg, M. 1938. Untersuchungen über den Aussagenkalkül von A. *Heyting. Wiadomości Matematyczne* 46: 429–435.
Wybraniec-Skardowska, U. 2016. On the mutual definability of the notions of entailment, rejection, and inconsistency. *Axioms* 5 (2): 15.
Wybraniec-Skardowska, U. 2018. Rejection in Łukasiewicz's and Słupecki's sense. In *Lvov-Warsaw School. Past and Present*, ed. A. Garrido, and U. Wybraniec-Skardowska, 575–598. Birkhäuser.
Wybraniec-Skardowska, U., and J. Waldmajer. 2011. On pairs of dual consequence operations. *Logica Universalis* 5 (2): 177–203.

Influence in Different Network Structures

Yunqi Xue, Rohit Parikh and Mihai Gociu

Abstract We are interested in revisable and actionable social knowledge/belief that leads to a large group action. Instead of centralized coordination, bottom-up approach is our focus. We explore multiple methods of belief revision in social networks. Such belief revision in groups represents social influence and power to some degree. Both influences from friends and from experts are explained. We define an intuitive concept of *expected influence* of a group. When different influence sources are suggesting conflicting actions, agents could make strategic decisions by analyzing expected influence of different subgroups. We then show some properties of expected influence in different network structures. We also simulate the strategic influence emerging in small-world networks which represents many real-world networks.

1 Introduction

Humans can sometimes coordinate a large scale event from a bottom-up fashion. There were over 100,000 people appeared in the streets at times of a social activity. You may think such group events must be well planned and communicated in a top-down fashion or something alike. The truth is that people had limited means of communication. The telecommunication networks and WiFi were both overloaded at the time of protests. People were using a phone app called FireChat, which is a one-to-one messaging phone app and relies on Bluetooth or WiFi link. This app allowed hundreds of people to coordinate their activities.

This chapter is in its final form and it is not submitted to publication anywhere else.

Y. Xue (✉) · R. Parikh
Graduate Center, CUNY, New York City, USA
e-mail: xue.yun.qi@gmail.com

R. Parikh
e-mail: rparikh@gc.cuny.edu

M. Gociu
Spyhce, Cluj-Napoca, Romania

F. Liu et al. (eds.), *Knowledge, Proof and Dynamics*,
Logic in Asia: Studia Logica Library,
https://doi.org/10.1007/978-981-15-2221-5_10

It is amazing to see such one-to-one communication can help hundreds of thousands of people coordinate an action in a swift manner. More importantly, the coordination of the action happened in a bottom-up fashion. In fact, many group activities or even social movements may be initialized centrally or through a few big hubs, but the actual actions taken are due to the bottom-up influence from our immediate social network. For example, you are an art dealer and got some information about a gallery opening from a website, or through an email invite. When you make final decision of going or not going, you usually check with your social network to see if any of the people you know would go as well. In addition, you may receive opinions that influence your decision from multiple sources, e.g., friends or your professional contacts such as other art dealers.

We are interested in studying the relationship between the movement of social knowledge within a network and the group actions. We are looking at a particular type of social knowledge that is actionable. Also we use a very loose term of knowledge because it is often just an unjustifiable belief that people have. Therefore, when we model it we treat the process as belief revision. There are different possible ways of influence that can play out in the network. They may result in different opinions about an event. We show that by using a simple indicator, *Expected Influence*, individuals in the network can treat each conflict as a coordination game and make a strategic decision. When we aggregate the influence in the network, we can actually show who has more social influence, hence social power. In different network structures, we can see that the potential social power will be distributed differently. For example, in a fully connected social network, the social power is evenly distributed.

In this paper, we first explain two types of influence, Expert Influence and Friends Influence. Next, we introduce the concept of expected social influence. We show how agents can make strategic decisions when different influences are in conflict. We also apply this process into different network structures such as a complete graph, star, ring, etc. At the end, we will show some simulations of the social influence in different networks.

2 Belief Updates and Social Influence

The latest logic systems have placed such active agents in a social network. The belief and knowledge updates depend not only on the initial state of an agent and her influencer, but also on the network structure of their community and their positions. This development, therefore, shows the changes of knowledge and belief on a community level.

We will explain two different processes of social influence, Friends Influence, (Liu et al. 2014) and Expert Influence (Xue and Parikh 2015).

2.1 Friends Influence

In the framework from Liu et al. (2014), friendships is taken to be a symmetric and irreflexive relation. So an agent is a friend of any friend of hers, but she is not her own friend. In addition, they do not assume that friendship is transitive. Therefore friends of friends are not necessarily friends. Friendships among a group of people create a social network or community. Agents start with some initial belief towards an issue p. They can be Bp (believe p), $B\neg p$ (believe the negation of p), or Up (no opinion on p). In the process of communicating with friends, a belief can be changed in two ways. When the agent is strongly influenced by friends (Sp), it leads to belief revision (Rp). Similarly, when she is weakly influenced (Wp), it leads only to belief contraction (Cp).

Symbol F means "all my friends". Therefore FBp means "all my friends believe p". In the framework, Liu et al. further define the dual operator $\langle F\rangle$ as "some of my friends". $\langle F\rangle Bp$ means some of my friends believe p. Then they add the aspect of a threshold. For instance, assuming my threshold is 100%, if I believe p but all of my friends believe $\neg p$, then I will change my own belief to $\neg p$. but if only some of my friends believe $\neg p$ then I will become undecided about p. When we set the threshold to $\frac{2}{3}$, we could have a Condorcet's paradox at hand (List and Pettit 2002).

Within a group of friends, beliefs tend to be adopted from one to the other. It creates a distribution of beliefs in the community. Liu et al. (2014) look at the influence one could gain from a socially connected agent.

The three possible doxastic states of a proposition p can be defined through following axioms:

- Strong influence: $S\varphi \leftrightarrow (FB\varphi \wedge \langle F\rangle B\varphi)$
- Weak influence: $W\varphi \leftrightarrow (F\neg B\neg\varphi \wedge \langle F\rangle B\varphi)$.

2.2 Language Splitting

In this subsection, we will introduce an alternative method for belief update through influence. We call it Expert Influence (Xue and Parikh 2015), which allows different sub-languages.[1] A sub-language used by some members of the community can be seen as a language for a particular topic or a sub-culture. For example, a community of people all speaks English. Among them, some of them are physicians. These physicians may share a sub-language that is filled with jargon and medical statements. A sub-language on a topic can also be shared among non-professionals. A good example can be a sub-language shared by foodies.

There are a few benefits of splitting a language into multiple sub-languages. First of all, we can express a more realistic process of day-to-day belief or knowledge updates. Due to the restrictions of the mathematical axioms in dynamic epistemic logic, when an agent revises her belief or knowledge on one particular statement, she

[1] Past work have assumed one language without any sub-languages for the whole community.

might disregard all her other beliefs in the same language.[2] With a splitting language scenario, the agent would only revise all her beliefs or knowledge in related topics. For example, if I learn, at some point, that there is a two-term limit for presidency in the US, then I only revise my belief in the topic of politics. I disregard my past belief that Obama will be the president of the US in 2017. However, I will not revise any of my beliefs about my dentist.

Second benefit is that we will be able to capture the multiple sub-cultures. Often people who are knowledgeable about politics are not necessarily the ones who are also into food. Therefore, knowledge influences may spread in different directions within a network of agents. A splitting language allows us to demonstrate this.

Last but not least, when we separate knowledge into different topics, we naturally have experts in each topic. These are the people who have more knowledge on a topic than most of the others in the network.

Therefore, we propose an alternative model that captures Expert Influence. In the this section, we will first briefly review AGM belief revision and the P Axioms which are proposed by Parikh (2011). Then we will explain how they are applied to Expert Influence.

2.2.1 AGM Axioms and P Axioms

Work on belief revision can be traced back to Peirce's classical papers, "the Fixation of Belief" in late nineteenth century and "how to make our ideas clear". A more recent, more technical, and a widely recognized work is by Alchourrón et al. (1985). However, none of the two papers discussed the influence on our beliefs by other people. They are interested in belief change but not on its source in other people. How do we adopt beliefs from people and adjust our beliefs accordingly?

Parikh (2011) shows that a person's beliefs can be uniquely divided into different *subject matters* and so if his beliefs are a theory T in L then L will split naturally into sub-languages $L_1 \cup L_2 \cup L_n$ and T will split into $T_1 \cup T_2 \cup \cdots T_n$, each T_i in L_i.

This idea of splitting language was naturally applied to belief revision. If I learn some political fact, it will not change my views about my teeth or the location of my children. Moreover, if have trouble with my car, I will not consult a dentist. Conversely, I will not consult a garage mechanic about my teeth.

Before we look into Parikh's axioms for language splitting, let us review the original AGM axioms which can lead to some forgetful updates.

AGM Axioms

1. $T * A$ is a theory
2. $A \in T * A$

[2]This point will be elaborated in the following two sections: Methodology—relevant literature and Projected results.

3. If $A \Leftrightarrow B$, then $T * A = T * B$.
4. $T * A \subseteq T + A$
5. If A is consistent with T, i.e., it is not the case that $\neg A \in T$, then $T * A = T + A$
6. $T * A$ is consistent if A is.
7. $T * (A \wedge B) \subseteq (T * A) + B$
8. If $\neg B \notin T * A$ then $(T * A) + B \subseteq T * (A \wedge B)$.

If A is consistent with T, then $T * A = T + A$, otherwise $T * A = Con(A)$. Given these axioms,[3] a forgetful update can happen in which case when A is inconsistent with T, all information in T is discarded.

Clearly, this kind of updates is unrealistic since in real life when we learn something about whether we retain our beliefs about politics. Parikh improves this update by using Craig's Interpolation Theorem.

Craig's Interpolation Theorem

Theorem 1 *Let L_1, L_2 be first order languages, $L = L_1 \cap L_2$ and T_1, T_2 be theories in L_1, L_2, respectively such that $T_1 \cup T_2$ has no model (is inconsistent). Then there is some formula ψ of L such that $T_1 \vdash \psi$ and $T_2 \vdash \neg\psi$. In particular, if ϕ is an L_1 formula and ξ is an L_2 formula and $T_1 \cup T_2$, $\phi \vdash \xi$ then there exists an L-formula ψ such that $T_1, \phi \vdash \psi$ and $T_2, \psi \vdash \xi$.*

P Axioms

Parikh proposed following axioms to improve the belief updates, which prevent the discarding all the information in the way we mentioned in a forgetful update.

1. **Axiom P1:** If T is split between L_1 and L_2, and A is an L_1 formula, then $T * A$ is also split between L_1 and L_2.
2. **Axiom P2:** If T is split between L_1 and L_2, A, B are in L_1 and L_2 respectively, then $T * A * B = T * B * A$.
3. **Axiom P2g:** If T split between L_1 and L_2, A, B are in L_1 and L_2 respectively, then $T * A * B = T * B * A = T * (A \wedge B)$.
4. **Axiom P3:** If T is confined to L_1 and A is in L_1 then $T * A$ is just the consequences in L of $T *' A$ where $*'$ is the update of T by A in the sub-language L_1.

All these axioms follow from axiom P below:
Axiom P: If $T = Con(A, B)$ where A, B are in L_1, L_2 respectively and C is in L_1, then $T * C = Con(A) *' C \dot{+} B$, where $*'$ is the update operator for the sub-language L_1.

[3] $T * A$ is the revision of theory T by A. $T + A$ is $Con(T \cup A)$ the logical closure of $Con(T \cup A)$.

2.3 Expert Influence

We are extending Friend Influence to show how influence on a subject travels in general. We call it *Expert Influence*. Before we discuss this type of belief updates, we are going to review some preconditions of the model.

Resolving differences of opinions among experts will bring us to the *judgment aggregation problem* which was investigated by List and Pettit (2002). Such issues are very difficult and this paper is not the right place to address it. We will start a simple assumption of having one expert in each area in a social network at the beginning of an influence process. In addition, we look at a connected social network only, i.e., there is no isolated person or groups in the network. We study how Expert Influence on a single proposition can travel through the network.

In this method, we employ all AGM axioms, P axioms from Parikh (2011), and two new additional axioms that instruct agents what to do when they are influenced by an expert or a non-expert.

- **Influence from Expert to Non-expert:** Before a non-expert talk to an expert on p, he can believe p, believe $\neg p$ or have no opinion about p. Once the non-expert talks to a neighboring expert about p, he adopts the beliefs of the expert and become an expert on p himself.
- **Influence between Non-experts:** Between two non-experts, if they have different beliefs about p, they both contract their beliefs on p since none of them are authoritative on the matter.

These conditions and axioms can be expressed formally in the following way:

1. Agents $N = \{1, 2, \dots, n\}$ in a social network of G
2. $R \subseteq N \times N$
3. R is symmetric and irreflexive
4. 3 possible doxastic states:
   a. believe p: Bp
   b. disbelieve p: $B\neg p$
   c. no belief about p: $Up = \neg Bp \wedge \neg B \neg p$.

Expert Influence Axioms

- Expert influence to a non-expert: $R(i, j) \wedge E(i, p, t) \wedge B(i, p, t) \wedge (B(j, p, t) \vee B(j, \neg p, t) \vee U(j, p, t)) \rightarrow B(j, p, t+1) \wedge E(j, p, t+1)$
  This can clearly be simplified to $R(i, j) \wedge E(i, p, t) \wedge B(i, p, t) \rightarrow B(j, p, t+1) \wedge E(j, p, t+1)$
- Influence between non-experts: $R(i, j) \wedge \neg E(i, p, t) \wedge \neg E(j, p, t) \wedge B(i, p, t) \wedge B(j, \neg p, t) \rightarrow U(i, p, t+1) \wedge U(j, p, t+1)$.

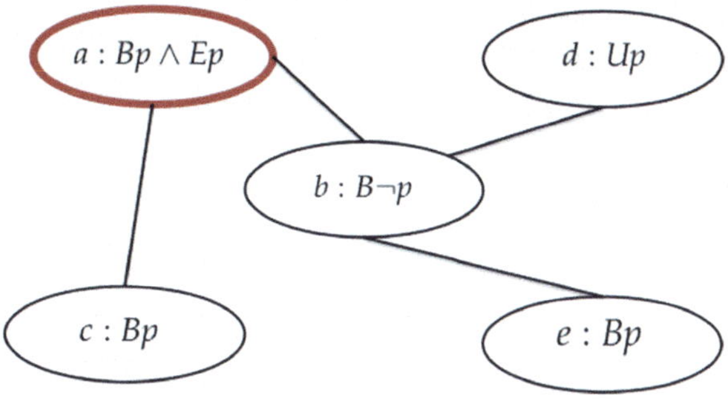

Fig. 1 Example 1: single expert—beginning

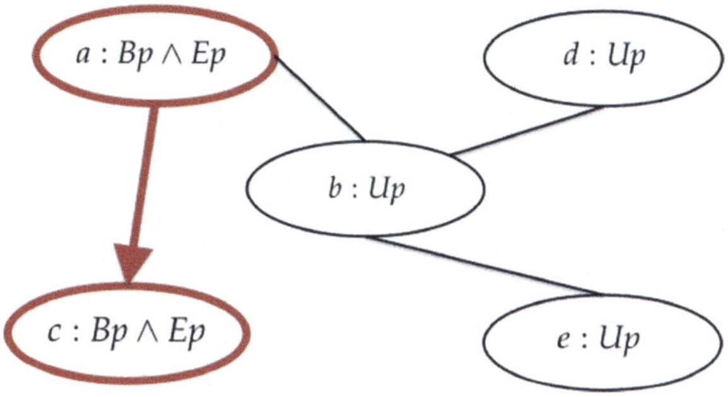

Fig. 2 Example 1: single expert—during

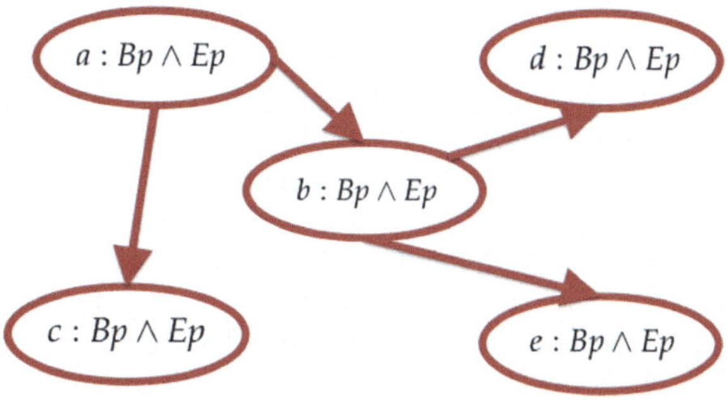

Fig. 3 Example 1: single expert—end

2.3.1 Examples of Expert Influence

Let us now look at two possible scenarios of Expert Influence.

In Example 1 (Figs. 1, 2 and 3), 5 agents are in the network with a as an expert on p (marked in with a bold circle) initially. Communication can happen anytime between two agents, however, Expert Influence starts to spread only when a starts communicating with her neighbors. The figures show one of the many possible influence order. Figure 2 shows that agent a influenced c who has already become an expert on p. At the same time, we can see that b and e also communicated. Since they had different beliefs on p and none of them is an expert, they contracted their

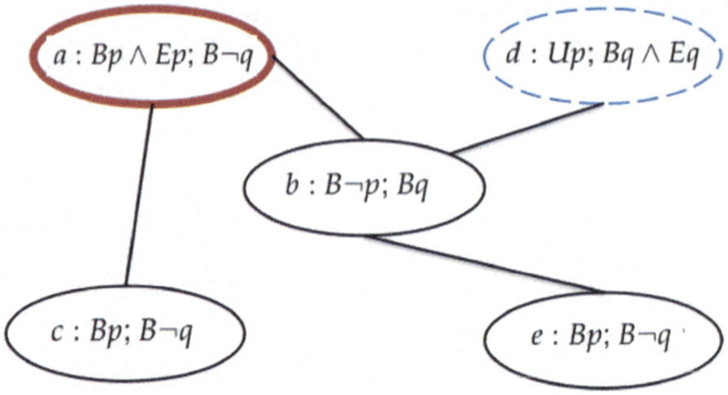

Fig. 4 Example 2 (two experts)—beginning

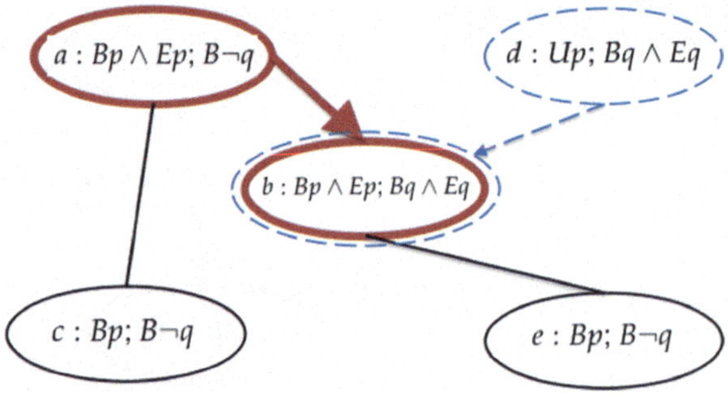

Fig. 5 Example 2—end the influence for agent b

beliefs on the matter. At the end of the influence process, we can see that every agent in the network adopt the opinion of the expert and become an expert on p (Fig. 3).

We have assumed that at the beginning of the influence process, there is only one expert in each area. However, it is possible to have two experts when we look at two areas of expertise. In Example 2 (Figs. 4 and 5), the 5 agents are in same network structure as in Example 1. Both a and d are experts, although in two areas (p and q) which are indicated by bold and dashed circles, respectively. The experts' influence spread out in a similar fashion as in Example 1. Agent b is at a special position in the network since he is connected to both experts a and expert d. By applying the axiom P2G, we can show that b will have $Ep \wedge Eq$. In addition, b is going to be the first person in the network to become an expert on both subject matters, regardless how the two different influence processes play out.

3 Influence and Coordination in Networks

There are many events in our daily life that require coordination with our friends and acquaintance. For example, we would like to go to a party only if enough interesting people are going. In some situations, such as a dinner party for a small group of people who we know, we can just contact our friends and ask. However, in some situations, such a gallery opening party which you have heard from some friends, you cannot contact everyone. In the second type of party, you may want to meet strangers who are interesting. On one hand, you cannot contact people you have not met. On the other hand, since your friends told you about the event, they may know some of the people who are going. Then coordination can happen through understanding influence in our social network.

In the past, we may still be able to find the information for deciding whether we would go to an event or take an action easily, since our social networks were generally smaller. In the era of infostorm, not only we are overwhelmed by information, but also having ever expending social network through all kinds of social media. How do we make good decision given different kinds of influence in a large social network?

We have introduced two different belief-update methods, namely, Friend Influence and Expert Influence. It is clear that both can be separately applied in different situations. In the case of a small dinner party, we probably get influenced through friends who are invited and coordinate with them. In the case of a gallery opening party, the influence may come naturally from an artist friend who is going and is an expert of such events. It is also possible that we get influenced both from our friends and experts at the same time. How do we choose which one to follow when the two are in conflict? More importantly, in an event that requires group coordination, can we choose strategically as an individual without global planning with all the participants?

In this section, we investigate how agents can be influenced differently in one event. We will first introduce the concept of *Expect Influence* which indicate potential influence of an agent or a group of agents. Then we will discuss how conflicting influence may raise between two different belief-update methods. At the end, we will discuss how network structure can affect people's choices of belief updates.

3.1 Expected Influence

During an event, such as a party, the people who are invited, are coordinating on an action which has two options either attend when there are enough interesting people going or not attend. Of course, they could contact each other and coordinate globally within a community. However, as we enter the era of infostorm, our network is growing ever larger. Such centralized coordination becomes more difficult and inefficient in some situations.

Every person in the network has her own knowledge about the local network she is in. She can then use that knowledge to decide how she wants to be influenced and take action. In order to make such decision, she needs to have a simple indicator, *Expected Influence*, which allows her to compare the influence from different groups of people in her network. This way she can maximize her benefit from taking the action under a particular influence.

Let us look back to the example of a gallery opening. Alice received an invitation to the opening. However, she is not too familiar with the crowd. She does know that a few people in her social network are invited too. Some of them actually encouraged her to go along with them while some told her that it is going to be a boring event.

Alice, being a mellow person, is happy to go if she believes that the majority of the invited guests are going. She would also be happy if she chooses to stay at home, and most people also decide not to go. Before she makes a decision, she gathers information from her social network. Then she could update her belief using the Friend Influence model which means her belief and action depend on her friends'. She could also update her belief using the Expert Influence model which means her belief and action depend on what she hears from the expert in her network, for example, a seasoned artist who knows the crowd and has extensive experience with this type of gallery openings. It is possible that the two methods lead to the same action which she would take. It is also possible that two methods are in conflict. In the second case, Alice would think about her goal, being happy in this case, and strategically choose a method that maximizes the happiness. This means she needs to pick an action which is going to be taken by most people. Therefore, she would assess whether her friends have more influence or the expert.

As we are influenced more and more through the social media platforms, it is harder for a community member to know how and when her acquaintances are influenced. Therefore, in this paper, we do not look at how an agent reaches to the point of being influenced and has to make a decision for her action. We are more interested in given the network, the influence she receives at a moment, and how she would strategically choose.

When we look at the influence of a person or a group of people, we do not look at how many people they have already influenced, but how many people they can potentially influence. Therefore we define the concept in the following way.

Definition 1 $I_{\mathcal{G}}, \mathcal{G} \subseteq Ag$ where Ag is the set of agents in the network, indicates the potential **influence** of the group $\mathcal{G}$. It is the number of agent in this set, $|\mathcal{G}|$, plus the total number of direct neighbors of each agent in group $\mathcal{G}$ removing the number of duplicated agents.

$$I_{\mathcal{G}} = |\mathcal{G} + \sum_{i=1}^{n} \mathcal{D}_i|$$

Figure 6 is a local network that is connected to the rest of the network through E. There are 5 agents, namely, A, B, C, D, and E. "Y" and "N" indicates if each agent believes that the opening will be an interesting one. "Y" means yes, and implies that the agent would probably go to the opening party. "N" means no. "u" in B means that B is undecided. The numbers in Fig. 6 indicates the degree of each node, i.e., the number of neighbors of each agent. E is the expert in this local network, who has a reputation for being the expert. Therefore, the number of her neighbors is known to everyone in the local network. Some agents' neighbors are completely observable in the local network, such as A, B, C, and D. Some agent's neighbors, such as the 10 neighbors of E, are not all on this local graph. Here we also assume that the rest of 9 neighbors of E are not in this local graph.

The influence of agent A is $I_{\{A\}} = 3$. The influence of agent B and C, $I_{\{B,C\}}$, is 4. Influence of agent E and C, $I_{\{E,C\}}$, is 13.

3.2 *Belief Conflict and Coordination Equilibria*

Now we continue to look at the gallery opening example in Fig. 6 and find out how an agent can choose between the two methods of belief update. Since agent B is undecided, we will focus on her and see how she can update her belief and choose the better update that maximizes her happiness.

Just as we mentioned earlier, there is only one expert in the local network. In this example, it is E. Furthermore, we will assume that the expert is always more actively talking about her expertise to the people in her network. In the example here, it means agent B get influenced by E first. The other agents, A and D are friends[4] of B. We can see that all of B's friends think it will be a great party and probably going, while the expert E believes that the party will not be so great. So how is B going to be influenced and how will she choose what to do?

B receives the information from E first and updates her belief to E's according to the expert method that was defined earlier in the paper. Then B communicates with her friends and finds out that her friends think otherwise. Now she has a choice between being influenced by her friends and being influenced by the expert.

B has a coordination problem to solve. She wants to make a choice that majority of the global network would choose. However, she does not have much knowledge about the actual global network. She does not know how many are invited. She certainly does not know who they are and how they are connected to each other. She knows her neighbors and the number of neighbors they each have. What she could do is to choose an update that gives her maximized influence in her local network. We further assume that the expert cannot be influenced by others since she believes that she has the authority on the matter. B can consider the pair-wise coordination with each of her friends.

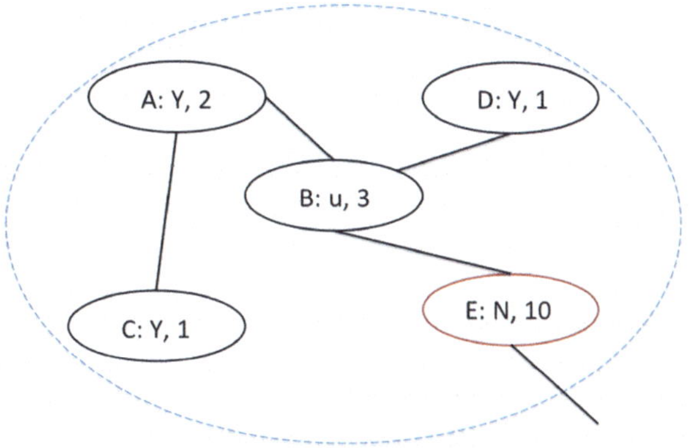

Fig. 6 Conflicting influence 1

[4]Experts are not friends.

Table 1 Influence matrix

		Agent D: Friends	Agent D: Expert
Agent B	Friends	4, 4	0, 0
	Expert	0, 0	13, 13

Let us look at the coordination between B and D. Once they communicate their beliefs with each other, they both know that the friends belief differs from the expert's belief. Since they each represent a different opinion, they are vague about the actions they take, going to the party or not, so that they will not hurt each others' feeling. They ponder what to do given the Influence Matrix from table 1. When they choose different beliefs to act upon ({Friends, Expert} and {Expert, Friends}), they are both not happy. When they are choosing the same belief to act upon, then their happiness depends on the expected influence.

$$I_{Friends} = I_{B,D} = |\mathcal{G}_{\{B,D\}} + \mathcal{D}_B + \mathcal{D}_D| = 4$$

$$I_{Expert} = I_{B,E} = I_{D,B,E} = |\mathcal{G}_{\{B,E\}} + \mathcal{D}_B + \mathcal{D}_E| = 13$$

Because agent B is the undecided one, therefore, when we estimate the influence we have B's degree in both methods. When if they choose expert method, it is I{B,E} for B, and I{D, B, E} for D (Table 1).

In this payoff matrix, we have two equilibria, with {Friends, Friends} as a risk-dominant equilibrium and {Expert, Expert} as a payoff-dominant equilibrium (Harsanyi and Selten 1988; Cassar 2007). It is reasonable to assume that they will pick the payoff-dominant equilibrium as they both know the influence of the expert is higher.

Gintis (2009) suggests that a payoff-dominant equilibrium is selected when players believe that the Principle of Honest Communication, according to which players keep their promises unless they can benefit by violating these promises and being believed.

We further make the natural assumption that regardless of the influence method, it is always better to pick the group that gives more expected influence. This can explain some social hypes Chwe (2013, p. 37) through a perspective of social networks, such as the famous El Farol Bar Problem (Arthur 1994). This also shows that people tend to believe and act on views that give them highest potential influence and ignore the source of the information.

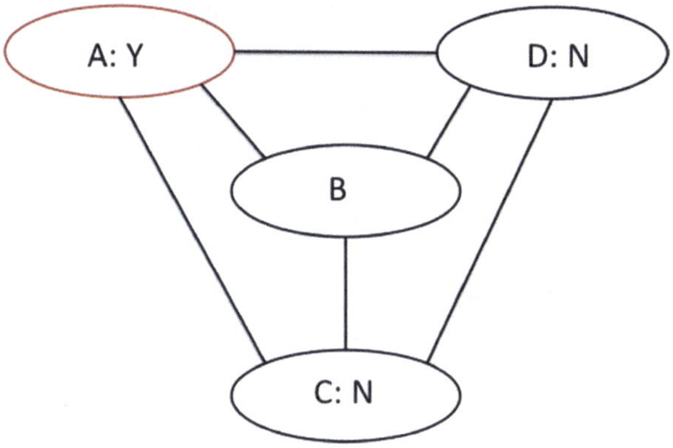

Fig. 7 Influence in a complete graph

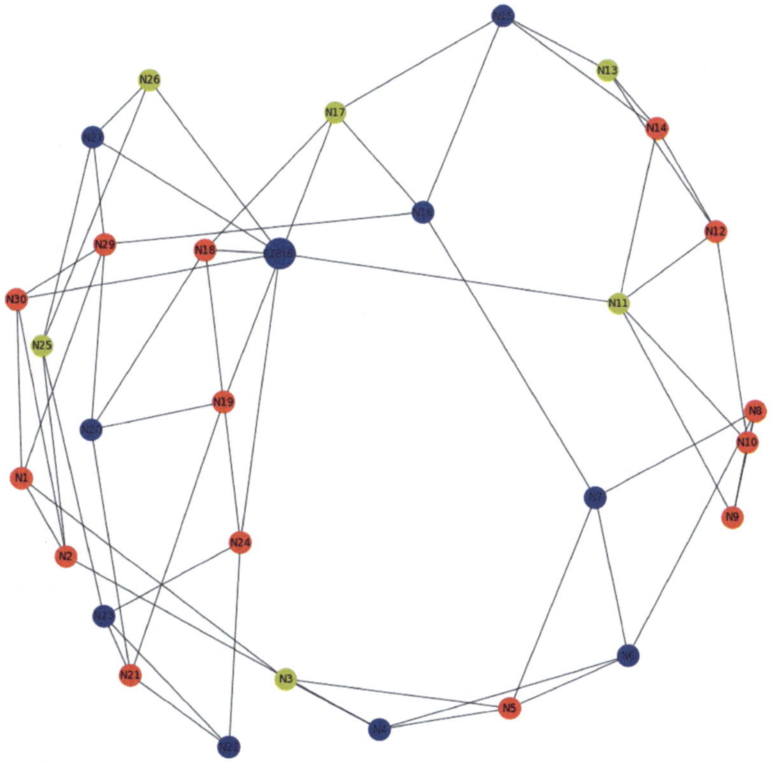

Fig. 8 Watts–Strogatz network example—initial

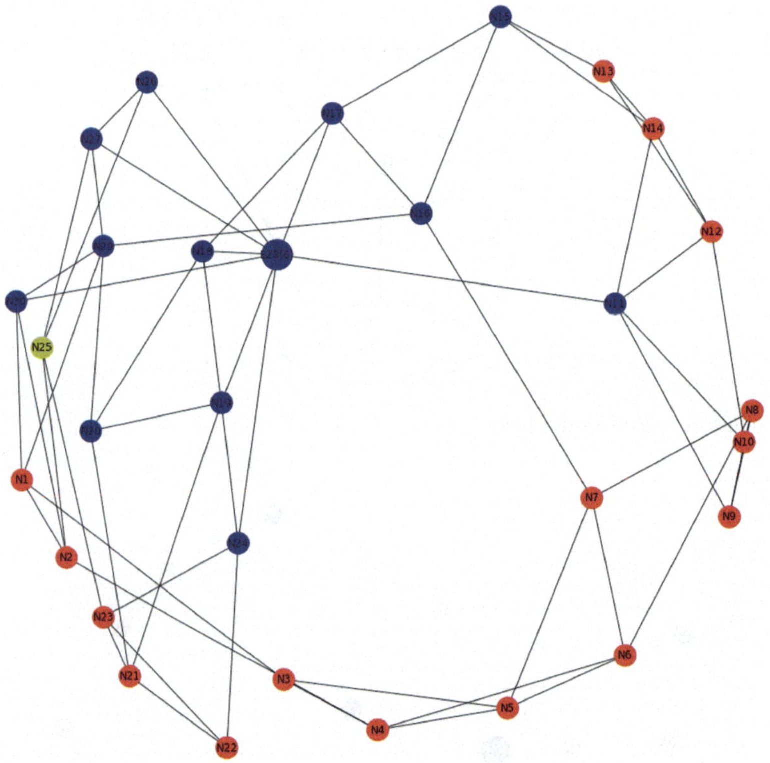

Fig. 9 Watts–Strogatz network example—final

3.3 Different Networks

With this simple influence indicator, we can investigate the relationship between a belief update method and the structure of the network. We will show a few properties in some basic network structures.

Proposition 1 *In a social network that is a* ***complete graph*** *(fully connected), two methods are the same. In fact, any group* $\mathcal{G} \in \wp(Ag)$ *has the same influence* $I^* = |Ag|$.

Proof By induction on $\mathcal{G}$. □

When a social network is so tightly connected, e.g., Fig. 7, each individual can potential influence everyone, which also means each of them has little to none influence. It is also natural to assume that such networks have few conflicts in general.

Proposition 2 *In a social network of* $|Ag|$ *number of agents, if the structure is a* ***ring****, friends* $I_{\{i,j\}} = 4$ *always have equal influence as the expert* $I_{\{i,e\}} = 4$.

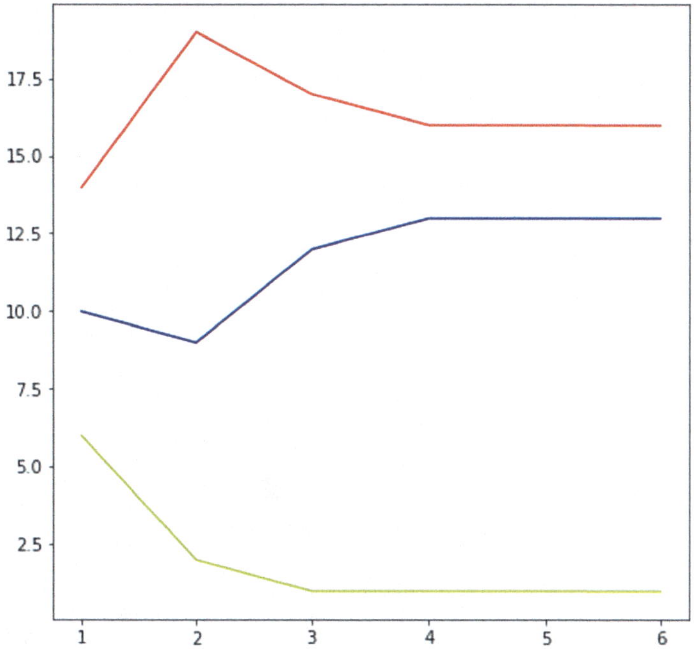

Fig. 10 Watts–Strogatz network example—history

Proof Any of the two neighboring agents have influence of 4. Induction on $|Ag|$. □

Proposition 3 *In a social network that is a structure of a* ***star*** *with* $|Ag| \geq 4$, $I_{\{i,e\}}$ *always has the most influence if e is at the center. Otherwise friend influence is always stronger.*

Proof (1) When the expert is at the center
$I_{\{i,e\}} = |Ag|$ since e is connected to everyone. $I_{\{i,j\}} = 2$ for all $i \neq j$ and $i, j \neq e$, since none of the two none-experts are neighboring each other.

(2) When a non-expert is at the center
$I_{\{c,i\}} = |Ag| = I_{\{c,e\}}$ with c as the agent who is at the center. $I_{\{i,e\}} = 2$ for the influence of any non-expert and expert (except the influence of the center agent and expert). □

We can see that the network structure has an effect on how agents would choose which type of belief updates as well. In a social network that is fully connected or has a ring structure, the agents are likely to treat the two methods equally if they are in conflict. However in a network that has a structure of a star, the choice depends on the position of the agents and the expert. An expert in the center knows that he has enormous power that can influence the whole network. It also shows that whoever has absolute control of the information flows has the control of behavior of the

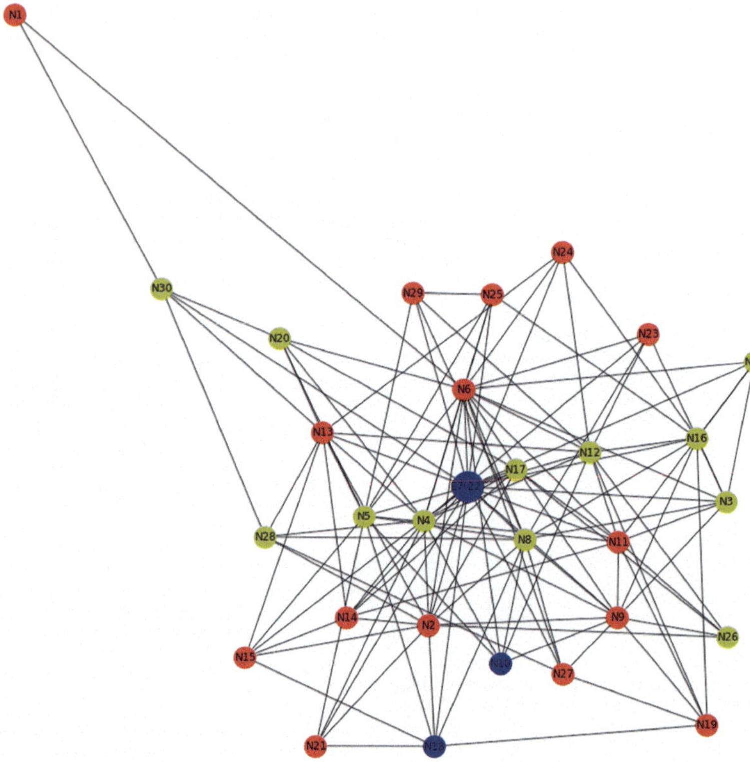

Fig. 11 Barabasi–Albert network example—initial

whole network. We should note that, comparing with the game theoretical approach we discussed earlier, these three propositions requires additional assumption that everyone knows the network structure.

4 Simulating Social Power

After studying some of the properties of influence indicator, we are interested in seeing how it could work on bigger network. Therefore, we simulated the process in two different types of network: Watts–Strogatz and Barabási–Albert (Figs. 8, 9, 10, 11, 12 and 13).

We discover that the expert[5] has different degrees of social power in different network structures. In particular, a well-connected expert has a lot of power in the Barabasi–Albert network, which is more similar to real-world social network.

[5]We have always assumed 1 expert in the network. In the simulation, we also set the number of expert to 1. This can be easily changed and extended in the future work.

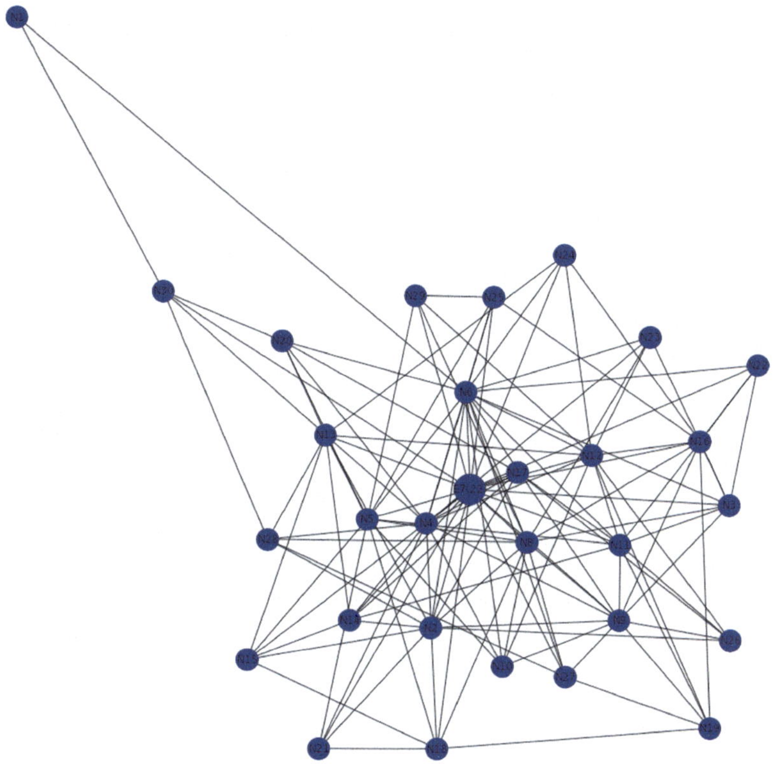

Fig. 12 Barabasi–Albert network example—final

5 Conclusion

In this paper, we are primarily interested decentralized social coordination. When people make decisions in participating social events, they often get influenced by different people in their social networks. We explained two possible belief update methods, namely Friend Influence and Expert Influence. Instead of assuming large amount of common knowledge, we assume limited local knowledge. We introduced a simple indicator for influence in a network. Sometimes two belief updates lead to two different suggestions. When this happens, each individual can use the influence indicator to assess the potential influence of various groups, e.g., her friends or the expert. Given the two options, each agent can make a strategic decision that maximizes her benefits. We also discussed how different network structures may affect agents' decisions. In some structures, the two methods always have the same influence. In some structures, such as a star, we can predicate the choice of agents by looking at their positions in the network.

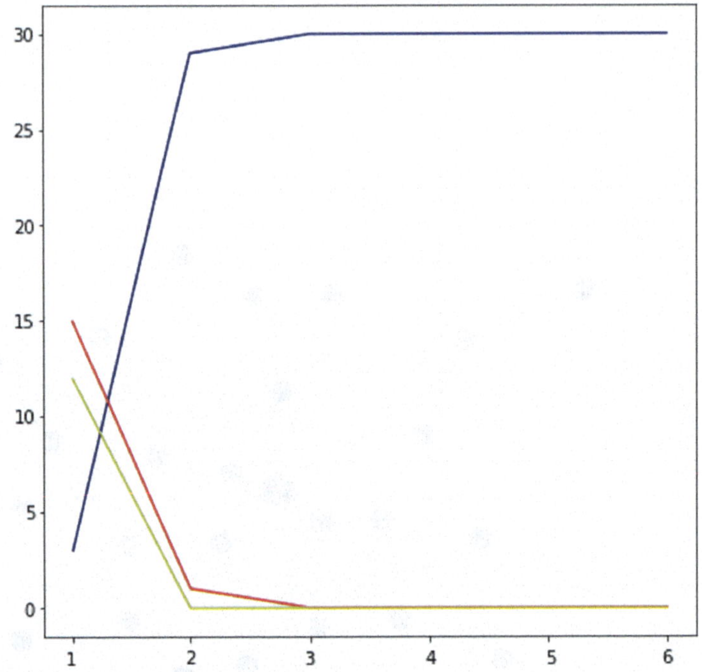

Fig. 13 Barabasi–Albert network example—history

Our simulation reflects our theoretical finding, which shows the importance of the network structure in spreading social powers.

References

Alchourrón, C., P. Gärdenfors, and D. Makinson. 1985. On the logic of theory change: Partial meet contraction and revision functions. *The Journal of Symbolic Logic* 50 (2): 510–530 (Cambridge University Press).

Arthur, B. 1994. Inductive reasoning and bounded rationality. *The American Economic Review* 84: 406–411 (JSTOR).

Cassar, A. 2007. Coordination and cooperation in local, random and small world networks: Experimental evidence. *Games and Economic Behavior* 58 (2): 209–230 (Elsevier).

Chwe, M. 2013. *Rational ritual: Culture, coordination, and common knowledge*. Princeton: Princeton University Press.

FireChat in Hong Kong: How an app tapped its way into the protests. CNN. 16 October 2014. http://www.cnn.com/2014/10/16/tech/mobile/tomorrow-transformed-firechat/.

Gintis, H. 2009. The local best response criterion: An epistemic approach to equilibrium refinement. *Journal of Economic Behavior and Organization* 71 (2): 89–97 (Elsevier).

Harsanyi, J., and R. Selten. 1988. *A general theory of equilibrium selection in games*. Cambridge: The MIT Press.

List, C., and P. Pettit. 2002. Aggregating sets of judgments: An impossibility result. *Economics and Philosophy* 18 (1): 89–110 (Cambridge University Press).
Liu, F., J. Seligman, and P. Girard. 2014. Logical dynamics of belief change in the community. *Synthese* 1–29.
Parikh, R. 2011. Beth definability, interpolation and language splitting. *Synthese* 179 (2): 211–221 (Springer).
Xue, Y., and R. Parikh. 2015. Strategic belief updates through influence in a community. *Studies in Logic* 4: 124–143.

The manufacturer's authorised representative in the EU is Springer
Nature Customer Service Centre GmbH, Europaplatz 3, 69115 Heidelberg,
Germany. If you have any concerns regarding our products, please
contact ProductSafety@springernature.com

Printed and bound by CPI Group (UK) Ltd, Croydon, CR0 4YY
05/10/2026
02228120-0001